Nuclear Physics: Advanced Principles and Applications

Nuclear Physics: Advanced Principles and Applications

Editor: Nadia Crawford

NY RESEARCH
P R E S S

New York

Published by NY Research Press
118-35 Queens Blvd., Suite 400,
Forest Hills, NY 11375, USA
www.nyresearchpress.com

Nuclear Physics: Advanced Principles and Applications
Edited by Nadia Crawford

International Standard Book Number: 978-1-63238-681-6 (Hardback)

Cataloging-in-Publication Data

  Nuclear physics : advanced principles and applications / edited by Nadia Crawford.
      p. cm.
  Includes bibliographical references and index.
  ISBN 978-1-63238-681-6
  1. Nuclear physics. 2. Physics. I. Crawford, Nadia.
QC776 .N83 2019
539.7--dc23

Contents

Preface

Nuclear physics is a field of physics that is concerned with the study of atomic nuclei, their constituents and interactions. The structure and properties of the nucleus can be understood from the classical and quantum mechanical approaches of liquid-drop model and the nuclear shell model respectively. Current research in nuclear physics studies nuclei under extreme conditions, such as under high excitation energy and high spin. Nuclei with extreme neutron-to-proton ratios can be artificially created, by the processes of nucleon transfer reactions or induced fusion reactions. Some of the significant areas of study in nuclear physics are nuclear decay processes, nuclear fusion and nuclear fission. Advances in nuclear physics have transformed the fields of medicine, warfare, energy, geology and archaeology. This book elucidates the advanced principles of nuclear physics and their applications in a multidisciplinary manner. It strives to provide a fair idea about this discipline and to help develop a better understanding of the latest advances within this field. It attempts to assist those with a goal of delving into the field of nuclear physics.

Significant researches are present in this book. Intensive efforts have been employed by authors to make this book an outstanding discourse. This book contains the enlightening chapters which have been written on the basis of significant researches done by the experts.

Finally, I would also like to thank all the members involved in this book for being a team and meeting all the deadlines for the submission of their respective works. I would also like to thank my friends and family for being supportive in my efforts.

Editor

Towards single-valued polylogarithms in two variables for the seven-point remainder function in multi-Regge kinematics

Johannes Broedel [a,*], Martin Sprenger [b], Alejandro Torres Orjuela [a,c]

[a] *Institut für Mathematik und Institut für Physik, Humboldt-Universität zu Berlin, IRIS Adlershof,*
Zum Großen Windkanal 6, 12489 Berlin, Germany
[b] *Institut für Theoretische Physik, Eidgenössische Technische Hochschule Zürich, Wolfgang-Pauli-Strasse 27,*
8093 Zürich, Switzerland
[c] *Institut für Mathematik, Technische Universität Berlin, Straße des 17. Juni 136, 10623 Berlin, Germany*

Editor: Stephan Stieberger

Abstract

We investigate single-valued polylogarithms in two complex variables, which are relevant for the seven-point remainder function in $\mathcal{N} = 4$ super-Yang–Mills theory in the multi-Regge regime. After constructing these two-dimensional polylogarithms, we determine the leading logarithmic approximation of the seven-point remainder function up to and including five loops.

1. Introduction

One of the most interesting results in the study of scattering amplitudes in planar $\mathcal{N} = 4$ super-Yang–Mills theory is that certain amplitudes can be constructed to high loop orders solely from understanding the space of functions describing the amplitude, its symmetries as well as its

---

* Corresponding author.
E-mail addresses: jbroedel@physik.hu-berlin.de (J. Broedel), sprengerm@itp.phys.ethz.ch (M. Sprenger), alejandro.torresorjuela@campus.tu-berlin.de (A. Torres Orjuela).

limiting behavior in special kinematic regimes. Indeed, following this program of bootstrapping the six-point amplitude, the remainder function is by now known up to four loops [1–4] and it is very likely that this program can be continued to higher orders.

Much less, however, is known about the seven-point amplitude. The MHV remainder function in general kinematics has been calculated up to two loops [5], while the symbol is known up to three loops [6,7]. It is therefore a sensible idea to consider specific kinematic configurations, hoping to obtain higher-order results in those special settings which can then be used as constraints on a potential ansatz for the full seven-point remainder function.

The special kinematic configuration considered in this paper is the multi-Regge limit. This limit has been studied in the seven-point case before: the Mandelstam regions have been classified [8,9], and the remainder function has been calculated in the most interesting Mandelstam region in the leading logarithmic approximation (LLA) [10]. Furthermore, the seven-point remainder function has been studied at strong coupling [11,12] as well as from the perspective of the symbol for two [13] and three loops [14].

In this paper, we follow the path of understanding for the six-point remainder function in the multi-Regge limit: while first expressed in Fourier–Mellin space up to next-to-leading logarithmic order (NLLA) [15–17], the identification of the relevant space of functions paved the way for an efficient evaluation of the remainder function in momentum space [18–20]. In the six-point case, these functions are single-valued harmonic polylogarithms (SVHPLs) which we will briefly review below.

Similarly, in ref. [10], the seven-point MHV remainder function was written down in Fourier–Mellin space. Due to the complicated nature of the integral, it was so far only evaluated up to two loops. In this paper we therefore set out to identify a suitable two-variable generalization of SVHPLs constituting the relevant space of functions in the seven-point case. We construct those functions from their differential behavior, which allows us to obtain results up to five loops.

The paper is organized as follows. In section 2 we review the six-point case and the construction of the SVHPLs describing the remainder function. In subsection 3.1 we then move on to seven gluons and highlight the differences to the six-point case and the two-dimensional generalization of HPLs in subsection 3.2 before constructing the two-dimensional analogue of SVHPLs in subsection 3.3. Using these functions, we obtain expressions for the remainder function in LLA up to five loops in subsection 3.4, before concluding in section 4.

2. Six-point remainder function in multi-Regge kinematics

2.1. Starting point/setup

The six-point remainder function R_6^{MHV} in $\mathcal{N} = 4$ super-Yang–Mills theory[1] describes the discrepancy between the full amplitude and the BDS ansatz [21],

$$A_6^{\mathrm{MHV}} = A_{\mathrm{BDS}}\, e^{R_6^{\mathrm{MHV}}}. \tag{2.1}$$

While the calculation of the six-point remainder function in general kinematics requires a multitude of different techniques, its determination in the so-called multi-Regge kinematics is simpler. The multi-Regge limit refers to the kinematical regime of the scattering amplitude in which the

---

[1] For simplicity of presentation, we confine ourselves to the MHV remainder function; the six-point remainder function for NMHV is described and evaluated in refs. [18,3,4,20].

rapidities of the outgoing particles are strongly ordered. In terms of external momenta k_i, the multi-Regge limit can be most easily described using dual variables $x_{ii+1} := x_i - x_{i+1} := k_i$. Expressed in terms of dual conformal cross ratios u_i, the multi-Regge behavior reads:

$$u_1 := \frac{x_{13}^2 x_{46}^2}{x_{14}^2 x_{36}^2} \to 1, \quad u_2 := \frac{x_{24}^2 x_{15}^2}{x_{25}^2 x_{14}^2} \to 0, \quad u_3 := \frac{x_{35}^2 x_{26}^2}{x_{36}^2 x_{25}^2} \to 0, \tag{2.2}$$

where the reduced cross ratios

$$\tilde{u}_2 := \frac{u_2}{1-u_1} =: \frac{1}{|1+w|^2}, \quad \tilde{u}_3 := \frac{u_3}{1-u_1} =: \frac{|w|^2}{|1+w|^2} \tag{2.3}$$

are kept finite. As visible from eq. (2.3), six-point multi-Regge kinematics is completely determined by a complex parameter w and the large cross ratio u_1.

While the naïve limit eq. (2.2) yields a vanishing remainder function, a non-trivial result can be obtained by an analytic continuation to the so-called Mandelstam region [15,16], which is implemented by a clockwise continuation of the large cross ratio u_1,

$$u_1 \to e^{-2i\pi} u_1 \tag{2.4}$$

before taking the limit (2.2). For this setup, the six-point remainder function in multi-Regge kinematics can be written as a Fourier–Mellin integral [16,22,17]:

$$e^{R_6^{MHV}+i\pi\delta}|_{MRK} = \cos\pi\omega_{ab} + i\frac{a}{2}\sum_{n=-\infty}^{\infty}(-1)^n\left(\frac{w}{w^*}\right)^{\frac{n}{2}}\int_{-\infty}^{\infty}\frac{dv}{v^2+\frac{n^2}{4}}|w|^{2iv}\Phi_{reg}^{MHV}(v,n)$$

$$\times \exp\left[-\omega(v,n)\left(\log(1-u_1)+i\pi+\frac{1}{2}\log\frac{|w|^2}{|1+w|^4}\right)\right]. \tag{2.5}$$

In the above equation, the first term originates from a Regge pole exchange, while the second term comes from the exchange of a two-Reggeon bound state, which gives rise to a Regge-cut contribution. The so-called impact factor $\Phi_{reg}^{MHV}(v,n)$ and the BFKL eigenvalue $\omega(v,n)$ appearing in the latter have an expansion in powers of the loop-counting parameter $a = \frac{g^2 N_c}{8\pi^2}$:

$$\omega(v,n) = -a(E_{v,n}+aE_{v,n}^{(1)}+a^2E_{v,n}^{(2)})+\mathcal{O}(a^4),$$
$$\Phi_{reg}^{MHV} = 1 + a\,\Phi^{(1),MHV}+a^2\,\Phi^{(2),MHV}+\mathcal{O}(a^3). \tag{2.6}$$

Physically, the BFKL eigenvalue $\omega(v,n)$ describes the evolution of the two-Reggeon bound state, while the impact factor describes the coupling of the two-Reggeon bound state to the physical gluons. While the first orders of these quantities were determined by direct calculation [16,22,17, 18,2], a general solution based on the Wilson-loop OPE [23–26] has been identified in ref. [27].

In the six-point (and seven-point) calculations below, we will only need the lowest-order term of the BFKL eigenvalue, which reads

$$E_{v,n} = -\frac{1}{2}\frac{|n|}{v^2+\frac{n^2}{4}}+\psi\left(1+iv+\frac{|n|}{2}\right)+\psi\left(1-iv+\frac{|n|}{2}\right)-2\psi(1), \tag{2.7}$$

where $\psi(x)$ is the digamma function. The two other quantities appearing in eq. (2.5), the phase δ and the Regge-pole contribution ω_{ab}, are related to the cusp anomalous dimension $\gamma_K(a)$ via

$$\omega_{ab} = \frac{1}{8}\gamma_K(a)\log|w|^2 \quad \text{and} \quad \delta = \frac{1}{8}\gamma_K(a)\log\frac{|w|^2}{|1+w|^4} \tag{2.8}$$

and are thus known to all orders [28].

Finally, the term $\log(1 - u_1)$ in the integrand of eq. (2.5) is large because of the behavior of the cross ratio u_1 in the multi-Regge limit eq. (2.2). This suggests to organize the remainder function in powers of $\log(1 - u_1)$ at each loop order:

$$R_6^{\text{MHV}}\Big|_{\text{MRK}} = 2\pi i \sum_{\ell=2}^{\infty}\sum_{n=0}^{\ell-1} a^\ell \log^n(1 - u_1)\Big[g_n^{(\ell)}(w, w^*) + 2\pi i\, h_n^{(\ell)}(w, w^*)\Big]. \tag{2.9}$$

In the above equation, all terms with $n = \ell - 1$ are referred to as the leading logarithmic approximation (LLA) and terms with $n = \ell - 1 - k$ belong to (Next-to)^k-LLA. Since the imaginary and real parts $g_n^{(\ell)}$ and $h_n^{(\ell)}$ are not independent [18], it is sufficient to calculate all imaginary parts $g_n^{(\ell)}$ in order to determine the full remainder function. For more details on the six-point remainder function in the multi-Regge limit we refer the reader to ref. [20] and continue with the description of the relevant functions for the evaluation of eq. (2.5).

2.2. Single-valued harmonic polylogarithms in one variable

Before describing the functions governing the integral eq. (2.5), let us introduce harmonic polylogarithms (or HPLs for short) [29], which are defined as iterated integrals

$$H_{a_1,a_2,\dots,a_n}(z) = \int_0^z dt f_{a_1} H_{a_2,\dots,a_n}(t), \tag{2.10}$$

where the integration weights f_a are given as

$$f_{-1} = \frac{1}{1+t}, \quad f_0 = \frac{1}{t}, \quad \text{and} \quad f_1 = \frac{1}{1-t}. \tag{2.11}$$

In eq. (2.10), the length of the index vector $\vec{a} = \{a_1, \dots, a_n\}$ is called the *weight* of a HPL, while z is referred to as the *argument*. For the six-point remainder function, only the latter two integration weights in eq. (2.11) will appear. Corresponding to the weights f_0 and f_1 we introduce two letters x_0 and x_1 which will be used as non-commutative bookkeeping variables in generating functions for polylogarithms below.

From their definition eq. (2.10) it is clear that HPLs satisfy the differential equation

$$\frac{\partial}{\partial z} H_{a_1,a_2,\dots,a_n} = f_{a_1}(z) H_{a_2,\dots,a_n}(z). \tag{2.12}$$

Harmonic polylogarithms of low weight can be conveniently expressed in terms of logarithms and dilogarithms, for example

$$\underbrace{H_{0,\dots,0}(z)}_{w} = \frac{1}{w!}\log^w(z), \quad \underbrace{H_{1,\dots,1}(z)}_{w} = \frac{1}{w!}(-\log(1-z))^w, \quad \underbrace{H_{0,\dots,0,1}(z)}_{(w-1)} = \text{Li}_w(z). \tag{2.13}$$

Furthermore, harmonic polylogarithms satisfy a scaling identity

$$H_{k\cdot a_1,\dots,k\cdot a_n}(k \cdot z) = H_{a_1,\dots,a_n}(z) \quad \text{for} \quad k \neq 0, z \neq 0, \tag{2.14}$$

which is valid whenever $a_n \neq 0$.

Given that the usual logarithm has a branch cut which is canonically chosen to lie along the negative real axis, its iterated and integrated versions have branch cuts, as well. However, it is possible to obtain single-valued harmonic polylogarithms (SVHPLs) by linearly combining products of the form $H_{s_1}(z)H_{s_2}(\bar{z})$ in a way that all branch cuts cancel. The combinations of HPLs leading to SVHPLs are unique and can be determined from demanding triviality of the monodromies around singular points of HPLs [30].

The lowest-weight SVHPLs[2] read

$$\mathcal{L}_0(z) = H_0(z) + H_0(\bar{z})$$
$$\mathcal{L}_1(z) = H_1(z) + H_1(\bar{z})$$
$$\mathcal{L}_{0,0}(z) = H_{0,0}(z) + H_{0,0}(\bar{z}) + H_0(z)H_0(\bar{z})$$
$$\mathcal{L}_{1,0}(z) = H_{1,0}(z) + H_{0,1}(\bar{z}) + H_1(z)H_0(\bar{z})$$
$$\mathcal{L}_{1,0,1}(z) = H_{1,0,1}(z) + H_{1,0,1}(\bar{z}) + H_{1,0}(z)H_1(\bar{z}) + H_1(z)H_{1,0}(\bar{z})$$

$$\vdots \tag{2.15}$$

While up to weight three the expressions follow an obvious pattern, ζ-values make an appearance starting at weight four, for example

$$\mathcal{L}_{1,0,1,0}(z) = H_{1,0,1,0}(z) + H_{0,1,0,1}(\bar{z}) + H_{1,0,1}(z)H_0(\bar{z}) + H_1(z)H_{0,1,0}(\bar{z})$$
$$+ H_{1,0}(z)H_{01}(\bar{z}) - 4\zeta_3 H_1(\bar{z}). \tag{2.16}$$

An elaborate introduction to SVHPLs in the context of the six-point remainder function in MRK in which the method for solving the single-valuedness condition is carefully explained can be found in section 3 of ref. [18].

In the remainder of this subsection, let us collect several properties of SVHPLs which will be useful below: Two SVHPLs labeled by words s_1 and s_2 satisfy the shuffle relation

$$\mathcal{L}_{s_1}(z)\,\mathcal{L}_{s_2}(z) = \sum_{s \in s_1 \shuffle s_2} \mathcal{L}_s(z), \tag{2.17}$$

where the shuffle $s_1 \shuffle s_2$ refers to all permutations of $s_1 \cup s_2$ which leave the order of elements in s_1 and s_2 unaltered. The generating functional for the SVHPLs,

$$\mathcal{L}^{\{0,1\}}(z) = \sum_{s \in X(\{x_0, x_1\})} \mathcal{L}_s(z)s = 1 + \mathcal{L}_0(z)\,x_0 + \mathcal{L}_1(z)\,x_1 + \mathcal{L}_{0,0}(z)\,x_0 x_0$$
$$+ \mathcal{L}_{0,1}(z)\,x_0 x_1 + \dots, \tag{2.18}$$

where $X(\{x_0, x_1\})$ are all words[3] in the alphabet $\{x_0, x_1\}$. This generating functional satisfies the differential equations

$$\frac{\partial}{\partial z}\mathcal{L}^{\{0,1\}}(z) = \left(\frac{x_0}{z} + \frac{x_1}{1-z}\right)\mathcal{L}^{\{0,1\}}(z), \quad \frac{\partial}{\partial \bar{z}}\mathcal{L}^{\{0,1\}}(z) = \mathcal{L}^{\{0,1\}}(z)\left(\frac{y_0}{\bar{z}} + \frac{y_1}{1-\bar{z}}\right), \tag{2.19}$$

---

[2] As will be clear from the examples given in eq. (2.15), the $\mathcal{L}_s(z)$ are functions of both z and $\bar{z}$. For simplicity, however, we will denote these functions as $\mathcal{L}_s(z)$.

[3] For convenience, SVHPLs will be labeled by the indices of the letters rather than by the letters themselves.

where $\{y_0, y_1\}$ is an additional alphabet, which appears in the construction of SVHPLs in ref. [30] and is related by the single-valuedness condition to the alphabet $\{x_0, x_1\}$ mentioned above. Solving this condition order by order, one finds

$$y_0 = x_0 \quad \text{and}$$

$$y_1 = x_1 - \zeta_3(2x_0x_0x_1x_1 - 4x_0x_1x_0x_1 + 2x_0x_1x_1x_1 + 4x_1x_0x_1x_0 + \cdots) + \cdots, \quad (2.20)$$

and can thus find the analogue of eq. (2.15) for $\mathcal{L}_s$ for an arbitrary label s constructed from the alphabet $\{x_0, x_1\}$. Note, however, that from eq. (2.19) the $\mathcal{L}_s(z)$ satisfy the same differential equation in z as the corresponding HPL $H_s(z)$, cf. eq. (2.12).

2.3. Calculation of the six-point remainder function

As pointed out at the end of subsection 2.1, the problem of calculating the remainder function $R_6^{\text{MHV}}|_{\text{MRK}}$ via eq. (2.5) boils down to evaluating the real part of the sum over the integral, which will yield the functions $g_n^{(\ell)}$. The crucial ingredients here are the loop expansions of the impact factor $\Phi_{\text{reg}}^{\text{MHV}}(\nu, n)$ and the BFKL eigenvalue $\omega(\nu, n)$ in eq. (2.6).

To calculate $g_n^{(\ell)}$ one would then expand the integral to the desired loop and logarithmic order, close the contour at infinity and sum up the residues. This, however, becomes cumbersome already beyond the lowest loop order.

As discussed in ref. [18], the functions $g_n^{(\ell)}$ can be expressed in terms of SVHPLs. This opens a natural and simpler way for the calculation of eq. (2.5): one starts from an ansatz in SVHPLs and compares the series expansions in (w, w^*) of both the ansatz and the integral. Following this approach the remainder function was calculated up to five loops, as well as for higher loop orders in LLA and NLLA [18,2,19].

A more direct evaluation of the remainder function eq. (2.5) was developed in ref. [20]. The key insight, first used in refs. [31,19], is that the leading term of any SVHPL is simply given by the harmonic polylogarithm with the same index structure

$$\mathcal{L}_s(w, w^*) = H_s(w) + \ldots, \quad (2.21)$$

as exemplified in eq. (2.15). Importantly, the term $H_w(w)$ is the only term in the expansion of the SVHPL which does not depend on w^*. Comparing with the dispersion relation (2.5), we see that the leading terms are encoded in the residues at $\nu = -\frac{in}{2}$ as only for those poles the residues will have no contribution from w^*.

Since the remainder function can be expressed in terms of SVHPLs exclusively, a viable approach consists of simply determining the leading terms, which will be a linear combination of HPLs and obtaining the full result by simply promoting HPLs to SVHPLs via

$$H_s(w) \rightarrow \mathcal{L}_s(w, w^*). \quad (2.22)$$

In performing the above replacement, the contributions from the omitted residues are restored automatically.

In ref. [20], we started from this observation and identified recursion relations between different integrals which hold on the locus of the poles $\nu = -\frac{in}{2}$, but which by using eq. (2.22) lift to relations of the full result. Employing these relations, we reduced all integrals to a set of trivial basis integrals. This allowed us to efficiently evaluate the remainder function up to very high loop- and logarithmic orders and to prove Pennington's formula [32] for the six-point remainder function in LLA.

Given the success of this approach for the six-point remainder function, it is natural to ask, whether a similar formalism can be established for seven points. This idea is going to be discussed in the next section.

3. Seven-point remainder function in multi-Regge kinematics

3.1. From six to seven gluons

We now move on to the seven-point MHV remainder function in multi-Regge kinematics, which is defined similarly to the six-point case,

$$A_7^{\text{MHV}} = A_{\text{BDS}} e^{R_7^{\text{MHV}}}. \tag{3.1}$$

The kinematics in this case is governed by seven conformal cross ratios,

$$u_{1,1} = \frac{x_{37}^2 x_{46}^2}{x_{47}^2 x_{36}^2}, \qquad u_{2,1} = \frac{x_{15}^2 x_{24}^2}{x_{14}^2 x_{25}^2}, \qquad u_{3,1} = \frac{x_{35}^2 x_{26}^2}{x_{25}^2 x_{36}^2},$$

$$u_{1,2} = \frac{x_{14}^2 x_{57}^2}{x_{15}^2 x_{47}^2}, \qquad u_{2,2} = \frac{x_{16}^2 x_{25}^2}{x_{15}^2 x_{26}^2}, \qquad u_{3,2} = \frac{x_{36}^2 x_{27}^2}{x_{26}^2 x_{37}^2}, \tag{3.2}$$

$$\tilde{u} = \frac{x_{13}^2 x_{47}^2}{x_{37}^2 x_{14}^2}.$$

In the multi-Regge limit the cross ratios $u_{1,s}$, $s = 1, 2$, and $\tilde{u}$ approach 1, while all other cross ratios tend to zero. Due to a conformal Gram relation, only six of the above cross ratios are independent. It is, however, advantageous for what follows to keep all seven cross ratios explicitly. The remaining kinematic freedom in the multi-Regge limit is again given by the reduced cross ratios

$$\tilde{u}_{a,s} := \frac{u_{a,s}}{1 - u_{1,s}}, \tag{3.3}$$

which are finite in the multi-Regge limit and which we again parameterize by two complex variables w_1, w_2 defined as

$$\tilde{u}_{2,1} =: \frac{1}{|1 + w_1|^2}, \quad \tilde{u}_{3,1} =: \frac{|w_1|^2}{|1 + w_1|^2}, \quad \tilde{u}_{2,2} =: \frac{1}{|1 + w_2|^2}, \quad \tilde{u}_{3,2} =: \frac{|w_2|^2}{|1 + w_2|^2}. \tag{3.4}$$

A key difference to the six-point case is that several interesting Mandelstam regions exist, in which Regge cuts appear. However, as is shown in [8,9], the seven-point remainder function can be written as a linear combination of three elementary building blocks in every Regge region. These building blocks are usually called the *short cuts* which describe a Regge cut in the s_{45}- and s_{56}-channel, respectively, as well as the *long cut* which describes a Regge cut which spans the s_{456}-channel, see [8,9] for details and Fig. 1 for a pictorial representation. As it turns out, the short cuts are fully determined by the BFKL eigenvalue and impact factor of the six-point amplitude. Therefore, we focus on the long cut in which a new ingredient appears. To study the long cut, one can take the analytic continuation[4]

---

[4] Subtleties in the choice of path that arise due to the conformal Gram relation are, for example, discussed in ref. [12], but do not play a role here.

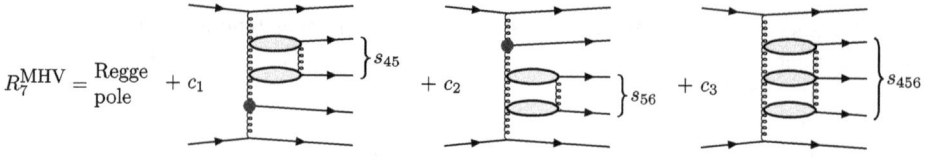

Fig. 1. In the seven-point case, the remainder function in every Regge region can be written as a linear combination of the Regge pole contribution, the *short cuts* in the s_{45}- and s_{56}-channel and the *long cut* in the combined s_{456}-channel. See [8,9] for details.

$$\tilde{u} \to e^{-2\pi i}\,\tilde{u} \tag{3.5}$$

of the remainder function, before going to the multi-Regge limit. In this Mandelstam region, the short cuts do not contribute and the remainder function is fully determined by the long cut. In LLA, the remainder function in this region was stated in ref. [9] and reads

$$R_7^{\text{MHV}} = 1 + i\pi \sum_{\ell=2}^{\infty} \sum_{k=0}^{\ell-1} \frac{a^\ell}{k!(\ell-1-k)!} \log^k(1-u_{1,1})$$
$$\times \log^{\ell-1-k}(1-u_{1,2}) \mathcal{I}_7\left[E_{\nu,n}^k E_{\mu,m}^{\ell-k-1}\right], \tag{3.6}$$

where we define

$$\mathcal{I}_7\left[\mathcal{F}(\nu,n,\mu,m)\right] := \sum_{n=-\infty}^{\infty} \sum_{m=-\infty}^{\infty} \int_{-\infty}^{+\infty} \frac{d\nu}{2\pi} \int_{-\infty}^{+\infty} \frac{d\mu}{2\pi} w_1^{i\nu+n/2}(w_1^*)^{i\nu-n/2}$$
$$\times w_2^{i\mu+m/2}(w_2^*)^{i\mu-m/2} C(\nu,n,\mu,m)\,\mathcal{F}(\nu,n,\mu,m), \tag{3.7}$$

and where

$$C(\nu,n,\mu,m) = (-1)^{n+m} \frac{\Gamma\left(-i\nu-\frac{n}{2}\right)\Gamma\left(i\mu+\frac{m}{2}\right)\Gamma\left(i(\nu-\mu)+\frac{m-n}{2}\right)}{\Gamma\left(1+i\nu-\frac{n}{2}\right)\Gamma\left(1-i\mu+\frac{m}{2}\right)\Gamma\left(1-i(\nu-\mu)+\frac{m-n}{2}\right)} \tag{3.8}$$

is the so-called central emission vertex. Comparing expression (3.7) with the corresponding equation for the six-point case (2.5), we see that this is a new ingredient. Like in the six-point case there is again a nice physical interpretation of all the terms in eqs. (3.7) and (3.6), with the central emission vertex $C(\nu,n,\mu,m)$ describing the emission of a physical gluon from a bound state of two reggeized gluons and the BFKL eigenvalue $E_{\nu,n}$ describing the evolution of the bound state of two reggeized gluons, of which we have two because of the appearance of the central emission vertex. The impact factor which describes the coupling of the bound state to the physical gluons does not appear in eq. (3.7) since it is trivial in LLA. We can represent eq. (3.6) graphically as shown in Fig. 2. Note that eq. (3.6) has a similar form to eq. (2.9), only with two distinct large logarithms. Furthermore, as always in LLA, the real part vanishes.

Let us now study the symmetry properties of eq. (3.7). Using the expressions for the LLA BFKL eigenvalue eq. (2.7) and the central emission vertex eq. (3.8) we see that the remainder function is invariant under exchange of

$$\nu \leftrightarrow -\mu,\ n \leftrightarrow -m, \tag{3.9}$$

which corresponds to a swap

$$w_1 \leftrightarrow \frac{1}{w_2}, \tag{3.10}$$

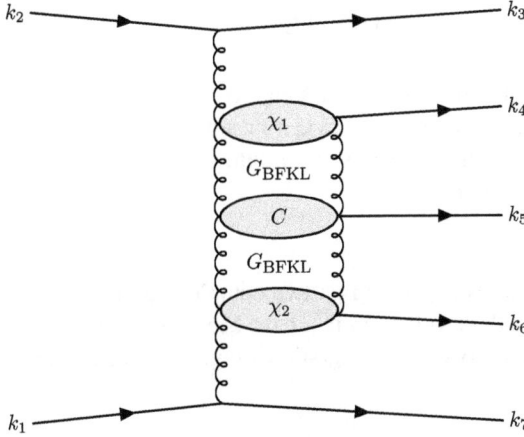

Fig. 2. Pictorial representation of eqs. (3.6) and (3.7), with C being the central emission vertex, G_{BFKL} describing the evolution of the two-Reggeon bound state and χ_1, χ_2 being the building blocks of the impact factor.

as well as under exchange of

$$n \leftrightarrow -n, \ m \leftrightarrow -m \tag{3.11}$$

which, in turn, corresponds to

$$w_1 \leftrightarrow w_1^*, \ w_2 \leftrightarrow w_2^*. \tag{3.12}$$

These symmetry properties will be very useful when evaluating the remainder function later on.

To evaluate eq. (3.7), one first closes the contours of the two integrals at infinity and then sums up the residues. A convenient choice is to close the contour of the μ-integration in the upper half-plane and the contour of the ν-integration in the lower half-plane. This corresponds to the choice $w_1 < 1$ and $w_2 > 1$, which is compatible with the symmetry $w_1 \leftrightarrow \frac{1}{w_2}$. In [10], this calculation was carried out for the integrals appearing at two loops, with the result

$$\mathcal{I}_7 \left[E_{\nu,n} \right] = \frac{1}{2} \left(\log \left| 1 + \frac{w_1}{1 + \frac{1}{w_2}} \right|^2 + \log \left| 1 + \frac{1 + \frac{1}{w_2}}{w_1} \right|^2 \right), \tag{3.13}$$

which takes the form of the six-point two-loop result with a rescaled variable, $w \to \frac{w_1}{1 + \frac{1}{w_2}}$. From this result, we can also immediately obtain $\mathcal{I}_7 \left[E_{\mu,m} \right]$ by making use of the symmetry (3.9) discussed before. Indeed, a special class of integrals is obtained when only one of the two energy eigenvalues $E_{\nu,n}$ or $E_{\mu,m}$ appears in the integrand, for definiteness let us choose $E_{\nu,n}$. In this case, the μ-integration can be carried out explicitly and results in a rescaling of the parameters (w_1, w_1^*). This reduces the integral to a six-point integral, that is, upon replacing

$$w \to \frac{w_1}{1 + \frac{1}{w_2}} \tag{3.14}$$

one can effectively obtain the simple two-loop solution eq. (3.13) from the corresponding two-loop LLA integral of eq. (2.5). Starting from three loops, integrals with both types of energy eigenvalues $E_{\nu,n}$ and $E_{\mu,m}$ appear causing a more complicated result. We therefore have to resort to other means of solving the integral. As in the six-point case, a sensible starting point is trying to understand the relevant functions describing the remainder function.

3.2. Harmonic polylogarithms in two variables

Contrary to the situation in the six-point scenario, where the kinematics is determined by one complex parameter w, in the seven-point case we need to find SVHPLs in two complex variables w_1 and w_2 (cf. eq. (3.13)). In the discussion to follow we will use variables y and z, which will be related to w_1 and w_2 later on.

3.2.1. Two new letters

Harmonic Polylogarithms depending on two complex parameters – or two-dimensional harmonic polylogarithms (2dHPLs) for short[5] – have been constructed in ref. [33]. The implementation relies on introducing two new integration weights accompanying the weights f_a defined in eq. (2.11),

$$f_z = \frac{1}{t+z} \quad \text{and} \quad f_{1-z} = \frac{1}{1-t-z}. \tag{3.15}$$

Similar to the six-point scenario, where the function f_{-1} does not appear, the function f_z is not needed for the seven-point remainder function in MRK. In accordance, we will introduce the additional letter x_{1-z} only.

Up to weight three, all 2dHPLs can be expressed in terms of generalized polylogarithms. For the simplest cases the relations read:

$$H_{1-z}(y) = H_1\left(\frac{y}{1-z}\right) = -\log\left(1 - \frac{y}{1-z}\right),$$

$$H_{1,1-z}(y) = \frac{1}{2}\log^2(1-y) - \log(1-y)\log(1-z) + \text{Li}_2\left(\frac{z}{1-y}\right) - \text{Li}_2(z), \tag{3.16}$$

where a complete set of relations leading to expressions for all possible labels up to weight three is presented in appendix A.2 of ref. [33].

While the labels y and z appear to be on unequal footing in the above formulæ, this is actually a choice of notation only: there are numerous relations between different representations of 2dHPLs. In particular, it is always possible to switch y and z in label and argument of a 2dHPL, for example:

$$H_{0,1,1-y}(z) = H_1(z)H_{0,1}(y) - H_1(y)H_{0,1}(z) - H_{0,1,1}(y) + H_{0,1,1}(z) + H_{0,1,1-z}(y). \tag{3.17}$$

This type of relation, which can be easily derived for every label by reverting to the integral representation of 2dHPls, will be a crucial ingredient in fixing ζ-terms in 2dSVHPLs below.

In order to have a canonical representation, we will choose 2dHPLs with labels from $\{0, 1, 1 - z\}$ for the argument y and 1dHPLs with labels from $\{0, 1\}$ for the argument z. Solving for shuffle relations by choosing a Lyndon basis [34] for the labels we will finally use

$$H_{\text{Lyndon}(\{0,1,1-z\})}(y) \quad \text{and} \quad H_{\text{Lyndon}(\{0,1\})}(z). \tag{3.18}$$

As pointed out in ref. [13], this choice is actually a basis for the 2dHPLs appearing in the seven-point remainder function in MRK.

---

[5] In the following sections we will sometimes refer to the HPLs and SVHPLs discussed in subsection 2.2 as 1dHPLs and 1dSVHPLs, respectively.

3.2.2. Differential equations for 2dHPLs

With a second complex parameter entering the definition of 2dHPLs, an obvious question is the one about the differential behavior of those functions. While one could use relations like eq. (3.17) and thus trace back derivatives with respect to one variable appearing in the label of the polylogarithm to a derivative with respect to the argument, it is far more efficient to consider derivatives with respect to the labels of a 2dHPL directly. The necessary formulæ for taking those derivatives are listed and explained in Appendix A.

In terms of a generating function of 2dHPLs[6] with argument y

$$H^{\{0,1,1-z\}}(y) = \sum_{s \in X(\{x_0, x_1, x_{1-z}\})} H_s(y)s$$

$$= 1 + H_0(y) x_0 + H_1(y) x_1 + H_{1-z}(y) x_{1-z}$$
$$+ H_{0,0}(y) x_0 x_0 + H_{0,1}(y) x_0 x_1 + H_{0,1-z}(y) x_0 x_{1-z}$$
$$+ H_{1,0}(y) x_1 x_0 + H_{1,1}(y) x_1 x_1 + H_{1,1-z}(y) x_1 x_{1-z}$$
$$+ H_{1-z,0}(y) x_{1-z} x_0 + H_{1-z,1}(y) x_{1-z} x_1 + H_{1-z,1-z}(y) x_{1-z} x_{1-z}$$
$$+ \ldots, \tag{3.19}$$

the y-derivative can be written down immediately after considering the defining equation (2.10) together with the additional integration weight f_{1-z}:

$$\frac{\partial}{\partial y} H^{\{0,1,1-z\}}(y) = \left(\frac{x_0}{y} + \frac{x_1}{1-y} + \frac{x_{1-z}}{1-y-z} \right) H^{\{0,1,1-z\}}(y). \tag{3.20}$$

Proceeding to the derivative with respect to z, it is no longer possible to write the derivative in multiplicative form as in eq. (3.20). Instead, one can describe the pattern of how letters and prefactors are attached to existing words depending on their particular letters. Writing

$$\frac{\partial}{\partial z} H^{\{0,1,1-z\}}(y) = \Xi\left(H^{\{0,1,1-z\}}(y) \right) \tag{3.21}$$

the operation Ξ acts as follows:

- for each sequence of letters $s = x_{1-z} \ldots x_{1-z}$, promote

$$s \quad \rightarrow \quad \frac{x_0 s}{1-z} + \frac{x_1 s}{z} - \frac{s x_0}{1-z} - \frac{s x_1}{z} \tag{3.22a}$$

- for each sequence of letters $s = x_1 \ldots x_1$, promote

$$s \quad \rightarrow \quad \frac{x_{1-z} s}{z(1-z)} - \frac{s x_{1-z}}{z(1-z)} \tag{3.22b}$$

- to any complete word s, add a leading $1-z$:

$$s \quad \rightarrow \quad \frac{y}{(1-z)(1-y-z)} x_{1-z} s. \tag{3.22c}$$

---

[6] Again, instead of noting the full word s in the subscript of a 2dHPL, we just write the indices of the letters.

In order to extract the derivative of a particular 2dHPL with label s one expands both sides of eq. (3.21) and compares the coefficients of the word s.

In practice, aiming to find the z-derivative of $H_{0,1-z,1}(y)$ for example, one has to browse through all words of length two, which upon adding one letter following the rules eq. (3.22) will yield the word $x_0 x_{1-z} x_1$. Starting from $H_{0,1}(y)x_0 x_1$, one can insert a letter x_{1-z} between x_0 and x_1, making use of the first part of rule (3.22b). Taking $H_{0,1-z}(y)x_0 x_{1-z}$, a letter x_1 can be appended to the right, which amounts to using the last part of rule (3.22a). Finally, the word $x_0 x_{1-z} x_1$ can be reached by prepending x_0 to the word $x_{1-z} x_1$ accompanying $H_{1-z,1}(y)$, thus using the first term in rule (3.22a):

$$\frac{\partial}{\partial z} H_{0,1-z,1}(y)x_0 x_{1-z} x_1 = \frac{H_{0,1}(y)}{z(1-z)} x_0 x_{1-z} x_1 - \frac{H_{0,1-z}(y)}{z} x_0 x_{1-z} x_1 + \frac{H_{1-z,1}(y)}{1-z} x_0 x_{1-z} x_1 .$$

(3.23)

Another example, where one has to make use of rules (3.22a) and (3.22c) reads:

$$\frac{\partial}{\partial z} H_{1-z,0,0}(y)x_{1-z} x_0 x_0 = \frac{y}{(1-z)(1-y-z)} H_{0,0}(y)\, x_{1-z} x_0 x_0 - \frac{1}{1-z} H_{1-z,0}(y)x_{1-z} x_0 x_0 .$$

(3.24)

3.3. Single-valued harmonic polylogarithms in two variables

The canonical way to identify single-valued versions of 2dHPLs would be to find a generalization of the single-valuedness condition formulated in ref. [30] for the alphabet $\{x_0, x_1\}$. However, although this generalization does most certainly exist, an explicit expression thereof is currently not known to us.[7] Therefore we continue on a different path: we postulate several constraints the single-valued versions of 2dHPLs should satisfy and later on argue that the functions thus constructed are indeed single-valued. In order to find those constraints, we take guidance by the properties of 1dHPLs reviewed in subsection 2.2:

- **Differential equations:** 1dSVHPLs satisfy the same differential equations as their 1dHPL counterpart (cf. eqs. (2.12) and (2.19)): therefore we require the generating functional of 2dSVHPLs with argument y

$$\mathcal{L}^{\{0,1,1-z\}}(y) = \sum_{s \in X(\{x_0, x_1, x_{1-z}\})} \mathcal{L}_s(y)s$$

$$= 1 + \mathcal{L}_0(y)\, x_0 + \mathcal{L}_1(y)\, x_1 + \mathcal{L}_{1-z}(y)\, x_{1-z}$$
$$+ \mathcal{L}_{0,0}(y)\, x_0 x_0 + \mathcal{L}_{0,1}(y)\, x_0 x_1 + \mathcal{L}_{0,1-z}(y)\, x_0 x_{1-z}$$
$$+ \mathcal{L}_{1,0}(y)\, x_1 x_0 + \mathcal{L}_{1,1}(y)\, x_1 x_1 + \mathcal{L}_{1,1-z}(y)\, x_1 x_{1-z}$$
$$+ \mathcal{L}_{1-z,0}(y)\, x_{1-z} x_0 + \mathcal{L}_{1-z,1}(y)\, x_{1-z} x_1 + \mathcal{L}_{1-z,1-z}(y)\, x_{1-z} x_{1-z}$$
$$+ \ldots ,$$

(3.25)

to satisfy

$$\frac{\partial}{\partial y}\mathcal{L}^{\{0,1,1-z\}}(y) = \left(\frac{x_0}{y} + \frac{x_1}{1-y} + \frac{x_{1-z}}{1-y-z}\right)\mathcal{L}^{\{0,1,1-z\}}(y)$$

(3.26)

---

[7] As pointed out in the conclusions, the paper [35] provides an explicit construction of those functions.

$$\frac{\partial}{\partial z}\mathcal{L}^{\{0,1,1-z\}}(y) = \Xi\left(\mathcal{L}^{\{0,1,1-z\}}(y)\right). \tag{3.27}$$

where the operation Ξ has been defined in eq. (3.22).

- **Limiting behavior:** In the limit $z \to 0$, the alphabet will shrink since $x_{1-z} \to x_1$. Thus we demand to recover the corresponding 1dSVHPLs in the limit $(1-z) \to 1$.
- **Switching variables:** Consistency with the relations switching the variable in the label and the argument. This means we require all relations like eq. (3.17) to hold upon replacing $H \to \mathcal{L}$.

In short, we require 2dSVHPLs to inherit the properties of 2dHPLs and in addition we demand a consistent reduction to 1dSVHPLs in the limit $z \to 0$. We will explicitly show in the following that those constraints are indeed sufficient to pin down 2dSVHPLs in two variables to at least weight four.

In practice, we can gain some experience regarding the structure of the 2dSVHPLs by studying a simple ad hoc construction: We start from the 2dHPLs as given by Gehrmann and Remiddi and reviewed in subsection 3.2. Since these functions can be expressed in terms of generalized polylogarithms up to weight three, we can promote each 1dHPL separately to its single-valued version using relations like eq. (2.15).

This is most easily explained using an example: a candidate for a single-valued 2dHPL can be obtained via

$$H_{1,1-z}(y) = \frac{1}{2}\log^2(1-y) - \log(1-y)\log(1-z) + \text{Li}_2\left(\frac{z}{1-y}\right) - \text{Li}_2(z)$$

$$= H_{1,1}(y) - H_1(y)H_1(z) + H_{0,1}\left(\frac{z}{1-y}\right) - H_{0,1}(z)$$

$$\to \mathcal{L}_{1,1}(y) - \mathcal{L}_1(y)\mathcal{L}_1(z) + \mathcal{L}_{0,1}\left(\frac{z}{1-y}\right) - \mathcal{L}_{0,1}(z). \tag{3.28}$$

Making use of the scaling relation

$$H_{a_1,a_2,\ldots,a_n}\left(\frac{y}{1-z}\right) = H_{(1-z)a_1,(1-z)a_2,\ldots,(1-z)a_n}(y), \tag{3.29}$$

where $a_i \in \{0,1\}$ and which holds whenever the last index is not 0 (see the discussion around eq. (2.14)), as well as

$$H_{\underbrace{0,\ldots,0}_{n}}\left(\frac{y}{1-z}\right) = \frac{1}{n!}\left(H_0(y) + H_1(z)\right)^n, \tag{3.30}$$

one can express the above candidate for a single-valued polylogarithm in terms of 2dHPLs:

$$\mathcal{L}_{1,1-z}(y) = H_{1,1-z}(y) + H_1(y)H_{1-\bar{z}}(\bar{y}) + H_{1-\bar{z},1}(\bar{y}) + \mathcal{L}_0(z)\left(H_{1-\bar{z}}(\bar{y}) - H_1(\bar{y})\right)$$

$$- \mathcal{L}_1(z)H_1(\bar{y}). \tag{3.31}$$

Note that in eq. (3.31) the variable z in the label is complex-conjugated whenever the argument of the HPL is $\bar{y}$. Following this ad hoc approach we find functions up to weight three which perfectly match the first orders of the integral eq. (3.6). Furthermore, note that we did not express 1dSVHPLs of z in terms of usual HPLs, as this relates to a feature of 2dSVHPLs to be elaborated on below: if expressed in the basis eq. (3.18), 2dSVHPLs split into a *canonical* part as well as a part in which 1dSVHPLs of argument z are multiplied by 2dHPLs of arguments y and $\bar{y}$ (with labels possibly containing $1-z$ and $1-\bar{z}$). By *canonical* we refer to the pattern

$$\mathcal{L}_{a_1,a_2,\ldots,a_n}(y)\big|_{\text{can.}} = \sum_{k=0}^{n} H_{a_1,\ldots,a_k}(y) H_{\bar{a}_n,\bar{a}_{n-1},\ldots,\bar{a}_{k+1}}(\bar{y}) \tag{3.32}$$

that is already present for 1dSVHPLs (see eq. (2.15)). Quite naturally, the additional terms one finds beyond the canonical part are exactly those needed to preserve the derivative rule eq. (3.27).

Since our ad hoc construction of 2dSVHPLs only works up to weight three, we would now like to turn these observations into a construction of higher-weight 2dSVHPLs by the following algorithm: We start from a known 2dSVHPL and add a letter to the left by integrating in y, thus making use of the differential equation (3.26). This fixes the 2dSVHPL of higher weight up to a function of $\bar{y}$, z and $\bar{z}$.

Based on the assumptions that HPLs of argument z only appear in single-valued combinations we make the most general ansatz of terms

$$\mathcal{L}_{a_1,\ldots,a_n}(z) H_{b_1,\ldots,b_m}(\bar{y}), \tag{3.33}$$

where $a_i \in \{0, 1\}$ and $b_i \in \{0, 1, 1-\bar{z}\}$, compatible with the overall weight. Demanding that this ansatz satisfies relations eq. (3.27) for differentiation in z then fixes the ansatz completely.

To clarify the procedure, let us consider an example. Starting from the obvious weight one expressions

$$\mathcal{L}_0(y) = H_0(y) + H_0(\bar{y}),$$
$$\mathcal{L}_1(y) = H_1(y) + H_1(\bar{y}),$$
$$\mathcal{L}_{1-z}(y) = H_{1-z}(y) + H_{1-\bar{z}}(\bar{y}), \tag{3.34}$$

we can, for example, write down an ansatz for $\mathcal{L}_{1,1-z}(y)$ as

$$\begin{aligned}
\mathcal{L}_{1,1-z}(y) = {} & H_{1,1-z}(y) + H_{1-\bar{z},1}(\bar{y}) + H_1(y) H_{1-\bar{z}}(\bar{y}) \\
& + \mathcal{L}_0(z)\left(c_1 H_0(\bar{y}) + c_2 H_1(\bar{y}) + c_3 H_{1-\bar{z}}(\bar{y})\right) \\
& + \mathcal{L}_1(z)\left(c_4 H_0(\bar{y}) + c_5 H_1(\bar{y}) + c_6 H_{1-\bar{z}}(\bar{y})\right), \tag{3.35}
\end{aligned}$$

which by differentiation with respect to z is then fixed to give eq. (3.31). Carrying this out for all functions up to weight three, we can compare the results of this algorithm with our ad hoc construction and find a perfect matching, as we should. Additional complications start from weight four where ζ-values are going to appear. As the 2dSVHPLs at weight three do not contain any zetas, those terms cannot be fixed by the y- and z-derivative and we have to add a term

$$\zeta_3\left(c_1 H_1(\bar{y}) + c_2 H_{1-\bar{z}}(\bar{y})\right) \tag{3.36}$$

to the ansatz of every 2dSVHPLs at weight four. Note that the two terms above are the only ones consistent with the reduction $z \to 0$, as well as vanishing in the limit $\bar{y} \to 0$. Demanding a consistent reduction in the limit $z \to 0$ and consistency with the relations exchanging argument and label fixes most of the ζ-terms, but not all. However, we can impose an additional constraint: as shown in [33], a 2dHPL evaluated at $y = 1 - z$ can be written as a combination of 1dHPLs of argument z. Similarly, setting $y = 1-z$, $\bar{y} = 1-\bar{z}$ in our ansatz, we expect to obtain a combination of 1dSVHPLs of weight four. As it turns out, this constraint is strong enough to fix all coefficients. As an example containing a ζ-value, we find

$$\begin{aligned}
\mathcal{L}_{0,0,1,1-z}(y) = {} & \mathcal{L}_{0,0,1,1-z}(y)\big|_{\text{can.}} + \mathcal{L}_{0,0,1}(z) H_{1-\bar{z}}(\bar{y}) \\
& + \mathcal{L}_{0,1}(z)\left(H_0(y) H_{1-\bar{z}}(\bar{y}) + H_{1-\bar{z},0}(\bar{y})\right)
\end{aligned}$$

$$+ \mathcal{L}_1(z)\Big(- H_{0,0}(y)H_1(\bar{y}) - H_0(y)H_{1,0}(\bar{y}) - H_{1,0,0}(\bar{y}) \Big)$$

$$+ \mathcal{L}_0(z)\Big(- H_{0,0}(y)H_1(\bar{y}) + H_{0,0}(y)H_{1-\bar{z}}(\bar{y}) - H_0(y)H_{1,0}(\bar{y})$$

$$+ H_0(y)H_{1-\bar{z},0}(\bar{y}) - H_{1,0,0}(\bar{y}) + H_{1-\bar{z},0,0}(\bar{y}) \Big) - 2\zeta_3 H_{1-\bar{z}}(\bar{y}). \tag{3.37}$$

The expressions for all other Lyndon basis elements up to weight four can be found in the file attached to the arXiv submission of this paper.

Going on to weight five, our constraints do not seem to be strong enough to fix all coefficients, which is due to the growth of both the number of Lyndon basis elements and the larger number of terms appearing in the ansatz for the ζ-terms, i.e. the analogue of eq. (3.36) at weight five. We are only able to fully fix those 2dSVHPLs at weight five whose label contains two different indices only.[8] The expressions for the weight-five 2dSVHPLs can also be found in the file attached to the arXiv submission, but note that those still contain fudge factors. It would be interesting to see if there are additional constraints which allow to completely fix the functions at weight five as well.

Up to weight three it is obvious from our ad hoc construction that the resulting functions are single-valued: they are composed from single-valued components by definition. For higher weights we can only argue that this is indeed the case: Starting from our ansatz and fixing all fudge coefficients does not only reproduce the 2dSVHPLs constructed naïvely, but yields functions, which including their ζ-parts perfectly match the analytical properties of the integral eq. (3.6). While this does not prove single-valuedness, the perfect matching with the explicit calculation of the integral strongly supports our conjecture.

3.4. Matching the results

Now that we have constructed a suitable class of functions, we want to generate expressions for the seven-point remainder function eq. (3.6) beyond two loops. We do this by simply writing down an ansatz with the correct weight and matching the series expansion of the ansatz to data generated from calculating residues of eq. (3.6). This also allows us to identify the variables in argument and label and we find that

$$(y, z) \rightarrow \left(-w_1, -\frac{1}{w_2} \right) \tag{3.38}$$

is the correct choice. In the following we will use the abbreviation $x := \frac{1}{w_2}$.

This leads to the following results at two loops,

$$\mathcal{I}_7\left[E_{v,n} \right] = \frac{1}{2}\mathcal{L}_1(-x)\mathcal{L}_{1+x}(-w_1) + \frac{1}{2}\mathcal{L}_{0,1+x}(-w_1) + \frac{1}{2}\mathcal{L}_{1+x,0}(-w_1)$$

$$+ \mathcal{L}_{1+x,1+x}(-w_1),$$

$$\mathcal{I}_7\left[E_{\mu,m} \right] = -\frac{1}{2}\mathcal{L}_1(-w_1)\mathcal{L}_0(-x) - \frac{1}{2}\mathcal{L}_1(-w_1)\mathcal{L}_1(-x) + \frac{1}{2}\mathcal{L}_0(-x)\mathcal{L}_{1+x}(-w_1)$$

$$+ \mathcal{L}_1(-x)\mathcal{L}_{1+x}(-w_1) - \frac{1}{2}\mathcal{L}_{1,1+x}(-w_1) - \frac{1}{2}\mathcal{L}_{1+x,1}(-w_1)$$

$$+ \mathcal{L}_{1+x,1+x}(-w_1) + \frac{1}{2}\mathcal{L}_{0,1}(-x) + \frac{1}{2}\mathcal{L}_{1,0}(-x) + \mathcal{L}_{1,1}(-x).$$

---

[8] This is not surprising, as those 2dSVHPLs can be constructed from 1dSVHPLs using the rescaling identity.

Note again that $\mathcal{I}_7\left[E_{\mu,m}\right]$ can be obtained from $\mathcal{I}_7\left[E_{\nu,n}\right]$ by using the symmetry (3.9) as well as the relations (3.17). At three loops we find

$$
\begin{aligned}
\mathcal{I}_7\left[E_{\nu,n}^2\right] = {} & \frac{1}{2}\mathcal{L}_1(-x)\mathcal{L}_{0,1+x}(-w_1) + \frac{1}{4}\mathcal{L}_{1,1}(-x)\mathcal{L}_{1+x}(-w_1) + \frac{1}{4}\mathcal{L}_1(-x)\mathcal{L}_{1+x,0}(-w_1) \\
& + \mathcal{L}_1(-x)\mathcal{L}_{1+x,1+x}(-w_1) + \frac{1}{4}\mathcal{L}_{0,0,1+x}(-w_1) + \frac{1}{2}\mathcal{L}_{0,1+x,0}(-w_1) \\
& + \mathcal{L}_{0,1+x,1+x}(-w_1) + \frac{1}{4}\mathcal{L}_{1+x,0,0}(-w_1) + \mathcal{L}_{1+x,0,1+x}(-w_1) \\
& + \mathcal{L}_{1+x,1+x,0}(-w_1) + 2\mathcal{L}_{1+x,1+x,1+x}(-w_1)
\end{aligned}
$$

as well as

$$
\begin{aligned}
\mathcal{I}_7 & \left[E_{\nu,n}\,E_{\mu,m}\right] \\
= {} & -\frac{1}{4}\mathcal{L}_{0,1}(-w_1)\mathcal{L}_0(-x) - \frac{1}{4}\mathcal{L}_{0,1}(-w_1)\mathcal{L}_1(-x) - \frac{3}{4}\mathcal{L}_{0,1}(-w_1)\mathcal{L}_{1+x}(-w_1) \\
& - \frac{1}{4}\mathcal{L}_1(-w_1)\mathcal{L}_{0,1}(-x) + \frac{1}{2}\mathcal{L}_{0,1}(-x)\mathcal{L}_{1+x}(-w_1) + \frac{1}{4}\mathcal{L}_0(-x)\mathcal{L}_{0,1+x}(-w_1) \\
& + \frac{1}{2}\mathcal{L}_1(-x)\mathcal{L}_{0,1+x}(-w_1) - \frac{1}{4}\mathcal{L}_{1,0}(-w_1)\mathcal{L}_0(-x) - \frac{1}{4}\mathcal{L}_{1,0}(-w_1)\mathcal{L}_1(-x) \\
& - \frac{3}{4}\mathcal{L}_{1,0}(-w_1)\mathcal{L}_{1+x}(-w_1) + \frac{1}{4}\mathcal{L}_{1,0}(-x)\mathcal{L}_{1+x}(-w_1) - \frac{1}{2}\mathcal{L}_{1,1}(-w_1)\mathcal{L}_0(-x) \\
& - \frac{1}{2}\mathcal{L}_{1,1}(-w_1)\mathcal{L}_1(-x) - \frac{1}{4}\mathcal{L}_1(-w_1)\mathcal{L}_{1,1}(-x) + \mathcal{L}_{1,1}(-x)\mathcal{L}_{1+x}(-w_1) \\
& - \frac{1}{4}\mathcal{L}_1(-x)\mathcal{L}_{1,1+x}(-w_1) + \frac{1}{4}\mathcal{L}_0(-x)\mathcal{L}_{1+x,0}(-w_1) + \frac{1}{2}\mathcal{L}_1(-x)\mathcal{L}_{1+x,0}(-w_1) \\
& + \frac{1}{2}\mathcal{L}_0(-x)\mathcal{L}_{1+x,1+x}(-w_1) + \frac{3}{2}\mathcal{L}_1(-x)\mathcal{L}_{1+x,1+x}(-w_1) + \frac{1}{2}\mathcal{L}_{0,1,1+x}(-w_1) \\
& + \frac{1}{2}\mathcal{L}_{0,1+x,1}(-w_1) + \frac{1}{2}\mathcal{L}_{0,1+x,1+x}(-w_1) + \frac{1}{4}\mathcal{L}_{1,0,1+x}(-w_1) - \frac{1}{2}\mathcal{L}_{1,1,1+x}(-w_1) \\
& + \frac{1}{2}\mathcal{L}_{1,1+x,0}(-w_1) - \frac{1}{2}\mathcal{L}_{1,1+x,1+x}(-w_1) + \frac{1}{4}\mathcal{L}_{1+x,0,1}(-w_1) + \mathcal{L}_{1+x,0,1+x}(-w_1) \\
& + \frac{1}{2}\mathcal{L}_{1+x,1,0}(-w_1) - \frac{1}{2}\mathcal{L}_{1+x,1,1}(-w_1) + \frac{1}{2}\mathcal{L}_{1+x,1+x,0}(-w_1) - \frac{1}{2}\mathcal{L}_{1+x,1+x,1}(-w_1) \\
& + 2\mathcal{L}_{1+x,1+x,1+x}(-w_1).
\end{aligned}
$$

As explained before, the remaining integral at three loops, $\mathcal{I}_7\left[E_{\mu,m}^2\right]$, can be obtained by symmetry. All further results up to five loops are too lengthy to be reproduced here and we refer the reader to the file accompanying the arXiv submission of this paper.

Before we close let us make one additional but important remark. Recall that the key property of the one-dimensional SVHPLs that allowed us to directly construct the full result from a small set of residues in the six-point was that the leading term of the 1dSVHPL was a HPL with the same index structure which only depends on w. However, comparing with the explicit expressions of our 2dSVHPLs, e.g. eq. (3.31), we see that a similar statement holds – the leading term of a given 2dSVHPL is the 2dHPL with the same index structure which only depends on y and z but not on the complex-conjugated variables. This means that we should be able to construct the full result solely from the residues at

$$\nu = -\frac{i\,n}{2}, \quad \text{and} \quad \mu = \frac{i\,m}{2}, \tag{3.39}$$

which leads to an expression in 2dHPLs, and then making the replacement

$$H_w(y) \to \mathcal{L}_w(y). \tag{3.40}$$

We have checked that this indeed reproduces the full result. This in turn means that it should be possible to follow a procedure similar to [20] to reconstruct the remainder function from a set of simple basis integrals.

It is because of this remark that we present the formula for the five-loop remainder function in terms of 2dSVHPLs in the attached file, even though we cannot fully fix the ζ-parts of all those functions yet. We have obtained the full result in 2dHPLs and checked that the prescription eq. (3.40) works for the ζ-free part. Furthermore, the integrals contributing to the five-loop remainder function which only contain one kind of energy eigenvalue,

$$\mathcal{I}_7\left[E_{\nu,n}^4\right], \, \mathcal{I}_7\left[E_{\mu,m}^4\right] \tag{3.41}$$

only contain 2dSVHPLs of weight five with two different indices, which we fully understand. Here, too, the prescription eq. (3.40) works. We are therefore convinced that the formula as written down in the file attached to the arXiv submission is correct.

4. Conclusions

Setting up an efficient approach for calculating the MHV remainder function for seven points in $\mathcal{N} = 4$ super-Yang–Mills requires the construction of single-valued harmonic polylogarithms in two variables, 2dSVHPLs. In this paper, we have started the investigation of their analytical properties and constructed those functions up to and including weight four. The analytical constraints we are using, however, are not strong enough for completely determining the single-valued version of 2dHPLs starting from weight five, since we cannot fix the coefficients of all terms proportional to zeta values.

Upon availability of expressions for 2dHPLs, it is possible to apply a similar concept as the one introduced in ref. [19]: by calculating a certain subset of the residues contributing to the seven-point MHV remainder function only, one can determine the leading term of the 2dSVH-PLs, which later on can be promoted to the full single-valued expression.

Using this method, we have expressed the remainder function in terms of 2dSVHPLs up to five loops. It would be interesting to see whether there are further constraints on the 2dSVHPLs which would allow us to go beyond weight five.

While we provide an ad-hoc construction of the 2dSVHPLs, in a recent paper [35] an explicit construction of those functions is provided to arbitrary weight in a different language. Naturally, it would be interesting to compare our results to the more general approach in ref. [35].

Furthermore, the pattern we find is simple enough to suggest that it should be possible to identify a formula similar to the formula derived in ref. [32] for the LLA of the six-point remainder function.

Acknowledgements

We would like to thank Claude Duhr, Falko Dulat, Matteo Rosso and Till Bargheer for helpful discussions. The research of JB is supported in part by the SFB 647 "Raum–Zeit–Materie. Analytische und Geometrische Strukturen" of the German Research Foundation and the Marie

Curie Network GATIS (gatis.desy.eu) of the European Union's Seventh Framework Programme FP7-2007-2013 under grant agreement No. 317089. The work of MS is partially supported by the Swiss National Science Foundation through the NCCR SwissMAP.

Appendix A. Derivatives of harmonic polylogarithms

Derivatives of HPLs can be most easily stated in general form using the language of Goncharov polylogarithms. Given their integral definition

$$G_{a_1,a_2,...,a_n}(y) = \int_0^y \frac{dt}{t-a_1} G_{a_2,...,a_n}(t),$$
(A.1)

and comparing with the integration weights f_0, f_1 and f_{1-z} leads to

$$G_{a_1,a_2,...,a_n}(y) = (-1)^{(\#(1)+\#(1-z))} H_{a_1,a_2,...,a_n}(y),$$
(A.2)

that is, the relative sign is determined by the number of 1's and $(1-z)$'s appearing in the label. In terms of Goncharov polylogarithms, derivatives with respect to label and argument read [36]:

$$\frac{\partial}{\partial y} G_{a_1,a_2,...,a_n}(y) = \frac{1}{y-a_1} G_{a_2,...,a_n}(y)$$

$$\frac{\partial}{\partial a_i} G_{a_1,a_2...,a_n}(y) = \frac{1}{a_{i-1}-a_i} G_{a_1,...,\hat{a}_{i-1},...,a_n}(y) + \frac{1}{a_i-a_{i+1}} G_{a_1,...,\hat{a}_{i+1},...,a_n}(y)$$
$$- \frac{a_{i-1}-a_{i+1}}{(a_{i-1}-a_i)(a_i-a_{i+1})} G_{a_1,...,\hat{a}_i,...,a_n}(y)$$

$$\frac{\partial}{\partial a_n} G_{a_1,...,a_n}(y) = \frac{1}{a_{n-1}-a_n} G_{a_1,...\hat{a}_{n-1},a_n}(y) - \frac{a_{n-1}}{(a_{n-1}-a_n)a_n} G_{a_1,...,a_{n-1}}(y),$$
(A.3)

where $\hat{a}$ denotes omission of the respective entry. Given the terms of the form $\frac{1}{a_{i-1}-a_i}$ in the derivatives, it is obvious that for neighboring elements of the same kind one will get divergent terms. However, by shifting the entries in the label by a small value ϵ one can safely determine the derivative and successively take $\epsilon \to 0$. For example one finds:

$$\frac{\partial}{\partial z} H_{1-z,1-z,1-z}(y) = -\frac{\partial}{\partial z} G_{1-z+\epsilon,1-z,1-z+\epsilon}(y)\Big|_{\epsilon\to 0}$$
$$= -\frac{y\,G_{1-z,1-z}(y)}{(1-z)(1-y-z)}$$
$$= -\frac{y\,H_{1-z,1-z}(y)}{(1-z)(1-y-z)}.$$
(A.4)

References

[1] L.J. Dixon, J.M. Drummond, M. von Hippel, J. Pennington, Hexagon functions and the three-loop remainder function, J. High Energy Phys. 1312 (2013) 049, arXiv:1308.2276.
[2] L.J. Dixon, J.M. Drummond, C. Duhr, J. Pennington, The four-loop remainder function and multi-Regge behavior at NNLLA in planar $N = 4$ super-Yang–Mills theory, J. High Energy Phys. 1406 (2014) 116, arXiv:1402.3300.
[3] L.J. Dixon, M. von Hippel, Bootstrapping an NMHV amplitude through three loops, J. High Energy Phys. 1410 (2014) 065, arXiv:1408.1505.
[4] L.J. Dixon, M. von Hippel, A.J. McLeod, The four-loop six-gluon NMHV ratio function, J. High Energy Phys. 1601 (2016) 053, arXiv:1509.08127.

[5] J. Golden, M. Spradlin, An analytic result for the two-loop seven-point MHV amplitude in $\mathcal{N} = 4$ SYM, J. High Energy Phys. 1408 (2014) 154, arXiv:1406.2055.

[6] S. Caron-Huot, Superconformal symmetry and two-loop amplitudes in planar $N = 4$ super Yang–Mills, J. High Energy Phys. 1112 (2011) 066, arXiv:1105.5606.

[7] J.M. Drummond, G. Papathanasiou, M. Spradlin, A symbol of uniqueness: the cluster bootstrap for the 3-loop MHV heptagon, J. High Energy Phys. 1503 (2015) 072, arXiv:1412.3763.

[8] J. Bartels, A. Kormilitzin, L. Lipatov, Analytic structure of the $n = 7$ scattering amplitude in $\mathcal{N} = 4$ SYM theory in the multi-Regge kinematics: conformal Regge pole contribution, Phys. Rev. D 89 (2014) 065002, arXiv:1311.2061.

[9] J. Bartels, A. Kormilitzin, L.N. Lipatov, Analytic structure of the $n = 7$ scattering amplitude in $\mathcal{N} = 4$ theory in multi-Regge kinematics: conformal Regge cut contribution, Phys. Rev. D 91 (2015) 045005, arXiv:1411.2294.

[10] J. Bartels, A. Kormilitzin, L.N. Lipatov, A. Prygarin, BFKL approach and $2 \to 5$ maximally helicity violating amplitude in $\mathcal{N} = 4$ super-Yang–Mills theory, Phys. Rev. D 86 (2012) 065026, arXiv:1112.6366.

[11] J. Bartels, V. Schomerus, M. Sprenger, Heptagon amplitude in the multi-Regge regime, J. High Energy Phys. 1410 (2014) 67, arXiv:1405.3658.

[12] J. Bartels, V. Schomerus, M. Sprenger, The Bethe roots of Regge cuts in strongly coupled $\mathcal{N} = 4$ SYM theory, J. High Energy Phys. 1507 (2015) 098, arXiv:1411.2594.

[13] T. Bargheer, G. Papathanasiou, V. Schomerus, The two-loop symbol of all multi-Regge regions, J. High Energy Phys. 1605 (2016) 012, arXiv:1512.07620.

[14] T. Bargheer, Systematics of the multi-Regge three-loop symbol, arXiv:1606.07640.

[15] J. Bartels, L.N. Lipatov, A. Sabio Vera, BFKL pomeron, Reggeized gluons and Bern–Dixon–Smirnov amplitudes, Phys. Rev. D 80 (2009) 045002, arXiv:0802.2065.

[16] J. Bartels, L.N. Lipatov, A. Sabio Vera, $N = 4$ supersymmetric Yang Mills scattering amplitudes at high energies: the Regge cut contribution, Eur. Phys. J. C 65 (2010) 587, arXiv:0807.0894.

[17] V.S. Fadin, L.N. Lipatov, BFKL equation for the adjoint representation of the gauge group in the next-to-leading approximation at $N = 4$ SUSY, Phys. Lett. B 706 (2012) 470, arXiv:1111.0782.

[18] L.J. Dixon, C. Duhr, J. Pennington, Single-valued harmonic polylogarithms and the multi-Regge limit, J. High Energy Phys. 1210 (2012) 074, arXiv:1207.0186.

[19] J.M. Drummond, G. Papathanasiou, Hexagon OPE resummation and multi-Regge kinematics, J. High Energy Phys. 1602 (2016) 185, arXiv:1507.08982.

[20] J. Broedel, M. Sprenger, Six-point remainder function in multi-Regge-kinematics: an efficient approach in momentum space, J. High Energy Phys. 1605 (2016) 055, arXiv:1512.04963.

[21] Z. Bern, L.J. Dixon, V.A. Smirnov, Iteration of planar amplitudes in maximally supersymmetric Yang–Mills theory at three loops and beyond, Phys. Rev. D 72 (2005) 085001, arXiv:hep-th/0505205.

[22] L.N. Lipatov, A. Prygarin, BFKL approach and six-particle MHV amplitude in $N = 4$ super Yang–Mills, Phys. Rev. D 83 (2011) 125001, arXiv:1011.2673.

[23] B. Basso, A. Sever, P. Vieira, Spacetime and flux tube S-matrices at finite coupling for $N = 4$ supersymmetric Yang–Mills theory, Phys. Rev. Lett. 111 (2013) 091602, arXiv:1303.1396.

[24] B. Basso, A. Sever, P. Vieira, Space–time S-matrix and flux tube S-matrix II. Extracting and matching data, J. High Energy Phys. 1401 (2014) 008, arXiv:1306.2058.

[25] B. Basso, A. Sever, P. Vieira, Space–time S-matrix and flux-tube S-matrix III. The two-particle contributions, J. High Energy Phys. 1408 (2014) 085, arXiv:1402.3307.

[26] B. Basso, A. Sever, P. Vieira, Space–time S-matrix and flux-tube S-matrix IV. Gluons and fusion, J. High Energy Phys. 1409 (2014) 149, arXiv:1407.1736.

[27] B. Basso, S. Caron-Huot, A. Sever, Adjoint BFKL at finite coupling: a short-cut from the collinear limit, J. High Energy Phys. 1501 (2015) 027, arXiv:1407.3766.

[28] N. Beisert, B. Eden, M. Staudacher, Transcendentality and crossing, J. Stat. Mech. 0701 (2007) P01021, arXiv:hep-th/0610251.

[29] E. Remiddi, J. Vermaseren, Harmonic polylogarithms, Int. J. Mod. Phys. A 15 (2000) 725, arXiv:hep-ph/9905237.

[30] F.C. Brown, Polylogarithmes multiples uniformes en une variable, C. R. Math. 338 (2004) 527.

[31] J. Drummond, C. Duhr, B. Eden, P. Heslop, J. Pennington, V.A. Smirnov, Leading singularities and off-shell conformal integrals, J. High Energy Phys. 1308 (2013) 133, arXiv:1303.6909.

[32] J. Pennington, The six-point remainder function to all loop orders in the multi-Regge limit, J. High Energy Phys. 1301 (2013) 059, arXiv:1209.5357.

[33] T. Gehrmann, E. Remiddi, Two loop master integrals for $\gamma^* \to 3$ jets: the planar topologies, Nucl. Phys. B 601 (2001) 248, arXiv:hep-ph/0008287.

[34] J. Blumlein, Algebraic relations between harmonic sums and associated quantities, Comput. Phys. Commun. 159 (2004) 19, arXiv:hep-ph/0311046.

2

The Heisenberg algebra as near horizon symmetry of the black flower solutions of Chern–Simons-like theories of gravity

M.R. Setare *, H. Adami

Department of Science, University of Kurdistan, Sanandaj, Iran

Editor: Stephan Stieberger

Abstract

In this paper we study the near horizon symmetry algebra of the non-extremal black hole solutions of the Chern–Simons-like theories of gravity, which are stationary but are not necessarily spherically symmetric. We define the extended off-shell ADT current which is an extension of the generalized ADT current. We use the extended off-shell ADT current to define quasi-local conserved charges such that they are conserved for Killing vectors and asymptotically Killing vectors which depend on dynamical fields of the considered theory. We apply this formalism to the Generalized Minimal Massive Gravity (GMMG) and obtain conserved charges of a spacetime which describes near horizon geometry of non-extremal black holes. Eventually, we find the algebra of conserved charges in Fourier modes. It is interesting that, similar to the Einstein gravity in the presence of negative cosmological constant, for the GMMG model also we obtain the Heisenberg algebra as the near horizon symmetry algebra of the black flower solutions. Also the vacuum state and all descendants of the vacuum have the same energy. Thus these zero energy excitations on the horizon appear as soft hairs on the black hole.

---

* Corresponding author.
 E-mail addresses: rezakord@ipm.ir (M.R. Setare), hamed.adami@yahoo.com (H. Adami).

1. Introduction

It is well known that the pure Einstein–Hilbert gravity in three dimensions exhibits no propagating physical degrees of freedom [1,2]. So choosing appropriate conditions at the boundary is crucial in this theory. Depending on the chosen boundary conditions, this theory can lead to completely different boundary theories. Adding the gravitational Chern–Simons term produces a propagating massive graviton [3]. The resulting theory is called topologically massive gravity (TMG). Including a negative cosmological constant, yields cosmological topologically massive gravity (CTMG). In this case the theory exhibits both gravitons and black holes. Unfortunately there is a problem in this model, with the usual sign for the gravitational constant, the massive excitations of CTMG carry negative energy. In the absence of a cosmological constant, one can change the sign of the gravitational constant, but if $\Lambda < 0$, this will give a negative mass to the BTZ black hole, so the existence of a stable ground state is in doubt in this model [4,5]. A few years ego a new theory of massive gravity (NMG) in three dimensions has been proposed [6]. This theory is equivalent to the three-dimensional Fierz–Pauli action for a massive spin-2 field at the linearized level. With the only Einstein–Hilbert term in the action there are no propagating degrees of freedom, but by adding the higher curvature terms in the action the situation becomes different. Usually the theories including the terms given by the square of the curvatures have the massive spin 2 mode and the massive scalar mode in addition to the massless graviton. Also the theory has ghosts due to negative energy excitations of the massive tensor. The unitarity of NMG was discussed in [7,8] (see also [9,10]) and this model is generalized to higher dimensions. It was found that there exists a choice of parameters for which these theories possess one AdS background on which neither massive fields, nor massless scalars propagate. By this special choice of the parameters, which is called a critical point, there appears a mode which behaves as a logarithmic function of the distance. The massive graviton modes obey Brown–Henneaux boundary conditions, at the critical point in parameter space, the massive gravitons become massless and are replaced by new modes, so-called logarithmic modes. Although, the compliance of the NMG with the holographic c-theorem has been shown [11,12], both TMG and NMG have a bulk-boundary unitarity conflict. In other terms, either the bulk or the boundary theory is non-unitary, so there is a clash between the positivity of the two Brown–Henneaux boundary c charges and the bulk energies [13]. Recently an interesting three dimensional massive gravity was introduced by Bergshoeff et al. [14], dubbed Minimal Massive Gravity (MMG), which has the same minimal local structure as TMG. The MMG model has the same gravitational degree of freedom as the TMG has and the linearization of the metric field equations for MMG yield a single propagating massive spin-2 field. It seems that the single massive degree of freedom of MMG is unitary in the bulk and gives rise to a unitary CFT on the boundary. More recently the author of [15] has introduced Generalized Minimal Massive Gravity (GMMG), an interesting modification of MMG. GMMG is a unification of MMG with NMG, so this model is realized by adding higher-derivative deformation term to the Lagrangian of MMG. As has been shown in [15], GMMG also avoids the aforementioned "bulk-boundary unitarity clash". Calculation of the GMMG action to quadratic order about AdS_3 space shows that the theory is free of negative-energy bulk modes. Also Hamiltonian analysis shows that the GMMG model has no Boulware–Deser ghosts and this model propagates only two physical modes. So these models are viable candidates for semi-classical limit of a unitary quantum $3D$ massive gravity.

Although the Chern–Simons-like theories of gravity (CSLTG) in $(2+1)$-dimension [16] (e.g. TMG, NMG, MMG, Zwei-dreibein gravity (ZDG) [17], GMMG, etc.), exhibit local physical degrees of freedom, but for these theories also, different boundary conditions can lead to completely different boundary theories. For matter-free Einstein–Hilbert gravity, the behavior of the three-dimensional metric at spatial infinity is given by the Brown–Henneaux boundary conditions [18]. But in the presence of matter, these boundary conditions can be modified [19]. This modification can occur in Topological massive gravity [20] and even in pure Einstein–Hilbert gravity [21].

Recently the authors of [22] have considered the black flower solution of the Einstein equations in $3d$ [23], then have proposed a new set of boundary conditions, which leads to a very simple near horizon symmetry algebra, the Heisenberg algebra.[1] In this paper we are going to study this near horizon symmetry in the framework of Chern–Simons-like theories of gravity. For this purpose, at first we should obtain boundary conserved charges. Here we use a formalism based on the concept of quasi-local conserved charges. We have obtained the quasi-local conserved charges of the Lorentz-diffeomorphism covariant theories of gravity in the first order formalism, in paper [30]. In previous paper [31] by introducing the total variation of a quantity due to the infinitesimal Lorentz-diffeomorphism transformation, we have obtained the conserved charges in the Lorentz-diffeomorphism non-covariant theories. Here we should find an expression for the quasi-local conserved charges of CSLTG associated with the field dependent Killing vector fields. So we need an extended version of the generalized off-shell ADT current such that it becomes conserved for the field dependent Killing vectors and the field dependent asymptotically Killing vectors. By this extension, we obtain the quasi-local conserved charge corresponding to a field dependent Killing vector field. After that we apply our formalism to the GMMG and obtain conserved charges of the non-extremal black hole solutions which are stationary but are not necessarily spherically symmetric. By writing conserved charges in Fourier modes, we find the Heisenberg algebra as the near horizon symmetry algebra of these black solutions. Similar to the Einstein gravity [22], for the GMMG also, we obtain the Hamiltonian as $H \equiv P_0$, where P_0 is a Casimir of the algebra, so the vacuum state and all descendants of the vacuum have the same energy. These zero energy excitations on the horizon appear as soft hairs on the black hole. By setting $\sigma = -1$, $\mu \to \infty$ and $m^2 \to \infty$, where the GMMG reduces to the Einstein gravity with negative cosmological constant, all our results for the GMMG are reduced to the results of [22] which have been obtained by a different way.

2. Quasi-local conserved charges associated with field dependent Killing vectors

In this section we consider Chern–Simons-like theories of gravity, then we find an expression for quasi-local conserved charge corresponding to a field dependent Killing vector which is admitted by a solution of considered theory. The Lagrangian 3-form of the CSLTG is given by [16]

$$L = \frac{1}{2}\tilde{g}_{rs}a^r \cdot da^s + \frac{1}{6}\tilde{f}_{rst}a^r \cdot a^s \times a^t. \qquad (1)$$

---

[1] Here we should mention that Donnay et al. [24], have shown that the asymptotic symmetries close to the horizon of the non-extremal black hole solution of the three-dimensional Einstein gravity in the presence of a negative cosmological term, are generated by an extension of supertranslations (see also [25]). The near horizon symmetries in three dimensions are related with the Bondi–van der Burg–Metzner–Sachs (BMS) algebra [26]. The authors of [24] have shown that for a special choice of boundary conditions, the near region to the horizon of a stationary black hole presents a generalization of supertranslation, including a semidirect sum with superrotations, represented by Virasoro algebra. From BMS supertranslation we know that the vacuum is not unique and infinite degenerate vacua are physically distinct and are related to each other by the BMS supertranslation (for more information see [27–29]).

In the above Lagrangian $a^{ra} = a^{ra}{}_\mu dx^\mu$ are the Lorentz vector valued one-forms, where $r = 1, \ldots, N$ and a indices refer to the flavor and the Lorentz indices, respectively. We should mention that, here the wedge products of the Lorentz-vector valued one-form fields are implicit. Also, $\tilde{g}_{rs}$ is a symmetric constant metric on the flavor space and $\tilde{f}_{rst}$ are the totally symmetric "flavor tensors" which are interpreted as the coupling constants. We use a 3D-vector algebra notation for the Lorentz vectors in which contractions with η_{ab} and ε^{abc} are denoted by dots and crosses, respectively.[2] It is worth saying that a^{ra} is a collection of the dreibein e^a, the dualized spin-connection ω^a, the auxiliary field $h^a{}_\mu = e^a{}_\nu h^\nu{}_\mu$ and so on. Also for all interesting CSLTG we have $\tilde{f}_{\omega rs} = \tilde{g}_{rs}$ [32].[3]

Let $\pounds_\xi$ denote the ordinary Lie derivative along ξ and the Lie–Lorentz derivative (L–L derivative) $\mathfrak{L}_\xi$ is defined by [33]

$$\mathfrak{L}_\xi A^{a\cdots}{}_{b\cdots} = \pounds_\xi A^{a\cdots}{}_{b\cdots} + \lambda^a_\xi{}_c A^{c\cdots}{}_{b\cdots} + \cdots - \lambda^c_\xi{}_b A^{a\cdots}{}_{c\cdots} - \cdots, \tag{2}$$

where λ^{ab}_ξ is generator of the Lorentz gauge transformations $SO(2,1)$. The total variation of a^{ra} due to a diffeomorphism generator ξ is [34]

$$\delta_\xi a^{ra} = \mathfrak{L}_\xi a^{ra} - \delta^r_\omega d\chi^a_\xi, \tag{3}$$

which is caused by a combination of variations due to the diffeomorphism and the infinitesimal Lorentz gauge transformation. In Eq. (3), $\chi^a_\xi = \frac{1}{2}\varepsilon^a{}_{bc}\lambda^{bc}_\xi$ and δ^r_s denotes the ordinary Kronecker delta. Also, χ^a_ξ is a general function of space-time coordinates and of the diffeomorphism generator ξ. It should be noted that χ^a_ξ is linear in ξ. One can find the total variation of the Lagrangian due to the diffeomorphism generator ξ as [34]

$$\delta_\xi L = \mathfrak{L}_\xi L + d\psi_\xi, \tag{4}$$

where ψ_ξ is given by

$$\psi_\xi = \frac{1}{2}\tilde{g}_{\omega r} d\chi_\xi \cdot a^r. \tag{5}$$

The variation of the Lagrangian (1) is given by

$$\delta L = \delta a^r \cdot E_r + d\Theta(a, \delta a), \tag{6}$$

where

$$E_r{}^a = \tilde{g}_{rs} da^{sa} + \frac{1}{2}\tilde{f}_{rst}(a^s \times a^t)^a, \tag{7}$$

so that $E_r{}^a = 0$ are the equations of motion, and

$$\Theta(a, \delta a) = \frac{1}{2}\tilde{g}_{rs}\delta a^r \cdot a^s, \tag{8}$$

is the surface term. The total variation of the surface term is

$$\delta_\xi \Theta(a, \delta a) = \mathfrak{L}_\xi \Theta(a, \delta a) + \Pi_\xi, \tag{9}$$

where

---

[2] Here we consider the notation used in [16].

[3] The Lagrangian of CSLTG contains combinations such as $f \cdot R = f \cdot d\omega + \frac{1}{2} f \cdot \omega \times \omega$, $f \cdot D(\omega)h = f \cdot dh + \omega \cdot f \times h$, $\omega \cdot d\omega + \frac{1}{2}\omega \cdot \omega \times \omega$ and so on. It can be seen that all of these combinations obey the equation $\tilde{f}_{\omega rs} = \tilde{g}_{rs}$.

$$\Pi_\xi = \frac{1}{2} g_{\omega r} d\chi_\xi \cdot \delta a^r. \tag{10}$$

Now, by considering that the variation in Eq. (6) is the total variation generated by ξ and by using the Bianchi identities, we find that [34]

$$dJ_\xi = 0, \tag{11}$$

where

$$J_\xi = \Theta(a, \delta_\xi a) - i_\xi L - \psi_\xi + i_\xi a^r \cdot E_r - \chi_\xi \cdot E_\omega, \tag{12}$$

here i_ξ denotes interior product in ξ. Strictly speaking, J_ξ is an off-shell conserved current, i.e. the equation (11) holds off-shell. By virtue of the Poincare lemma, one can write $J_\xi = dK_\xi$. It is easy to show that

$$K_\xi = \frac{1}{2} \tilde{g}_{rs} i_\xi a^r \cdot a^s - \tilde{g}_{\omega s} \chi_\xi \cdot a^s. \tag{13}$$

Let $\hat{\delta}$ denote variation due to dynamical fields. By varying Eq. (12) with respect to dynamical fields we will have

$$d\left(\hat{\delta} K_\xi - K_{\hat{\delta}\xi} - i_\xi \Theta(a, \hat{\delta}a)\right) = \hat{\delta}\Theta(a, \delta_\xi a) - \delta_\xi \Theta(a, \hat{\delta}a) - \Theta(a, \delta_{\hat{\delta}\xi} a) \\ + \hat{\delta}a^r \cdot i_\xi E_r + i_\xi a^r \cdot \hat{\delta} E_r - \chi_\xi \cdot \hat{\delta} E_\omega. \tag{14}$$

In the calculation of the above equation, we assumed that ξ is a function of dynamical fields and we used the fact that $\hat{\delta}\chi_\xi = \chi_{\hat{\delta}\xi}$, because χ_ξ is linear in ξ. We define the right hand side of Eq. (14) as extended off-shell ADT current, namely

$$\mathfrak{J}_{ADT}(a, \hat{\delta}a, \delta_\xi a) = \hat{\delta}a^r \cdot i_\xi E_r + i_\xi a^r \cdot \hat{\delta} E_r - \chi_\xi \cdot \hat{\delta} E_\omega \\ + \hat{\delta}\Theta(a, \delta_\xi a) - \delta_\xi \Theta(a, \hat{\delta}a) - \Theta(a, \delta_{\hat{\delta}\xi} a). \tag{15}$$

The extended off-shell ADT current will be reduced to the generalized off-shell ADT current [34] when ξ is independent of the dynamical fields, that is $\hat{\delta}\xi = 0$. The extended off-shell ADT current $\mathfrak{J}_{ADT}$ reduces to

$$\Omega_{Symp}(a, \hat{\delta}a, \delta_\xi a) = \hat{\delta}\Theta(a, \delta_\xi a) - \delta_\xi \Theta(a, \hat{\delta}a) - \Theta(a, \delta_{\hat{\delta}\xi} a), \tag{16}$$

when the equations of motion and the linearized equations of motion both are satisfied. The equation (16) is just the ordinary symplectic current [35–38] when ξ is independent of dynamical fields, that is $\hat{\delta}\xi = 0$. So it seems sensible that Eq. (16) is an extension of the symplectic current. By substituting Eq. (8) into Eq. (16) we have

$$\Omega_{Symp}(a, \hat{\delta}a, \delta_\xi a) = \tilde{g}_{rs} \delta_\xi a^r \cdot \hat{\delta} a^s. \tag{17}$$

By replacing $\hat{\delta} = \delta_1$ and $\delta_\xi = \delta_2$, the equation (17) becomes

$$\Omega_{Symp}(a, \delta_1 a, \delta_2 a) = \tilde{g}_{rs} \delta_2 a^r \cdot \delta_1 a^s. \tag{18}$$

It is clear that Ω_{Symp} is closed, skew-symmetric and non-degenerate, also it explicitly vanishes when ξ is a Killing vector field, namely $\delta_\xi a^r = 0$, then Ω_{Symp} has all properties of a symplectic current.

On the other hand, in the off-shell case, if we assume that ξ is a Killing vector field, then

$$\hat{\delta}\Theta(a, \delta_\xi a) - \delta_\xi \Theta(a, \hat{\delta}a) - \Theta(a, \delta_{\hat{\delta}\xi} a) = 0. \tag{19}$$

Thus, in this case, the extended off-shell ADT current reduces to the ordinary one [34]. So, by the above discussion, the definition of the extended off-shell ADT current as Eq. (15) makes sense.

Now we can write Eq. (14) as follows:

$$\mathfrak{J}_{ADT}(a, \hat{\delta}a, \delta_\xi a) = d\mathfrak{Q}_{ADT}(a, \hat{\delta}a; \xi), \tag{20}$$

where $\mathfrak{Q}_{ADT}$ is extended off-shell ADT conserved charge and it is defined as

$$\mathfrak{Q}_{ADT}(a, \hat{\delta}a; \xi) = \hat{\delta}K_\xi - K_{\hat{\delta}\xi} - i_\xi \Theta(a, \hat{\delta}a). \tag{21}$$

It should be noted that, the first term in the right hand side of the above equation, is just the Komar expression for the charge perturbation [52]. Second term comes from the fact that, it is assumed that ξ is dependent on the dynamical fields and the third term is the contribution of surface term in charge perturbation [35–38]. In this way, we can define quasi-local conserved charge perturbation associated with a field dependent vector field ξ as

$$\hat{\delta}Q(\xi) = \frac{1}{8\pi G} \int_\Sigma \mathfrak{Q}_{ADT}(a, \hat{\delta}a; \xi), \tag{22}$$

where G denotes the Newtonian gravitational constant and Σ is a space-like codimension two surface. Due to the definition (15), the quasi-local conserved charge (22) is not only conserved for Killing vectors which are admitted by spacetime everywhere but also it is conserved for the asymptotic Killing vectors. By substituting Eq. (8) and Eq. (13) into Eq. (22) we find that

$$Q(\xi) = \frac{1}{8\pi G} \int_0^1 ds \int_\Sigma \left(g_{rs} i_\xi a^r - g_{\omega s} \chi_\xi \right) \cdot \hat{\delta}a^s, \tag{23}$$

where we took an integration from (22) over the one-parameter path on the solution space [39,40, 34]. This has exactly the form of case in which the Killing vector field is independent of dynamical fields [34]. However, we argued that it is usable for the case in which ξ is field dependent.

The symplectic current (16) vanishes when ξ is a Killing vector field[4] admitted by spacetime everywhere, then it is easy to see from (14) that, the generalized off-shell ADT current becomes conserved for this case. However, if we assume ξ to be an asymptotically Killing vector field, the generalized off-shell ADT current is no longer a conserved quantity, instead the extended off-shell ADT current (15) is a conserved current (see Eq. (20)). Since we have $\delta_\xi a^r = 0$ asymptotically, then the symplectic current vanishes asymptotically. Hence, the extended off-shell ADT current asymptotically reduces to the generalized off-shell ADT current. Therefore the extended off-shell ADT current is appropriate to obtain conserved charges associated with asymptotically Killing vectors.

3. Extended near horizon geometry

In the paper [22], the authors have proposed following metric as a new fall-off condition for near horizon of a non-extremal black hole in three dimensions

---

[4] In this paragraph, we drop "field dependent" phrase for simplicity.

$$ds^2 = \left[l\rho \left(f_+ \zeta^+ + f_- \zeta^- \right) + \frac{l^2}{4} \left(\zeta^+ - \zeta^- \right)^2 \right] dv^2 + 2ldvd\rho$$

$$+ l \left(\frac{\mathcal{J}^+}{\zeta^+} - \frac{\mathcal{J}^-}{\zeta^-} \right) d\rho d\phi + l\rho \left(\frac{\mathcal{J}^+}{\zeta^+} - \frac{\mathcal{J}^-}{\zeta^-} \right) \left(f_+ \zeta^+ + f_- \zeta^- \right) dvd\phi \qquad (24)$$

$$+ \left[\frac{l^2}{4} \left(\mathcal{J}^+ + \mathcal{J}^- \right)^2 - \frac{l\rho}{\zeta^+ \zeta^-} \left(f_+ \zeta^+ + f_- \zeta^- \right) \mathcal{J}^+ \mathcal{J}^- \right] d\phi^2,$$

where l is AdS radii, $\zeta^\pm$ are constant parameters, $\mathcal{J}^\pm = \mathcal{J}^\pm(\phi)$ are arbitrary functions of ϕ and $f_\pm = f_\pm(\rho)$ are given as

$$f_\pm(\rho) = 1 - \frac{\rho}{2l\zeta^\pm}. \qquad (25)$$

The line-element (24) is written in ingoing Eddington–Finkelstein coordinates, also v, ρ and ϕ are the advanced time, the radial coordinate and the angular coordinate, respectively. In the particular case of $\zeta^\pm = -a$, where the constant a is the Rindler acceleration, the line-element (24) will be reduced to

$$ds^2 = -2al\rho f(\rho)dv^2 + 2ldvd\rho - 2a^{-1}\theta(\phi)d\phi d\rho + 4\rho\theta(\phi)f(\rho)dvd\phi$$

$$+ \left[\gamma(\phi)^2 + \frac{2\rho}{al} f(\rho) \left(\gamma(\phi)^2 - \theta(\phi)^2 \right) \right] d\phi^2, \qquad (26)$$

where $l\mathcal{J}^\pm = \gamma \pm \theta$ and $f(\rho) = 1 + \frac{\rho}{2la}$. The line-element (26) describes a spacetime which possesses an event horizon located at $\rho = 0$. The line-element (24) solves the Einstein equations with negative cosmological constant

$$R(\Omega) + \frac{1}{2l^2} e \times e = 0, \qquad T(\Omega) = 0, \qquad (27)$$

where $R(\Omega) = d\Omega + \frac{1}{2}\Omega \times \Omega$ is curvature 2-form, $T(\Omega) = D(\Omega)e$ is torsion 2-form and Ω is torsion free spin-connection. Also, $D(\Omega)$ denotes exterior covariant derivative with respect to Ω.

The following Killing vector

$$\xi^v = \frac{1}{2} \left\{ -\left(\frac{1}{\zeta^+} - \frac{1}{\zeta^-} \right) \left(\frac{\mathcal{J}^+}{\zeta^+} - \frac{\mathcal{J}^-}{\zeta^-} \right) \left(\frac{\mathcal{J}^+}{\zeta^+} + \frac{\mathcal{J}^-}{\zeta^-} \right)^{-1} + \left(\frac{1}{\zeta^+} + \frac{1}{\zeta^-} \right) \right\} \Xi(\phi)$$

$$\xi^\rho = 0 \qquad (28)$$

$$\xi^\phi = \left(\frac{1}{\zeta^+} - \frac{1}{\zeta^-} \right) \left(\frac{\mathcal{J}^+}{\zeta^+} + \frac{\mathcal{J}^-}{\zeta^-} \right)^{-1} \Xi(\phi)$$

preserves the fall-off conditions (24), up to terms that involve powers of $\delta\mathcal{J}$ higher than the order one, i.e. we ignore the terms of order $\mathcal{O}(\delta\mathcal{J}^2)$. In the Eq. (28), $\Xi(\phi)$ is an arbitrary function of ϕ. Under the transformation generated by the Killing vector field (28) the arbitrary functions $\mathcal{J}^\pm(\phi)$, which have appeared in the metric, transform as

$$\hat{\delta}_\xi \mathcal{J}^\pm = \pm \Xi', \qquad (29)$$

where the prime denotes differentiation with respect to ϕ. We introduce a modified version of Lie brackets [41]

$$[\xi_1, \xi_2] = \pounds_{\xi_1} \xi_2 - \hat{\delta}_{\xi_1} \xi_2 + \hat{\delta}_{\xi_2} \xi_1, \tag{30}$$

so that the algebra of the Killing vector fields to be closed. In the equation (30), $\hat{\delta}_{\xi_1} \xi_2$ denotes the change induced in ξ_2 due to the variation of metric $\delta_{\xi_1} g_{\mu\nu} = \pounds_{\xi_1} g_{\mu\nu}$ [36]. Thus, we have

$$[\xi_1, \xi_2] = 0. \tag{31}$$

Therefore, the Killing vectors $\xi_1 = \xi(\Xi_1)$ and $\xi_2 = \xi(\Xi_2)$ commute. The relation between metric tensor and dreibein is given by $g_{\mu\nu} = \eta_{ab} e^a{}_\mu e^b{}_\nu$, so we conclude this section by writing down dreibein corresponding to the line-element (24)

$$
\begin{aligned}
e^0 &= -\frac{1}{2}\left[2 - \frac{l\rho}{2}\left(f_+\varsigma^+ + f_-\varsigma^-\right)\right] dv + \frac{l}{2} d\rho \\
&\quad + \frac{1}{2}\left[-\left(\frac{\mathcal{J}^+}{\varsigma^+} - \frac{\mathcal{J}^-}{\varsigma^-}\right) + \frac{l\rho}{2}\left(f_+\mathcal{J}^+ - f_-\mathcal{J}^-\right)\right] d\phi \\
e^1 &= \frac{l}{2}\left(\varsigma^+ - \varsigma^-\right) dv + \frac{l}{2}\left[\left(\mathcal{J}^+ + \mathcal{J}^-\right) - \frac{\rho}{l}\left(\frac{\mathcal{J}^+}{\varsigma^+} + \frac{\mathcal{J}^-}{\varsigma^-}\right)\right] d\phi \\
e^2 &= -\frac{1}{2}\left[2 + \frac{l\rho}{2}\left(f_+\varsigma^+ + f_-\varsigma^-\right)\right] dv - \frac{l}{2} d\rho \\
&\quad - \frac{1}{2}\left[\left(\frac{\mathcal{J}^+}{\varsigma^+} - \frac{\mathcal{J}^-}{\varsigma^-}\right) + \frac{l\rho}{2}\left(f_+\mathcal{J}^+ - f_-\mathcal{J}^-\right)\right] d\phi.
\end{aligned}
\tag{32}
$$

4. Application to the generalized minimal massive gravity

Generalized minimal massive gravity (GMMG) is an example of the Chern–Simons-like theories of gravity [15]. This model is realized by adding the CS deformation term, the higher derivative deformation term, and an extra term to pure Einstein gravity with a negative cosmological constant. In [15] it is discussed that this theory is free of negative-energy bulk modes, and also avoids the aforementioned "bulk-boundary unitarity clash". By a Hamiltonian analysis one can show that the GMMG model has no the Boulware–Deser ghosts and this model propagates only two physical modes. In the GMMG, there are four flavors of one-form, $a^r = \{e, \omega, h, f\}$ and the non-zero components of the flavor metric and the flavor tensor are

$$
\begin{aligned}
&\tilde{g}_{e\omega} = -\sigma, &&\tilde{g}_{eh} = 1, &&\tilde{g}_{\omega f} = -\frac{1}{m^2}, &&\tilde{g}_{\omega\omega} = \frac{1}{\mu}, \\
&\tilde{f}_{e\omega\omega} = -\sigma, &&\tilde{f}_{eh\omega} = 1, &&\tilde{f}_{f\omega\omega} = -\frac{1}{m^2}, &&\tilde{f}_{\omega\omega\omega} = \frac{1}{\mu}, \\
&\tilde{f}_{eff} = -\frac{1}{m^2}, &&\tilde{f}_{eee} = \Lambda_0, &&\tilde{f}_{ehh} = \alpha,
\end{aligned}
\tag{33}
$$

where σ, Λ_0, μ, m and α are a sign, cosmological parameter with dimension of mass squared, mass parameter of the Lorentz Chern–Simons term, mass parameter of the new massive gravity term and a dimensionless parameter, respectively. In this case, the equations of motion (7) are reduced to the following equations

$$-\sigma R(\omega) + \frac{\Lambda_0}{2} e \times e + D(\omega) h - \frac{1}{2m^2} f \times f + \frac{\alpha}{2} h \times h = 0, \tag{34}$$

$$-\sigma T(\omega) + \frac{1}{\mu} R(\omega) - \frac{1}{m^2} D(\omega) f + e \times h = 0, \tag{35}$$

$$R(\omega) + e \times f = 0, \tag{36}$$

$$T(\omega) + \alpha e \times h = 0. \tag{37}$$

Dreibein (32) solves the equations of motion (34)–(37) when the following equations are satisfied [25]

$$f^a = F e^a, \qquad h^a = H e^a, \tag{38}$$

$$\frac{\sigma}{l^2} - \alpha(1 + \sigma\alpha)H^2 + \Lambda_0 - \frac{F^2}{m^2} = 0, \tag{39}$$

$$-\frac{1}{\mu l^2} + 2(1 + \sigma\alpha)H + \frac{2\alpha}{m^2}FH + \frac{\alpha^2}{\mu}H^2 = 0, \tag{40}$$

$$-F + \mu(1 + \sigma\alpha)H + \frac{\mu\alpha}{m^2}FH = 0, \tag{41}$$

where F and H are constant parameters. It should be noted that one can decompose the spin-connection in two independent parts $\omega = \Omega + \kappa$, where Ω is the torsion-free part which is known as the Riemannian spin-connection and κ is the contorsion 1-form. It is easy to check that (using Eq. (37)) the contorsion 1-form for this case is given as $\kappa = \alpha h$.

By using equations (38)–(41) and $\omega = \Omega + \alpha h$, one can simplify Eq. (22) in the context of GMMG as

$$\hat{\delta}Q(\xi) = \frac{1}{8\pi G} \int_\Sigma \left\{ -\left(\sigma + \frac{\alpha H}{\mu} + \frac{F}{m^2} \right) \left((i_\xi\Omega - \chi_\xi) \cdot \hat{\delta}e + i_\xi e \cdot \hat{\delta}\Omega \right) \right.$$
$$\left. + \frac{1}{\mu} \left((i_\xi\Omega - \chi_\xi) \cdot \hat{\delta}\Omega + \frac{1}{l^2} i_\xi e \cdot \hat{\delta}e \right) \right\}. \tag{42}$$

Parameters that appear in the Eq. (42) (F and H), satisfy equations (39)–(41). On one hand, since the torsion free spin-connection is given as

$$\Omega^a{}_\mu = \frac{1}{2}\varepsilon^{abc}e_b{}^a \overset{\bullet}{\nabla}_\mu e_{c\alpha} \tag{43}$$

where $\overset{\bullet}{\nabla}$ denotes covariant derivative with respect to the Levi-Civita connection, then by substituting Eq. (32) into Eq. (43) we find that

$$\Omega^0 = -\frac{1}{4}\left(\zeta^+ - \zeta^- \right)\rho dv + \frac{1}{2l}\left[\left(\frac{\mathcal{J}^+}{\zeta^+} + \frac{\mathcal{J}^-}{\zeta^-} \right) - \frac{l\rho}{2}\left(f_+\mathcal{J}^+ + f_-\mathcal{J}^- \right) \right]d\phi$$
$$\Omega^1 = -\frac{1}{2}\left[\left(\zeta^+ + \zeta^- \right) - \frac{2\rho}{l} \right]dv - \frac{1}{2}\left[\left(1 - \frac{\rho}{l\zeta^+} \right)\mathcal{J}^+ - \left(1 - \frac{\rho}{l\zeta^-} \right)\mathcal{J}^- \right]d\phi \tag{44}$$
$$\Omega^2 = \frac{1}{4}\left(\zeta^+ - \zeta^- \right)\rho dv + \frac{1}{2l}\left[\left(\frac{\mathcal{J}^+}{\zeta^+} + \frac{\mathcal{J}^-}{\zeta^-} \right) + \frac{l\rho}{2}\left(f_+\mathcal{J}^+ + f_-\mathcal{J}^- \right) \right]d\phi.$$

On the other hand, by demanding that the Lie–Lorentz derivative of e^a becomes zero explicitly when ξ is a Killing vector field, we find the following expression for χ_ξ [31,33]

$$\chi_\xi^a = i_\xi\omega^a + \frac{1}{2}\varepsilon^a{}_{bc}e^{vb}(i_\xi T^c)_v + \frac{1}{2}\varepsilon^a{}_{bc}e^{b\mu}e^{cv}\overset{\bullet}{\nabla}_\mu\xi_v. \tag{45}$$

It has been shown that this expression can be rewritten as [42]

$$i_\xi \Omega - \chi_\xi = -\frac{1}{2}\varepsilon^a_{\ bc}e^{b\mu}e^{c\nu}\overset{\bullet}{\underset{\mu}{\nabla}}\xi_\nu. \tag{46}$$

Thus, using equations (28), (32), (44) and (46), we find that

$$
\begin{aligned}
(i_\xi\Omega - \chi_\xi)\cdot\hat{\delta}e + i_\xi e\cdot\hat{\delta}\Omega &= -\frac{l}{2}\left(\Xi\hat{\delta}\mathcal{J}^+ + \Xi\hat{\delta}\mathcal{J}^-\right)d\phi + \mathcal{O}(\hat{\delta}\mathcal{J}^2),\\
(i_\xi\Omega - \chi_\xi)\cdot\hat{\delta}\Omega + \frac{1}{l^2}i_\xi e\cdot\hat{\delta}e &= \frac{1}{2}\left(\Xi\hat{\delta}\mathcal{J}^+ - \Xi\hat{\delta}\mathcal{J}^-\right)d\phi + \mathcal{O}(\hat{\delta}\mathcal{J}^2).
\end{aligned}
\tag{47}
$$

By substituting Eq. (47) into Eq. (42), then by taking an integration over the one-parameter path on the solution space, we obtain

$$Q(\xi) = Q(\tau^+) + Q(\tau^-) \tag{48}$$

where $\tau^\pm = \pm\Xi(\phi)$ and $Q(\tau^\pm)$ are given as

$$Q(\tau^\pm) = \pm\frac{k}{4\pi}\left(\sigma \pm \frac{1}{\mu l} + \frac{\alpha H}{\mu} + \frac{F}{m^2}\right)\int_0^{2\pi}\tau^\pm(\phi)\mathcal{J}^\pm(\phi)d\phi. \tag{49}$$

In the equation (49) we set $k = l/(4G)$. The algebra of conserved charges can be written as [43]

$$\{Q(\xi_1), Q(\xi_2)\} = Q([\xi_1, \xi_2]) + \mathcal{C}(\xi_1, \xi_2) \tag{50}$$

where $\mathcal{C}(\xi_1, \xi_2)$ is central extension term. Also, the left hand side of the equation (50) can be defined by

$$\{Q(\xi_1), Q(\xi_2)\} = \hat{\delta}_{\xi_2}Q(\xi_1). \tag{51}$$

Due to the Eq. (31) one can deduce that $\hat{\delta}_{\xi_2}Q(\xi_1) = \mathcal{C}(\xi_1, \xi_2)$. By varying Eq. (49) with respect to the dynamical fields so that the variation is generated by a Killing vector, we have

$$
\hat{\delta}_{\tau_2^\pm}Q(\tau_1^\pm) = \pm\frac{k}{8\pi}\left(\sigma \pm \frac{1}{\mu l} + \frac{\alpha H}{\mu} + \frac{F}{m^2}\right)\int_0^{2\pi}\Xi_{12}(\phi)d\phi,
\tag{52}
$$

$$\hat{\delta}_{\tau_2^\pm}Q(\tau_1^\mp) = 0,$$

where

$$\Xi_{12} = \Xi_1\Xi_2' - \Xi_2\Xi_1'. \tag{53}$$

By setting $\tau^\pm = \pm\Xi(\phi) = \pm e^{in\phi}$, one can expand $Q(\tau^\pm)$ in Fourier modes

$$J_n^\pm = \frac{k}{4\pi}\left(\sigma \pm \frac{1}{\mu l} + \frac{\alpha H}{\mu} + \frac{F}{m^2}\right)\int_0^{2\pi}e^{in\phi}\mathcal{J}^\pm(\phi)d\phi. \tag{54}$$

Also, by substituting $\Xi_1 = e^{in\phi}$, $\Xi_2 = e^{im\phi}$ into Eq. (52) and replacement of Dirac brackets by commutators $\{,\} \rightarrow i[,]$, we have

$$
\left[J_n^\pm, J_m^\pm\right] = \mp\frac{k}{2}\left(\sigma \pm \frac{1}{\mu l} + \frac{\alpha H}{\mu} + \frac{F}{m^2}\right)n\delta_{m+n,0},
\tag{55}
$$

$$\left[J_n^\pm, J_m^\mp\right] = 0.$$

Similar to the near horizon symmetry algebra of the black flower solutions of the Einstein gravity in the presence of negative cosmological constant, the above algebra consists of two $U(1)$ current algebras, but instead with levels $\pm\frac{k}{2}$, here the level of algebra is given by $\mp\frac{k}{2}\left(\sigma\pm\frac{1}{\mu l}+\frac{\alpha H}{\mu}+\frac{F}{m^2}\right)$.

One can change the basis according to following definitions

$$X_n = \frac{1}{\sqrt{2u_+}}J_n^+ - \frac{i}{\sqrt{2u_-}}J_n^- \quad \text{for} \quad n \in \mathbb{Z}$$

$$P_n = \frac{i}{n\sqrt{2u_+}}J_{-n}^+ - \frac{1}{n\sqrt{2u_-}}J_{-n}^- \quad \text{for} \quad n \neq 0 \tag{56}$$

$$P_0 = J_0^+ + J_0^- \quad \text{for} \quad n = 0,$$

where

$$u_\pm = \mp\frac{k}{2}\left(\sigma\pm\frac{1}{\mu l}+\frac{\alpha H}{\mu}+\frac{F}{m^2}\right). \tag{57}$$

By using the above equations, the algebra (55), takes following form

$$[X_n, X_m] = [P_n, P_m] = [X_0, P_n] = [P_0, X_n] = 0 \tag{58}$$

$$[X_n, P_m] = i\delta_{nm} \quad \text{for} \quad n, m \neq 0. \tag{59}$$

It is clear that X_0 and P_0 are the two Casimirs and Eq. (59) is the Heisenberg algebra. It is interesting that, for the GMMG model also we obtain the Heisenberg algebra as the near horizon symmetry algebra of the black flower solutions. By comparing the definition of P_0 and Eq. (48), one can deduce that P_0 is just the Hamiltonian, i.e. $H \equiv P_0$.

By setting $\sigma = -1$, $\mu \to \infty$ and $m^2 \to \infty$, the results of this work, namely Eq. (49), Eq. (54) and Eq. (55), which we obtained for the Chern–Simons-like theories of gravity, are reduced to the results of the Einstein gravity with negative cosmological constant case which have obtained in [22] by a different way.

5. Soft hair and the soft hairy black hole entropy

We know that the Hamiltonian $H \equiv P_0$ gives us the dynamics of the system near the horizon. Let us consider all vacuum descendants [22]

$$|\psi(q)\rangle = N(q) \prod_{i=1}^{N^+} \left(J_{-n_i^+}^+\right)^{m_i^+} \prod_{i=1}^{N^-} \left(J_{-n_i^-}^-\right)^{m_i^-} |0\rangle \tag{60}$$

where q is a set of arbitrary non-negative integer quantum numbers $N^\pm$, $n_i^\pm$ and $m_i^\pm$. Also, $N(q)$ is a normalization constant such that $\langle\psi(q)|\psi(q)\rangle = 1$. The Hamiltonian $H \equiv P_0 = J_0^+ + J_0^-$ commutes with all generators $J_n^\pm$, so the energy of all states are the same. The energy of the vacuum state is given by the following eigenvalue equation

$$H|0\rangle = E_{\text{vac}}|0\rangle. \tag{61}$$

Also, for all descendants, we have

$$E_\psi = \langle\psi(q)|H|\psi(q)\rangle. \tag{62}$$

Due to the mentioned property of the Hamiltonian, we find that all descendants of the vacuum have the same energy as the vacuum,

$$E_\psi = E_{\text{vac}}, \tag{63}$$

in other words, they are soft hairs in the sense of being zero-energy excitations [44,22].

For the case of the BTZ black hole, we have

$$\mathcal{J}^\pm = \frac{1}{l}(r_+ \pm r_-), \qquad \zeta^\pm = -\frac{r_+^2 - r_-^2}{l^2 r_+}, \tag{64}$$

where r_- and r_+ are inner and outer horizon radiuses of the BTZ black hole [45,46]. By substituting Eq. (64) into Eq. (54), we find the eigenvalues of $J_n^\pm$ as follows:

$$J_n^\pm = \frac{1}{8G}\left(\sigma \pm \frac{1}{\mu l} + \frac{\alpha H}{\mu} + \frac{F}{m^2}\right)(r_+ \pm r_-)\,\delta_{n,0}. \tag{65}$$

The entropy of a soft hairy black hole is related to the zero mode charges $J_0^\pm$ by the following formula [47–50,22]

$$S = 2\pi\left(J_0^+ + J_0^-\right). \tag{66}$$

Hence, by substituting Eq. (65) into Eq. (66), we find the entropy of the BTZ black hole solution of GMMG as

$$S = -\frac{\pi}{2G}\left\{\left(\sigma + \frac{\alpha H}{\mu} + \frac{F}{m^2}\right)r_+ + \frac{r_-}{\mu l}\right\} \tag{67}$$

which is exactly matched with the results of the paper [51]. As we know, $J_0^+ + J_0^- = P_0$ is one of two Casimirs of the algebra, i.e. P_0 is a constant of motion. Therefore, one expects that the zero mode eigenvalue of P_0 should correspond to a conserved charge of considered spacetime. We have shown that entropy is the intended conserved charge in the context of GMMG, as the pure-gravity case.

6. Conclusion

Our aim in this paper was to study the near horizon symmetry algebra of the black hole solutions of the Chern–Simons-like theories of gravity, which are stationary but are not necessarily spherically symmetric. The Lagrangian of such theories is given by Eq. (1) in the first order formalism. We have tried to find an expression for the quasi-local conserved charges of CSLTG associated with the field dependent Killing vector fields. For this purpose, we have used the concept of total variation (3) to define an off-shell conserved current (12). We took a variation from Eq. (12) with respect to dynamical fields and then we defined the extended off-shell ADT current (15). We have shown that the extended off-shell ADT current is an extension of the generalized off-shell ADT current, i.e. we have extended the generalized off-shell ADT current such that it becomes conserved for the field dependent Killing vectors and the field dependent asymptotically Killing vectors. So this expression reduced to the generalized off-shell ADT current [34] when ξ is independent of dynamical fields, i.e., where $\hat{\delta}\xi = 0$. Then, we have found extended off-shell ADT conserved charge associated with the field dependent Killing vector field (21). Consequently, we have defined the quasi-local conserved charge corresponding to a field dependent Killing vector field as Eq. (23) which is conserved for the field dependent asymptotically Killing vectors as well. In section 3, we have considered the extended near horizon geometry which has been proposed in [22]. In the paper [22], the metric (24) was introduced as new fall-off conditions for the near horizon of a non-extremal black hole in 3D. This geometry is not spherically symmetric, and generically describes a "black flower" [23]. We have shown that Killing vectors of the form (28) preserve the fall-off conditions up to terms that involve powers of perturbations of dynamical fields higher than the one. In section 4, we have applied the provided formalism to the generalized minimal massive gravity as an example of the Chern–Simons-like theories of gravity. We have found the conserved charges correspond to the near horizon symmetry of a non-extremal non-spherically symmetric black hole solution of GMMG, see Eq. (48) and Eq. (49). Then, we have obtained the algebra of conserved charges in Fourier modes, see Eq. (55) or Eq. (58) and Eq. (59). It is interesting that, similar to the Einstein gravity in the presence of negative cosmological constant, for the GMMG model also we obtain the Heisenberg algebra as the near horizon symmetry algebra of the black flower solutions. In the section 5, we have summarized the concept of soft hair presented in [22] and we have argued that it is also valid in GMMG. In other words, since the Hamiltonian is given by $H \equiv P_0$, and P_0 is a Casimir of the algebra, the vacuum state and all descendants of the vacuum have the same energy. So these zero energy excitations on horizon appear as soft hairs on the black hole. Then by finding the eigenvalues of $J_n^{\pm}$ for the BTZ black hole, see Eq. (65), we have checked that the formula for the entropy of a soft hairy black hole gives us the correct value of the entropy of the BTZ black hole solution of the GMMG. It should be mentioned that, as one expected, by setting $\sigma = -1$, $\mu \to \infty$ and $m^2 \to \infty$, where the GMMG reduces to the Einstein gravity with negative cosmological constant, the results of this paper, namely Eq. (49), Eq. (54) and Eq. (55), reduced to the results of [22] which have been obtained by a different way.

Acknowledgements

M.R. Setare thanks Stephane Detournay for helpful comments and discussions.

References

[1] L.F. Abbott, S. Deser, Nucl. Phys. B 195 (1982) 76;
    L.F. Abbott, S. Deser, Phys. Lett. B 116 (1982) 259.
[2] S. Deser, B. Tekin, Phys. Rev. D 67 (2003) 084009;
    S. Deser, B. Tekin, Phys. Rev. Lett. 89 (2002) 101101.
[3] S. Deser, R. Jackiw, S. Templeton, Ann. Phys. 140 (1982) 372; Erratum: Ann. Phys. 185 (1988) 406; Ann. Phys. 281 (2000) 409–449.
[4] K.A. Moussa, G. Clement, C. Leygnac, Class. Quantum Gravity 20 (2003) L277.
[5] W. Li, W. Song, A. Strominger, J. High Energy Phys. 0804 (2008) 082.
[6] E.A. Bergshoeff, O. Hohm, P.K. Townsend, Phys. Rev. Lett. 102 (2009) 201301.
[7] M. Nakasone, I. Oda, Prog. Theor. Phys. 121 (2009) 1389.
[8] M. Nakasone, I. Oda, Phys. Rev. D 79 (2009) 104012.
[9] D. Grumiller, N. Johansson, J. High Energy Phys. 0807 (2008) 134.
[10] S. Ertl, D. Grumiller, N. Johansson, arXiv:0910.1706 [hep-th], 2009.
[11] A. Sinha, J. High Energy Phys. 1006 (2010) 061.
[12] R.C. Myers, A. Sinha, J. High Energy Phys. 1101 (2011) 125.
[13] S. Deser, B. Tekin, Class. Quantum Gravity 20 (2003) L259.
[14] E. Bergshoeff, O. Hohm, W. Merbis, A.J. Routh, P.K. Townsend, Class. Quantum Gravity 31 (2014) 145008.
[15] M.R. Setare, Nucl. Phys. B 898 (2015) 259.
[16] E.A. Bergshoeff, O. Hohm, W. Merbis, A.J. Routh, P.K. Townsend, Lect. Notes Phys. 892 (2015) 181.
[17] E.A. Bergshoeff, S. de Haan, O. Hohm, W. Merbis, P.K. Townsend, Phys. Rev. Lett. 111 (2013) 111102.
[18] J.D. Brown, M. Henneaux, Commun. Math. Phys. 104 (1986) 207.
[19] M. Henneaux, C. Martinez, R. Troncoso, J. Zanelli, Phys. Rev. D 65 (2002) 104007.

[20] D. Grumiller, N. Johansson, Int. J. Mod. Phys. D 17 (2009) 2367;
     M. Henneaux, C. Martinez, R. Troncoso, Phys. Rev. D 79 (2009) 081502R.
[21] G. Compere, W. Song, A. Strominger, J. High Energy Phys. 1305 (2013) 152;
     C. Troessaert, J. High Energy Phys. 1308 (2013) 044.
[22] H. Afshar, S. Detournay, D. Grumiller, W. Merbis, A. Perez, D. Tempo, R. Troncoso, Phys. Rev. D 93 (2016) 101503.
[23] G. Barnich, C. Troessaert, D. Tempo, R. Troncoso, Phys. Rev. D 93 (2016) 084001.
[24] L. Donnay, G. Giribet, H.A. Gonzalez, M. Pino, Phys. Rev. Lett. 116 (2016) 091101.
[25] M.R. Setare, H. Adami, Phys. Lett. B 760 (2016) 411.
[26] H. Bondi, M. van der Burg, A. Metzner, Proc. R. Soc. Lond. A 269 (1962) 21;
     R. Sachs, Phys. Rev. 128 (1962) 2851.
[27] L. Donnay, G. Giribet, H.A. Gonzalez, M. Pino, J. High Energy Phys. 09 (2016) 100.
[28] R.G. Cai, S.M. Ruan, Y.L. Zhang, J. High Energy Phys. 09 (2016) 163.
[29] A. Strominger, A. Zhiboedov, J. High Energy Phys. 01 (2016) 086.
[30] M.R. Setare, H. Adami, Eur. Phys. J. C 76 (2016) 187.
[31] M.R. Setare, H. Adami, Nucl. Phys. B 902 (2016) 115.
[32] W. Merbis, arXiv:1411.6888 [gr-qc], 2014.
[33] T. Jacobson, A. Mohd, Phys. Rev. D 92 (2015) 124010.
[34] M.R. Setare, H. Adami, Nucl. Phys. B 909 (2016) 345.
[35] J. Lee, R. Wald, J. Math. Phys. (N.Y.) 31 (1990) 725.
[36] R. Wald, J. Math. Phys. (N.Y.) 31 (1990) 2378.
[37] R.M. Wald, Phys. Rev. D 48 (1993) 3427.
[38] R.M. Wald, A. Zoupas, Phys. Rev. D 61 (2000) 084027.
[39] W. Kim, S. Kulkarni, S.H. Yi, Phys. Rev. Lett. 111 (2013) 081101.
[40] W. Kim, S. Kulkarni, S.H. Yi, Phys. Rev. D 88 (2013) 124004.
[41] G. Barnich, C. Troessaert, J. High Energy Phys. 1005 (2010) 062.
[42] M.R. Setare, H. Adami, arXiv:1604.07837 [hep-th], 2016.
[43] G. Barnich, F. Brandt, Nucl. Phys. B 633 (2002) 3.
[44] S.W. Hawking, M.J. Perry, A. Strominger, Phys. Rev. Lett. 116 (2016) 231301.
[45] M. Banados, C. Teitelboim, J. Zanelli, Phys. Rev. Lett. 69 (1992) 1849.
[46] M. Banados, M. Henneaux, C. Teitelboim, J. Zanelli, Phys. Rev. D 48 (1993) 1506.
[47] A. Perez, D. Tempo, R. Troncoso, Phys. Lett. B 726 (2013) 444.
[48] A. Perez, D. Tempo, R. Troncoso, arXiv:1301.0847 [hep-th], 2013.
[49] J. de Boer, J.I. Jottar, J. High Energy Phys. 1401 (2014) 023.
[50] C. Bunster, M. Henneaux, A. Perez, D. Tempo, R. Troncoso, J. High Energy Phys. 1405 (2014) 031.
[51] M.R. Setare, H. Adami, Phys. Lett. B 744 (2015) 280.
[52] A. Komar, Phys. Rev. 113 (1959) 934.

3

Holographic paramagnetism–ferromagnetism phase transition with the nonlinear electrodynamics

Cheng-Yuan Zhang, Ya-Bo Wu *, Ya-Nan Zhang, Huan-Yu Wang,
Meng-Meng Wu

Department of Physics, Liaoning Normal University, Dalian, 116029, China

Editor: Stephan Stieberger

Abstract

In the probe limit, we investigate the nonlinear electrodynamical effects of the both exponential form and the logarithmic form on the holographic paramagnetism–ferromagnetism phase transition in the background of a Schwarzschild-AdS black hole spacetime. Moreover, by comparing the exponential form of nonlinear electrodynamics with the logarithmic form of nonlinear electrodynamics and the Born–Infeld nonlinear electrodynamics which has been presented in Ref. [55], we find that the higher nonlinear electrodynamics correction makes the critical temperature smaller and the magnetic moment harder form in the case without external field. Furthermore, the increase of nonlinear parameter b will result in extending the period of the external magnetic field. Especially, the effect of the exponential form of nonlinear electrodynamics on the periodicity of hysteresis loop is more noticeable.

1. Introduction

In the early part of the last century, the phenomenon of superconductivity that the electrical resistivity of a material suddenly drops to zero below a critical temperature T_c was discovered and then the most successful microscopic theory of superconductivity was proposed by Bardeen,

---

* Corresponding author.
E-mail address: ybwu61@163.com (Y.-B. Wu).

Cooper and Schrieffer (BCS) to describe various properties of usual (low temperature) superconducting materials with great accuracy. However, in the modern condensed matter physics, some materials of significant theoretical and practical interest, such as the high-temperature cuprates and heavy fermion compounds, are beyond BCS theory. Fortunately, the anti-de Sitter (AdS)/conformal field theory (CFT) correspondence [1–4] provides a window into the strongly coupled condensed matter system, especially the construction of the holographic superconductor, by relating a weak coupling gravity theory in a d-dimensional AdS spacetime to a strong CFT on the $(d-1)$-dimensional boundary. It was suggested that the instability of the bulk black hole corresponds to a second order phase transition from normal state to superconducting state which brings the spontaneous U(1) symmetry breaking [5]. The authors of Ref. [6] introduced the first superconductor that can indeed be reproduced in this simple model. And following this, a variety of the holographic dual models have been shown in Refs. [7–16].

Recently, some efforts have been made to generalize the AdS/CFT correspondence to magnetism. The authors of Ref. [17] realized the holographic description of the paramagnetism–ferromagnetism phase transition in a dyonic Reissner–Nordström-AdS black brane. In that model, the magnetic moment is realized by the condensation of a real antisymmetric tensor field which couples to the background gauge field strength in the bulk. In the case without an external magnetic field, the time reversal symmetry is spontaneously broken, and the spontaneous magnetization happens in low temperatures. The critical exponents are in agreement with the ones from mean field theory. In the case of a nonzero magnetic field, the model realizes the hysteresis loop of the single magnetic domain, and the magnetic susceptibility satisfies the Curie–Weiss law. Obviously, this model in Ref. [17] gives a good starting point to explore more complicated magnetic phenomena and quantum phase transitions. Since then, a large number of the holographic dual models have been constructed and some interesting behaviors have been found, for reviews, see Refs. [18–25] and references therein.

All of the above mentioned models are carried out in the framework of usual Maxwell electrodynamics. However, besides the conventional framework of Maxwell electrodynamics, there is always a provision for non-linear electrodynamics, which correspond to the high derivative corrections to the gauge fields in various aspects of gravity theories. Moreover, it turned out that the higher derivative corrections of the gauge field carry more plentiful information than the usual Maxwell electrodynamics [26–31], and has been a focus for these years since most physical systems are inherently nonlinear to some extent. Among the various theories with non-linear electrodynamics, the Born–Infeld (BI) theory [32] has attained renewed attentions due to its several remarkable features. One of the interesting properties of the BI theory is that the electric field is regular for a point-like particle. The regular BI theory with finite energy gives the non-singular solutions of the field equations. In fact the BI electrodynamics is the only non-linear electrodynamic theory with a sensible weak field limit [33]. Another fascinating feature of the BI theory is that it remains invariant under electromagnetic duality. Therefore, considering Born–Infeld electrodynamics, Jing and Chen firstly introduced holographic dual model and observed that the nonlinear Born–Infeld corrections make it harder for the scalar condensation to form [34]. Subsequently, some rich physics in the phase transition of the holographic superconductor with Born–Infeld electrodynamics in Gauss–Bonnet gravity has been observed [35]. Along this direction, there has been accumulated interest to study various holographic dual models with the nonlinear electrodynamics [36–48]. At the same time, similar to the case of Born–Infeld nonlinear electrodynamics, other types of nonlinear electrodynamics in the context of gravitational field have been introduced, which can also remove the divergence arising in Maxwell theory at the origin. Two well known nonlinear Lagrangian for electrodynamics are logarithmic (LEN) [49,50] and exponential (ENE) [51–53] Lagrangian. The authors of Ref. [54] observed that the exponential form of nonlinear electrodynamics has stronger effect on the condensation formation and conductivity for the holographic conductors in the backgrounds of AdS black hole by considering three types of typical nonlinear electrodynamics. However, in the AdS soliton background the critical chemical potentials are independent of the explicit form of the nonlinear electrodynamics, i.e., the ENE, BINE and LNE correction do not effect on the critical potentials. So based on the research about holographic superconductor with the nonlinear electrodynamics, we have studied the effect of BI coupling parameter on the paramagnetism–ferromagnetism phase transition [55]. And now it is interesting to investigate how the other two types of nonlinear electrodynamics influence the paramagnetism–ferromagnetism phase transition.

The structure of this work is as follows. In section 2, we introduce the basic field equations of holographic ferromagnetism model with ENE and LEN in the Schwarzschild-AdS black hole which have not been studied as far as we know, and compare them with the BINE holographic paramagnetism model. In section 3 by the numerical method we obtain the critical temperature and study the magnetic moment in the presence of the three kinds of typical nonlinear electrodynamics. Magnetic susceptibility density and hysteresis loop will be shown in section 4. Finally in the last section we will include our summary and discussion.

2. Holographic model

In this paper, the model we are considering is just general relativity with a negative cosmological constant $\Lambda = -3/L^2$, a U(1) field A_μ and a massive 2-form field $M_{\mu\nu}$ in 4-dimension space-time. The ghost free action

$$S = \frac{1}{2\kappa^2} \int d^4x \sqrt{-g}(L_1 + \lambda^2 L_2) \tag{1}$$

$$L_1 = R + 6/L^2 + L(F), \tag{2}$$

$$L_2 = -\frac{1}{12}(dM)^2 - \frac{m^2}{4}M_{\mu\nu}M^{\mu\nu} - \frac{1}{2}M^{\mu\nu}F_{\mu\nu} - \frac{J}{8}V(M) \tag{3}$$

where dM is the exterior differential of 2-form field $M_{\mu\nu}$, m^2 is the squared mass of 2-form field $M_{\mu\nu}$ being greater than zero (see Ref. [21] for detail), λ and J are two real model parameters with $J < 0$ for producing the spontaneous magnetization, λ^2 characterizes the back reaction of the 2-form field $M_{\mu\nu}$ to the background geometry and to the Maxwell field strength, and $V(M)$ is a nonlinear potential of the 2-form field describing the self-interaction of the polarization tensor. For simplicity, we take the form of $V(M)$ as follows,

$$V(M) = (^*M_{\mu\nu}M^{\mu\nu})^2 = [^*(M \wedge M)]^2, \tag{4}$$

where * is the Hodge-star operator. As shown in Ref. [21], this potential shows a global minimum at some nonzero value of ρ. Meanwhile, $L(F)$ is the Lagrangian of three classes of Born–Infeld-like nonlinear electrodynamics

$$L(F) = \begin{cases} \dfrac{1}{b}\ln(1 + bF), & \text{LNE} \\ \dfrac{1}{b}(1 - \sqrt{1 + 2bF}), & \text{BINE} \\ -\dfrac{1}{4b}(e^{-4bF} - 1). & \text{ENE} \end{cases} \tag{5}$$

Here $F \equiv F_{\mu\nu}F^{\mu\nu}$ and $F_{\mu\nu}$ is the nonlinear electromagnetic tensor. As the nonlinear parameter b tends to zero, the Lagrangian $L(F)$ approaches to $F_{\mu\nu}F^{\mu\nu}$. Note that the higher order terms in the parameter b essentially correspond to the higher derivative corrections of the gauge fields. With the same value of b, we can discuss the difference in the three types of the holographic dual models with the nonlinear electrodynamics quantitatively. It should be noted that the horizon geometry of nonlinear charged black holes is closed to the horizon of uncharged (Schwarzschild) black hole solution for very large values of b [51], so in this case $L(F)$ can be neglected. By varying action (1), we can get the equations of motion for 2-form field

$$\nabla^\tau (dM)_{\tau\mu\nu} - m^2 M_{\mu\nu} - J(^*M_{\tau\sigma}M^{\tau\sigma})(^*M_{\mu\nu}) = F_{\mu\nu}, \tag{6}$$

and gauge field

$$\nabla^\mu(\frac{F_{\mu\nu}}{1+bF} + \frac{\lambda^2}{4}M_{\mu\nu}) = 0, \quad \text{LNE}$$

$$\nabla^\mu(\frac{F_{\mu\nu}}{\sqrt{1+2bF}} + \frac{\lambda^2}{4}M_{\mu\nu}) = 0, \quad \text{BINE} \tag{7}$$

$$\nabla^\mu(\frac{F_{\mu\nu}}{e^{4bF}} + \frac{\lambda^2}{4}M_{\mu\nu}) = 0. \quad \text{ENE}$$

In what follows, we will work in the probe limit and the background is a 4-dimensional planner Schwarzschild-AdS black hole

$$ds^2 = L^2(-r^2 f(r)dt^2 + \frac{dr^2}{r^2 f(r)} + r^2(dx^2 + dy^2)), \tag{8}$$

with

$$f(r) = 1 - \frac{r_+^3}{r^3}, \tag{9}$$

where the r_+ is the event horizon of the black hole and the Hawking temperature is

$$T = \frac{3r_+}{4\pi}. \tag{10}$$

In order to study systematically the effects of the b on the holographic ferromagnetic phase transition, we take the following self-consistent ansatz with matter fields,

$$M_{\mu\nu} = -p(r)dt \wedge dr + \rho(r)dx \wedge dy,$$
$$A_\mu = \phi(r)dt + Bxdy, \tag{11}$$

where B is a constant magnetic field viewed as the external magnetic field in the boundary field theory. Thus nontrivial equations of motion read,

$$\rho'' + \frac{f'}{f}\rho' - \frac{1}{r^2 f}[m^2 + 4Jp^2]\rho + \frac{B}{r^2 f} = 0,$$

$$(m^2 - \frac{4J\rho^2}{r^4})p - \phi' = 0, \tag{12}$$

which are the same form for the three types of nonlinear electrodynamics (7). For the gauge field ϕ, however, we obtain the following equations of motion

$$(4 + 8b(\frac{B^2}{r^4} + \phi'^2))\phi'' + \frac{8\phi'}{r}(1 + (\frac{6B^2}{r^4} - 2\phi'^2)b) - \lambda^2$$

$$(p' + \frac{2p}{r})[\frac{4B^2b}{r^4}(1 - 2\phi'^2b + \frac{B^2b}{r^4}) + (1 - 2\phi'^2b)^2] = 0, \quad \text{LNE} \qquad (13)$$

$$\phi'' + \frac{1}{16B^2br + 4r^5}(16B^2b + 8r^4 - 32br^4\phi'^2)\phi'$$

$$- \frac{r^3\lambda^2(2p + p'r)}{16B^2b + 4r^4}(1 + \frac{4B^2b}{r^4} - 4b\phi'^2)^{3/2} = 0, \quad \text{BINE}$$

$$(4 + 64b\phi'^2)\phi'' - (p' + \frac{2p}{r})\lambda^2 e^{(\frac{8bB^2}{r^4} - 8b\phi'^2)} + \frac{8\phi'}{r}(1 + \frac{16B^2b}{r^4}) = 0, \quad \text{ENE}$$

here a prime denotes the derivative with respect to r. Obviously, Eqs. (12) and (13) reduce to the standard holographic paramagnetism–ferromagnetism phase transition models discussed in Ref. [21] when $b \to 0$. In order to solve the nonlinear Eqs. (12) and (13) numerically, we should first solve the equation of ϕ' and put it into Eq. (13) and get the equation of p. And then we need to seek the boundary condition for ρ, ϕ and p near the black hole horizon $r \to r_+$ and at the spatial infinite $r \to \infty$. The regularity condition for $\rho(r_+)$ at the horizon gives the boundary condition $\phi(r_+) = 0$. Near the boundary $r \to \infty$, the nonlinear equations give the following asymptotic solution for matter fields

$$\rho = \rho_+ r^{\Delta_+} + \rho_- r^{\Delta_-} + \cdots + \frac{B}{m^2},$$

$$\phi = \mu - \frac{\sigma}{r} + \cdots, \quad p = \frac{\sigma}{m^2 r^2} + \cdots, \qquad (14)$$

with

$$\Delta_\pm = \frac{1}{2} \pm \frac{1}{2}\sqrt{1 + 4m^2}, \qquad (15)$$

where $\rho_\pm$, μ and σ are all constants, and μ and σ are interpreted as the chemical potential and charge density in the dual field theory respectively. The coefficient ρ_+ and ρ_- correspond to the source and vacuum expectation value of dual operator in the boundary field theory when $B = 0$. Therefore one should set $\rho_+ = 0$ since one wants the condensation to happen spontaneously below a critical temperature. When $B \neq 0$, the asymptotic behavior is governed by external magnetic field B.

3. Spontaneous magnetization

In this paper we work in the grand canonical ensemble where the chemical potential μ will be fixed. And the expression of magnetic moment as

$$N = -\lambda^2 \int \frac{\rho}{2r^2}dr. \qquad (16)$$

Here, we take $J = -1/8$, $m^2 = 1/8$ and $\lambda = 1/2$ as a typical example, which can capture the basic features of the model. In other words, the other choices of the parameters will not qualitatively modify our results. Using the shooting method, we can solve the Eqs. (12) and (13) numerically and then discuss the effect of the nonlinear electrodynamics on the magnetic moment.

Varying the nonlinearity parameter b, we present in the upper half plane of Fig. 1 the magnetic moment with the LNE (left two panels), BINE (middle two panels) and ENE (right two panels)

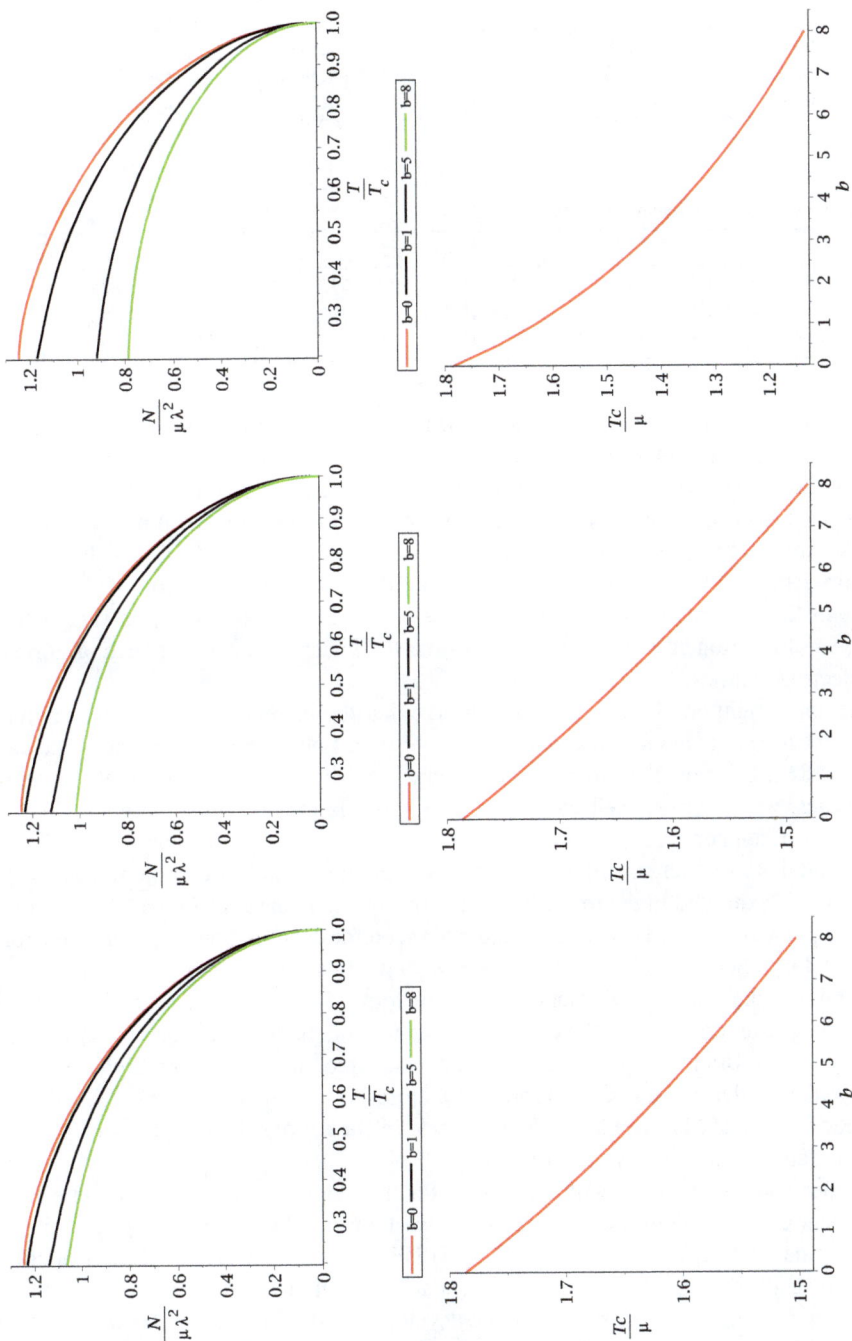

Fig. 1. The variety of the magnetic moment N and the critical temperature T_c with the LNE (left two panels), BINE (middle two panels) and ENE (right two panels) in the presence of nonlinear parameter b.

Table 1

The critical temperature T_c with different values of b.

b	0	1	5	8
LNE	1.7870	1.7422	1.5937	1.5026
BINE	1.7870	1.7417	1.5834	1.4806
ENE	1.7870	1.6337	1.2901	1.1349

Table 2

The magnetic moment N with different values of b.

b	0	1	5	8
LNE	$2.9409(1 - T/T_c)^{1/2}$	$2.8529(1 - T/T_c)^{1/2}$	$2.5177(1 - T/T_c)^{1/2}$	$2.2908(1 - T/T_c)^{1/2}$
BINE	$2.9409(1 - T/T_c)^{1/2}$	$2.8514(1 - T/T_c)^{1/2}$	$2.4835(1 - T/T_c)^{1/2}$	$2.2130(1 - T/T_c)^{1/2}$
ENE	$2.9409(1 - T/T_c)^{1/2}$	$2.6186(1 - T/T_c)^{1/2}$	$1.7843(1 - T/T_c)^{1/2}$	$1.4090(1 - T/T_c)^{1/2}$

as a function of temperature in $d = 4$ dimension. It is found that the spontaneous condensate of ρ (corresponding to the magnetic moment) in the bulk in the absence of external magnetic field appears and has similar behavior for different b when the temperature is lower than critical temperature T_c. Meanwhile, by fitting this curve in the vicinity of critical temperature, we find that the phase transition is a second order one with behavior $N \propto \sqrt{1 - T/T_c}$ for all cases calculated above. The results are still consistent with ones in the mean field theory and have been shown in Table 2. In other words, similar to the case of BINE, the holographic paramagnetism–ferromagnetism transition still exist even we consider the logarithmic and exponential forms of nonlinear electrodynamics.

From the upper right corner of Fig. 1, we observe that the increasing value of the nonlinear parameter b makes the magnetic moment smaller with the ENE, which is similar to the cases of BINE and LNE. It means that the magnetic moment is harder to be formed in the nonlinear electrodynamics, which agrees well with the results given in [55]. In Table 1 and Table 2, we present the critical temperature T_c and the behaviors of these condensation curves near $T \sim T_c$. It is easy to find that as b increases the critical temperature decreases for each nonlinear electrodynamic, which is exhibited in the lower half plane of Fig. 1 and agrees well with the finding in the upper half plane of Fig. 1. This behavior has been seen for the holographic superconductor in the background of a Schwarzschild–AdS black hole, where the three types of typical nonlinear electrodynamics make scalar condensation harder to form [54]. At the same time, the dependence of the magnetic moment and the critical temperature on the nonlinear parameter is similar to that on the Gauss–Bonnet term in the holographic superconductor, i.e., the higher curvature corrections make condensation harder to form. Therefore, we conclude that the ENE, BINE and LNE corrections to usual Maxwell field and the curvature corrections share some similar features for the condensation of the massive 2-form field ρ.

On the other hand, comparing with the curves for the magnetic moment in the three types of the nonlinear electrodynamics considered here, we find that the value of magnetic moment with ENE is smaller than ones in the BINE and LNE cases for the fixed value of nonlinear parameter b (except the case of $b = 0$, i.e., the usual Maxwell electrodynamics), which means that the magnetic moment is more difficult to be developed in the exponential form of nonlinear electrodynamics. This is also in good agreement with the results shown in Table 1 and in lower half plane of Fig. 1, where the critical temperature T_c for the condensate of ρ with the ENE is smaller than ones in the BINE and LNE cases for the fixed value of b.

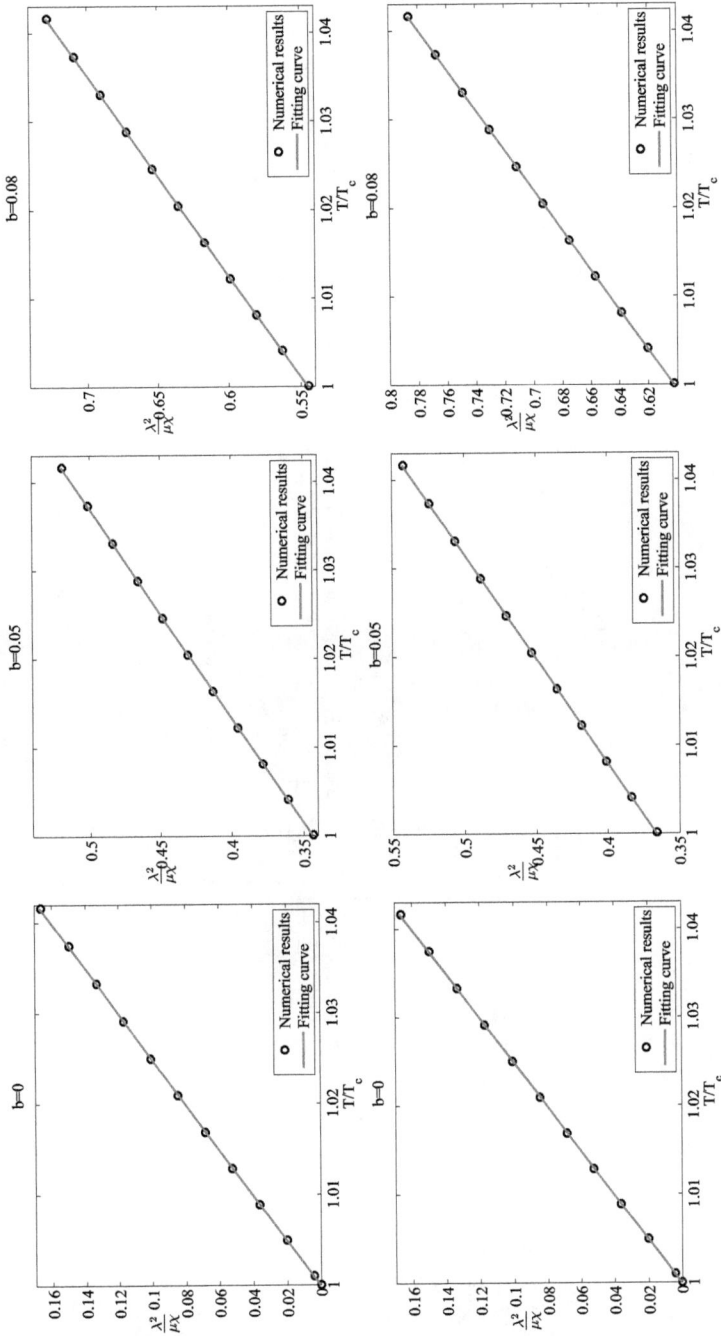

Fig. 2. The magnetic susceptibility as a function of temperature with the BINE (top three panels), LNE (middle three panels), and ENE (bottom three panels) in the presence of nonlinear parameter b.

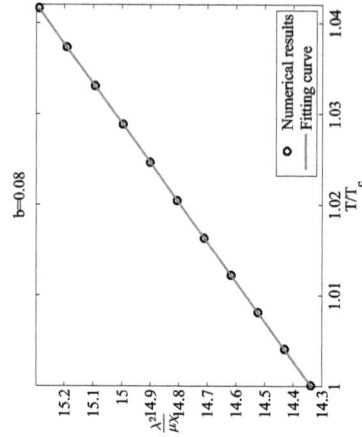

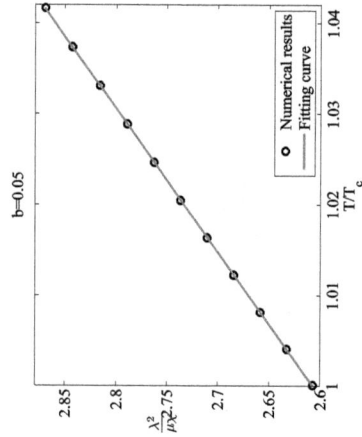

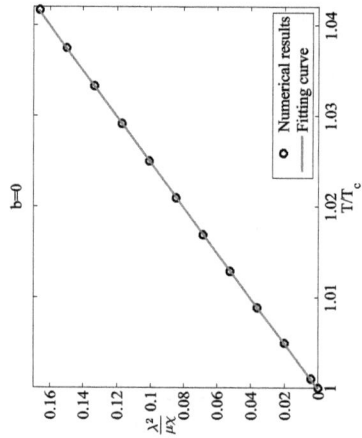

Fig. 2. (*continued*)

Table 3
The magnetic susceptibility χ with different values of b.

	b	0	0.05	0.08
LNE	$\lambda^2/\chi\mu$	$3.9906(T/T_c - 1)$	$4.2328(T/T_c - 0.91)$	$4.4227(T/T_c - 0.86)$
	θ/μ	-1.7870	-1.6303	-1.5404
BINE	$\lambda^2/\chi\mu$	$3.9906(T/T_c - 1)$	$4.2227(T/T_c - 0.92)$	$4.3911(T/T_c - 0.88)$
	θ/μ	-1.7870	-1.6395	-1.5620
ENE	$\lambda^2/\chi\mu$	$3.9906(T/T_c - 1)$	$6.3220(T/T_c - 0.5877)$	$22.8769(T/T_c - 0.3733)$
	θ/μ	-1.7870	-1.0448	-0.6616

4. The response to the external magnetic field

Let us turn on the external field to examine the response to magnetic moment N. This can be described by magnetic susceptibility density χ, defined as

$$\chi = \lim_{B \to 0} \frac{\partial N}{\partial B}. \tag{17}$$

In the high temperature region $T > T_c$, the ferromagnetic material is in a paramagnetic phase whose magnetic moments are randomly distributed. So the susceptibility obeys the Curie–Weiss law

$$\chi = \frac{C}{T + \theta}, T > T_c, \theta < 0, \tag{18}$$

where C and θ are two constants. Note that a significant difference between the antiferromagnetism and paramagnetism can be seen from the magnetic susceptibility. In the paramagnetic phase of antiferromagnetic material and paramagnetic material, the magnetic susceptibility also obeys the Curie–Weiss law, but the constant θ in Eq. (18) is positive and zero, respectively. For the three types of nonlinear electrodynamics, Fig. 2 shows the magnetic susceptibility as a function of temperature by solving Eq. (17) with $b = 0$, 0.05, 0.08. In the paramagnetic phase for all cases considered here, we observe that the magnetic susceptibility increases when the temperature is lowered for the fixed nonlinearly parameter b. Moreover, the magnetic susceptibility satisfies the Curie–Weiss law of the ferromagnetism near the critical temperature whether $b = 0$ or not. Concretely, the results have been presented in Table 3 for the chosen model parameters. It is easy to see that coefficient in front of $\frac{T}{T_c}$ for $\frac{1}{\chi}$ increases with the increasing b, which meets well with the discovery in Fig. 2. However, the absolute value of $\frac{\theta}{\mu}$ will decrease when the Born–Infeld scale parameter b increases. On the other hand, from Fig. 2 and Table 3 we can see the value of coefficient in front of $\frac{T}{T_c}$ for $\frac{1}{\chi}$ of the ENE is larger than that of BINE and LNE for the fixed value of b (except the case of $b = 0$, i.e., the usual Maxwell electrodynamics). Comparing the cases of BINE and LNE, however, the absolute value of $\frac{\theta}{\mu}$ for ENE is smaller. In the plot of Fig. 3, we show that the magnetic moment with respect to external field B in region of $T < T_c$ (i.e., $T = 0.89T_c$) with different parameter b. And from the each line in Fig. 3, we see that the magnetic moment is not single valued when the external magnetic field continuously changes between $-B_{max}$ and B_{max} periodically. Thus a hysteresis loop in the single magnetic domain will be obtained and the nonlinear parameter b has an effect on it quantitatively. Along the horizontal direction (the magnetic moment has been taken a same value), one need a larger external field as the nonlinear parameter b increases. In other words, the nonlinear electrodynamics makes the periodicity of hysteresis loop bigger which is different from the effect of Lifshitz dynamical

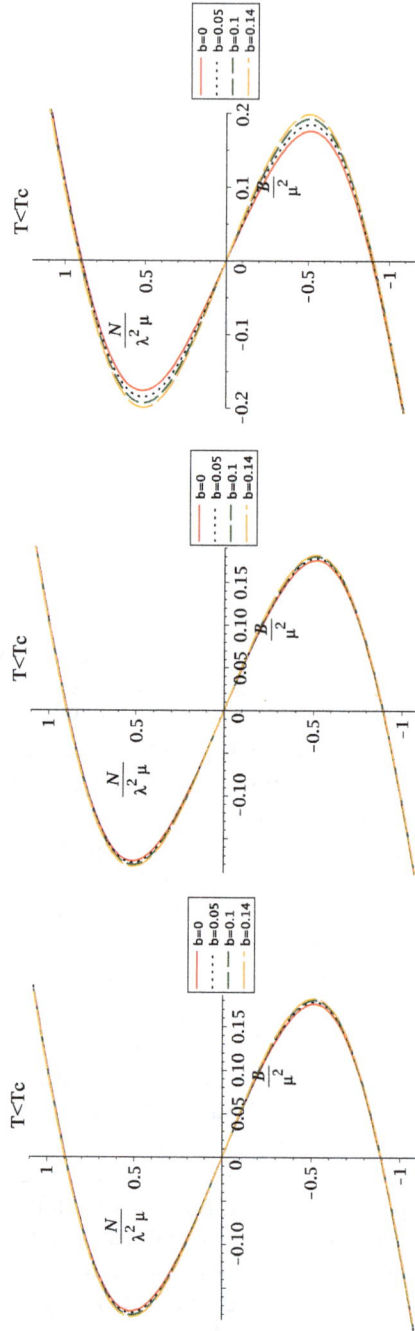

Fig. 3. The magnetic moment with respect to external magnetic field B in lower temperature with the LNE (left panel), BINE (middle panel) and ENE (right panel) in the presence of nonlinear parameter b.

exponent z on it. Particularly, for the case of ENE, whose effect on the periodicity of hysteresis loop is more noticeable. However, all the curves will overlap once the value of the magnetic field exceeds the maximum that corresponding to the case of $b = 0.14$, which can be seen from Fig. 3.

5. Summary and discussion

Sum up, we have investigated systematically holographic paramagnetism–ferromagnetism phase transitions in the presence of three kinds of typical Born–Infeld-like nonlinear electrodynamics correction to the Maxwell electrodynamics in 4-dimension Schwarzschild-AdS black hole spacetime, and obtained the effect of the nonlinear parameter b on the holographic paramagnetism–ferromagnetism phase transition. Considering that these nonlinear generalizations essentially imply the higher derivative corrections of the gauge fields, this study may help to understand the influences of the $1/N$ or $1/\lambda$ corrections on the holographic dual model. In the probe limit, comparing the exponential form of nonlinear electrodynamics (ENE) with the Born–Infeld nonlinear electrodynamics (BINE) and logarithmic form of nonlinear electrodynamics (LNE), in the black hole background, we found it has stronger effects on critical temperature and the magnetic moment. Furthermore, we observed that for all three types of the nonlinear electrodynamics considered here the higher nonlinear electrodynamics correction term can make the condensation harder form and result in the decreases of critical temperature and magnetic moment in the absence of magnetic field. This behavior is similar to that of the holographic superconductor in the background of a Schwarzschild-AdS black hole, where the three kinds of nonlinear electrodynamics make scalar condensation harder form and result in the larger deviations from the universal value $\omega_g/T_c \approx 8$ for the gap frequency. In the vicinity of the critical point, however, the behavior of the magnetic moment is always as $(1 - T/T_c)^{1/2}$, which is independent of the explicit form of the nonlinear electrodynamics, i.e., the ENE, BINE, LNE correction terms do not have any effect on the relationship. Meanwhile, it is in agreement with the result from mean field theory. Moreover, in the presence of the external magnetic field, the inverse magnetic susceptibility as $T \sim T_c$ behaves as $C/(T + \theta)$, $(\theta < 0)$ in all cases, which satisfies the Cure-Weiss law. Yet both the constant C and the absolute value of θ decrease with the increasing nonlinear parameter b. Furthermore, we have observed the hysteresis loop in the single magnetic domain when the external field continuously changes between the maximum and minimum values periodically with b. The increase of the nonlinear parameter b could result in extending the period of the external magnetic field. Especially, the effect of the exponential form of nonlinear electrodynamics on the periodicity of hysteresis loop is more noticeable.

Note that in this paper we just investigate the influences of the three kinds of nonlinear electrodynamics on paramagnetism–ferromagnetism phase transition. It would be of interest to generalize our study to holographic paramagnetism–antiferromagnetism model. Work in this direction will be reported in the future.

Acknowledgements

We would like to thank Prof. R. G. Cai and Dr. R. Q. Yang for their helpful discussions and comments. This work is supported by the National Natural Science Foundation of China (Grant Nos. 11175077, 11575075).

References

[1] J.M. Maldacena, The large N limit of superconformal field theories and supergravity, Int. J. Theor. Phys. 38 (1999) 1113, Adv. Theor. Math. Phys. 2 (1998) 231, arXiv:hep-th/9711200.

[2] S.S. Gubser, I.R. Klebanov, A.M. Polyakov, Gauge theory correlators from noncritical string theory, Phys. Lett. B 428 (1998) 105, arXiv:hep-th/9802109.

[3] E. Witten, Anti-de Sitter space and holography, Adv. Theor. Math. Phys. 2 (1998) 253, arXiv:hep-th/9802150.

[4] E. Witten, Anti-de Sitter space, thermal phase transition, and confinement in gauge theories, Adv. Theor. Math. Phys. 2 (1998) 505, arXiv:hep-th/9803131.

[5] S.S. Gubser, Breaking an Abelian gauge symmetry near a black hole horizon, Phys. Rev. D 78 (2008) 065034, arXiv:0801.2977 [hep-th].

[6] S.A. Hartnoll, C.P. Herzog, G.T. Horowitz, Building a holographic superconductor, Phys. Rev. Lett. 101 (2008) 031601, arXiv:0803.3295 [hep-th].

[7] S.A. Hartnoll, C.P. Herzog, G.T. Horowitz, Holographic superconductors, J. High Energy Phys. 0812 (2008) 015, arXiv:0810.1563 [hep-th];
S.A. Hartnoll, Lectures on holographic methods for condensed matter physics, Class. Quantum Gravity 26 (2009) 224002, arXiv:0903.3246 [hep-th].

[8] C.P. Herzog, Lectures on holographic superfluidity and superconductivity, J. Phys. A 42 (2009) 343001, arXiv:0904.1975 [hep-th];
C.P. Herzog, An analytic holographic superconductor, Phys. Rev. D 81 (2010) 126009, arXiv:1003.3278 [hep-th].

[9] G.T. Horowitz, Introduction to holographic superconductors, Lect. Notes Phys. 828 (2011) 313, arXiv:1002.1722 [hep-th].

[10] Z.Y. Nie, R.G. Cai, X. Gao, H. Zeng, Competition between the s-wave and p-wave superconductivity phases in a holographic model, J. High Energy Phys. 1311 (2013) 087, arXiv:1309.2204 [hep-th];
Z.Y. Nie, R.G. Cai, X. Gao, L. Li, H. Zeng, Phase transitions in a holographic s + p model with back-reaction, Eur. Phys. J. C 75 (2015) 559, arXiv:1501.00004 [hep-th];
Z.Y. Nie, H. Zeng, P–T phase diagram of a holographic s+p model from Gauss–Bonnet gravity, J. High Energy Phys. 1510 (2015) 047, arXiv:1505.02289 [hep-th].

[11] R.G. Cai, L. Li, L.F. Li, R.Q. Yang, Introduction to holographic superconductor models, Sci. China, Phys. Mech. Astron. 58 (6) (2015) 060401, arXiv:1502.00437 [hep-th].

[12] Y. Ling, P. Liu, C. Niu, J.P. Wu, Z.Y. Xian, Holographic superconductor on Q-lattice, J. High Energy Phys. 1502 (2015) 059, arXiv:1410.6761 [hep-th];
Y. Ling, P. Liu, C. Niu, J.P. Wu, Z.Y. Xian, Holographic entanglement entropy close to quantum phase transitions, J. High Energy Phys. 1604 (2016) 114, arXiv:1502.03661 [hep-th].

[13] S.A.H. Mansoori, B. Mirza, A. Mokhtari, F.L. Dezaki, Z. Sherkatghanad, Weyl holographic superconductor in the Lifshitz black hole background, J. High Energy Phys. 1607 (2016) 111, arXiv:1602.07245 [hep-th].

[14] X.M. Kuang, E. Papantonopoulos, Building a holographic superconductor with a scalar field coupled kinematically to Einstein tensor, J. High Energy Phys. 1608 (2016) 161, arXiv:1607.04928 [hep-th].

[15] J.W. Lu, Y.B. Wu, P. Qian, Y.Y. Zhao, X. Zhang, Lifshitz scaling effects on holographic superconductors, Nucl. Phys. B 887 (2014) 112, arXiv:1311.2699 [hep-th];
Y.B. Wu, J.W. Lu, M.L. Liu, J.B. Lu, C.Y. Zhang, Z.Q. Yang, Lifshitz effects on vector condensate induced by a magnetic field, Phys. Rev. D 89 (10) (2014) 106006, arXiv:1403.5649 [hep-th];
J.W. Lu, Y.B. Wu, J. Xiao, C.J. Lu, M.L. Liu, Holographic superconductors in IR modified Hořava–Lifshitz gravity, Int. J. Mod. Phys. A 31 (19) (2016) 1650110.

[16] C. Lai, Q. Pan, J. Jing, Y. Wang, Analytical study on holographic superfluid in AdS soliton background, Phys. Lett. B 757 (2016) 65, arXiv:1601.00134 [hep-th].

[17] R.G. Cai, R.Q. Yang, Paramagnetism—ferromagnetism phase transition in a dyonic black hole, Phys. Rev. D 90 (8) (2014) 081901, arXiv:1404.2856 [hep-th].

[18] R.G. Cai, R.Q. Yang, Holographic model for the paramagnetism/antiferromagnetism phase transition, Phys. Rev. D 91 (8) (2015) 086001, arXiv:1404.7737 [hep-th].

[19] R.G. Cai, R.Q. Yang, Coexistence and competition of ferromagnetism and p-wave superconductivity in holographic model, Phys. Rev. D 91 (2) (2015) 026001, arXiv:1410.5080 [hep-th].

[20] N. Yokoi, M. Ishihara, K. Sato, E. Saitoh, Holographic realization of ferromagnets, Phys. Rev. D 93 (2) (2016) 026002, arXiv:1508.01626 [hep-th].

[21] R.G. Cai, R.Q. Yang, Antisymmetric tensor field and spontaneous magnetization in holographic duality, Phys. Rev. D 92 (4) (2015) 046001, arXiv:1504.00855 [hep-th].

[22] R.G. Cai, R.Q. Yang, Y.B. Wu, C.Y. Zhang, Massive 2-form field and holographic ferromagnetic phase transition, J. High Energy Phys. 1511 (2015) 021, arXiv:1507.00546 [hep-th].

[23] R.G. Cai, R.Q. Yang, Insulator/metal phase transition and colossal magnetoresistance in holographic model, Phys. Rev. D 92 (10) (2015) 106002, arXiv:1507.03105 [hep-th].

[24] R.G. Cai, R.Q. Yang, Understanding strongly coupling magnetism from holographic duality, arXiv:1601.02936 [hep-th].

[25] C.Y. Zhang, Y.B. Wu, Y.Y. Jin, Y.T. Chai, M.H. Hu, Z. Zhang, Lifshitz scaling effects on the holographic paramagnetism–ferromagnetism phase transition, Phys. Rev. D 93 (2016) 126001, arXiv:1603.04149 [gr-qc].

[26] D. Anninos, G. Pastras, Thermodynamics of the Maxwell–Gauss–Bonnet anti-de Sitter black hole with higher derivative gauge corrections, J. High Energy Phys. 0907 (2009) 030, arXiv:0807.3478 [hep-th].

[27] S.H. Hendi, B.E. Panah, Thermodynamics of rotating black branes in Gauss–Bonnet-nonlinear Maxwell gravity, Phys. Lett. B 684 (2010) 77, arXiv:1008.0102 [hep-th].

[28] O. Miskovic, R. Olea, Conserved charges for black holes in Einstein–Gauss–Bonnet gravity coupled to nonlinear electrodynamics in AdS space, Phys. Rev. D 83 (2011) 024011, arXiv:1009.5763 [hep-th].

[29] O. Miskovic, R. Olea, Quantum statistical relation for black holes in nonlinear electrodynamics coupled to Einstein–Gauss–Bonnet AdS gravity, Phys. Rev. D 83 (2011) 064017, arXiv:1012.4867 [hep-th].

[30] O. Gurtug, S.H. Mazharimousavi, M. Halilsoy, $2 + 1$-dimensional electrically charged black holes in Einstein-power-Maxwell theory, Phys. Rev. D 85 (2012) 104004, arXiv:1010.2340 [gr-qc].

[31] A.E. Shabad, V.V. Usov, Effective Lagrangian in nonlinear electrodynamics and its properties of causality and unitarity, Phys. Rev. D 83 (2011) 105006, arXiv:1101.2343 [hep-th].

[32] M. Born, L. Infeld, Foundations of the new field theory, Proc. R. Soc. Lond. Ser. A, Math. Phys. Sci. 144 (1934) 425.

[33] G. Boillat, Nonlinear electrodynamics – lagrangians and equations of motion, J. Math. Phys. 11 (1970) 941.

[34] J. Jing, S. Chen, Holographic superconductors in the Born–Infeld electrodynamics, Phys. Lett. B 686 (2010) 68, arXiv:1001.4227 [gr-qc].

[35] J. Jing, L. Wang, Q. Pan, S. Chen, Holographic superconductors in Gauss–Bonnet gravity with Born–Infeld electrodynamics, Phys. Rev. D 83 (2011) 066010, arXiv:1012.0644 [gr-qc].

[36] J. Jing, Q. Pan, S. Chen, Holographic superconductors with Power-Maxwell field, J. High Energy Phys. 1111 (2011) 045, arXiv:1106.5181 [hep-th].

[37] J. Jing, L. Jiang, Q. Pan, Holographic superconductors for the Power-Maxwell field with backreactions, Class. Quantum Gravity 33 (2) (2016) 025001.

[38] A. Sheykhi, H.R. Salahi, A. Montakhab, Analytical and numerical study of Gauss–Bonnet holographic superconductors with power-Maxwell field, J. High Energy Phys. 1604 (2016) 058, arXiv:1603.00075 [gr-qc].

[39] Q. Pan, J. Jing, B. Wang, Holographic superconductor models with the Maxwell field strength corrections, Phys. Rev. D 84 (2011) 126020, arXiv:1111.0714 [gr-qc].

[40] S. Gangopadhyay, D. Roychowdhury, Analytic study of Gauss–Bonnet holographic superconductors in Born–Infeld electrodynamics, J. High Energy Phys. 1205 (2012) 156, arXiv:1204.0673 [hep-th].

[41] C.O. Lee, The holographic superconductors in higher-dimensional AdS soliton, Eur. Phys. J. C 72 (2012) 2092, arXiv:1202.5146 [gr-qc].

[42] D. Roychowdhury, Effect of external magnetic field on holographic superconductors in presence of nonlinear corrections, Phys. Rev. D 86 (2012) 106009, arXiv:1211.0904 [hep-th].

[43] S. Gangopadhyay, D. Roychowdhury, Analytic study of properties of holographic superconductors in Born–Infeld electrodynamics, J. High Energy Phys. 1205 (2012) 002, arXiv:1201.6520 [hep-th].

[44] W. Yao, J. Jing, Analytical study on holographic superconductors for Born–Infeld electrodynamics in Gauss–Bonnet gravity with backreactions, J. High Energy Phys. 1305 (2013) 101, arXiv:1306.0064 [gr-qc].

[45] W. Yao, J. Jing, Holographic entanglement entropy in metal/superconductor phase transition with Born–Infeld electrodynamics, Nucl. Phys. B 889 (2014) 109, arXiv:1408.1171 [hep-th].

[46] C. Lai, Q. Pan, J. Jing, Y. Wang, On analytical study of holographic superconductors with Born–Infeld electrodynamics, Phys. Lett. B 749 (2015) 437, arXiv:1508.05926 [hep-th].

[47] D. Ghorai, S. Gangopadhyay, Higher dimensional holographic superconductors in Born–Infeld electrodynamics with back-reaction, Eur. Phys. J. C 76 (3) (2016) 146, arXiv:1511.02444 [hep-th].

[48] A. Sheykhi, F. Shaker, Analytical study of holographic superconductor in Born–Infeld electrodynamics with back-reaction, Phys. Lett. B 754 (2016) 281, arXiv:1601.04035 [hep-th].

[49] H.H. Soleng, Charged black points in general relativity coupled to the logarithmic U(1) gauge theory, Phys. Rev. D 52 (1995) 6178, arXiv:hep-th/9509033.

[50] J. Jing, Q. Pan, S. Chen, Holographic superconductor/insulator transition with logarithmic electromagnetic field in Gauss–Bonnet gravity, Phys. Lett. B 716 (2012) 385, arXiv:1209.0893 [hep-th].

[51] S.H. Hendi, Asymptotic charged BTZ black hole solutions, J. High Energy Phys. 1203 (2012) 065, arXiv:1405.4941 [hep-th].

[52] S.H. Hendi, A. Sheykhi, Charged rotating black string in gravitating nonlinear electromagnetic fields, Phys. Rev. D 88 (4) (2013) 044044, arXiv:1405.6998 [gr-qc].

[53] W. Yao, J. Jing, Holographic entanglement entropy in metal/superconductor phase transition with exponential non-linear electrodynamics, Phys. Lett. B 759 (2016) 533, arXiv:1603.04516 [gr-qc].

[54] Z. Zhao, Q. Pan, S. Chen, J. Jing, Notes on holographic superconductor models with the nonlinear electrodynamics, Nucl. Phys. B 871 (2013) 98, arXiv:1212.6693.

[55] Y.B. Wu, C.Y. Zhang, J.W. Lu, B. Fan, S. Shu, Y.C. Liu, Holographic paramagnetism–ferromagnetism phase transition in the Born–Infeld electrodynamics, Phys. Lett. B 760 (2016) 469.

4

Fermionic T-duality in fermionic double space [☆]

B. Nikolić [*], B. Sazdović

Institute of Physics Belgrade, University of Belgrade, Pregrevica 118, 11080 Belgrade, Serbia

Editor: Stephan Stieberger

Abstract

In this article we offer the interpretation of the fermionic T-duality of the type II superstring theory in double space. We generalize the idea of double space doubling the fermionic sector of the superspace. In such doubled space fermionic T-duality is represented as permutation of the fermionic coordinates θ^α and $\bar{\theta}^\alpha$ with the corresponding fermionic T-dual ones, ϑ_α and $\bar{\vartheta}_\alpha$, respectively. Demanding that T-dual transformation law has the same form as initial one, we obtain the known form of the fermionic T-dual NS–R and R–R background fields. Fermionic T-dual NS–NS background fields are obtained under some assumptions. We conclude that only symmetric part of R–R field strength and symmetric part of its fermionic T-dual contribute to the fermionic T-duality transformation of dilaton field and analyze the dilaton field in fermionic double space. As a model we use the ghost free action of type II superstring in pure spinor formulation in approximation of constant background fields up to the quadratic terms.

1. Introduction

Two theories T-dual to one another can be viewed as being physically identical [1,2]. T-duality presents an important tool which shows the equivalence of different geometries and topologies. The useful T-duality procedure was first introduced by Buscher [3].

[☆] Work supported in part by the Serbian Ministry of Education, Science and Technological Development, under contract No. 171031.

[*] Corresponding author.
   E-mail addresses: bnikolic@ipb.ac.rs (B. Nikolić), sazdovic@ipb.ac.rs (B. Sazdović).

Mathematical realization of T-duality is given by Buscher T-dualization procedure [3], which is considered as standard one. There are also other frameworks in which we can represent T-dualization which should agree with the Buscher procedure. It is double space formalism which was the subject of the articles about twenty years ago [4–8]. Double space is spanned by coordinates $Z^M = (x^\mu \quad y_\mu)^T$ ($\mu = 0, 1, 2, \ldots, D - 1$), where x^μ and y_μ are the coordinates of the D-dimensional initial and T-dual space–time, respectively. Interest for this subject emerged recently with papers [9–13], where T-duality along some subset of d coordinates is considered as $O(d, d)$ symmetry transformation and [14,15], where it is considered as permutation of d initial with corresponding d T-dual coordinates.

Until recently only T-duality along bosonic coordinates has been considered. Analyzing the gluon scattering amplitudes in $N = 4$ super Yang–Mills theory, a new kind of T-dual symmetry, fermionic T-duality, was discovered [16,17]. It is a part of the dual superconformal symmetry which should be connected to integrability and it is valid just at string tree level. Mathematically, fermionic T-duality is realized within the same procedure as bosonic one, except that dualization is performed along fermionic variables. So, it can be considered as a generalization of Buscher T-duality. Fermionic T-duality consists in certain non-local redefinitions of the fermionic variables of the superstring mapping a supersymmetric background to another supersymmetric background. In Refs. [16,17] it was shown that fermionic T-duality maps gluon scattering amplitudes in the original theory to an object very close to Wilson loops in the dual one. Calculation of gluon scattering amplitudes in the initial theory is equivalent to the calculation of Wilson loops in fermionic T-dual theory. Generalizing the idea of double space to the fermionic case we would get fermionic double space in which fermionic T-duality is a symmetry [18] which exchanges scattering amplitudes and Wilson loops. Fermionic double space can be also successfully applied in random lattice [19], where doubling of the supercoordinate was done. Relation between fermionic T-duality and open string noncommutativity was considered in Ref. [20].

Let us explain our motivation for fermionic T-duality. It is well known that T-duality is important feature in understanding the M-theory. In fact, five consistent superstring theories are connected by web of T and S dualities. We are going to pay attention to the T-duality, hoping that S-duality (which can be understood as transformation of dilaton background field also) can be later successfully incorporated into our procedure. If we start with arbitrary (of five consistent superstring) theory and find all corresponding T-dual theories we can achieve any of other four consistent superstring theories. But to obtain formulation of M-theory it is not enough. We must construct one theory which contains the initial theory and all corresponding T-dual ones.

In the bosonic case (which is substantially simpler that supersymmetric one) we have succeeded to realize such program. In Refs. [14,15] we doubled all bosonic coordinates and showed that such theory contained the initial and all corresponding T-dual theories. We can connect arbitrary two of these theories just replacing some initial coordinates x^a with corresponding T-dual ones y_a. This is equivalent with T-dualization along coordinates x^a. So, introducing double space T-duality ceases to be transformation which connects two physically equivalent theories but it becomes symmetry transformation in extended space with respect to permutation group. We proved this in the bosonic string case both for constant and for weakly curved background with linear dependence on coordinates.

Unfortunately, this is not enough for construction of M-theory, because the T-duality for superstrings is much more complicated then in the bosonic case [21]. In Ref. [22] we have tried to extend such approach to the type II theories. In fact, doubling all bosonic coordinates we have unified types IIA, IIB as well as type $II^\star$ [23] (obtained by T-dualization along time-like direction) theories. There is an incompleteness in such approach. Doubling all bosonic coordinates, by simple permutations of initial with corresponding T-dual coordinates, we obtained all T-dual background fields except T-dual R–R field strength $F^{\alpha\beta}$. To obtain ${}_aF^{\alpha\beta}$ (the field strength after T-dualization along coordinates x^a) we need to introduce some additional assumptions. The explanation is that R–R field strength $F^{\alpha\beta}$ appears coupled with fermionic momenta π_α and $\bar{\pi}_\alpha$ along which we did not performed T-dualization and consequently we did not double these variables. It is an analogue of ij-term in approach of Refs. [9,10] where x^i coordinates are not doubled.

Therefore, in the first step of our approach to the formulation of M-theory (unification of types II theories) we must include T-dualization along fermionic variables (π_α and $\bar{\pi}_\alpha$ in particular case). It means that we should doubled these fermionic variables, also. The present article represents a necessary step for understanding T-dualization along all fermionic coordinates in fermionic double space. We expect that final step in construction of M-theory will be unification of all theories obtained after T-dualization along all bosonic and all fermionic variables [18,19]. In that case we should double all coordinates in superspace, anticipating that some superpermutation will connect arbitrary two of our five consistent supersymmetric string theories.

In this article we are going to double fermionic sector of type II theories adding to the coordinates θ^α and $\bar{\theta}^\alpha$ their fermionic T-duals, ϑ_α and $\bar{\vartheta}_\alpha$, where index α counts independent real components of the spinors, $\alpha = 1, 2, \ldots, 16$. Rewriting T-dual transformation laws in terms of the double coordinates, $\Theta^A = (\theta^\alpha, \vartheta_\alpha)$ and $\bar{\Theta}^A = (\bar{\theta}^\alpha, \bar{\vartheta}_\alpha)$, we define the "fermionic generalized metric" $\mathcal{F}_{AB}$ and the generalized currents $\bar{\mathcal{J}}_{+A}$ and $\mathcal{J}_{-A}$. The permutation matrix $\mathcal{T}^A{}_B$ exchanges $\bar{\theta}^\alpha$ and θ^α with their T-dual partners, $\bar{\vartheta}_\alpha$ and ϑ_α, respectively. From the requirement that fermionic T-dual coordinates, ${}^\star\Theta^A = \mathcal{T}^A{}_B\Theta^B$ and ${}^\star\bar{\Theta}^A = \mathcal{T}^A{}_B\bar{\Theta}^B$, have the same transformation law as initial ones, Θ^A and $\bar{\Theta}^A$, we obtain the expressions for fermionic T-dual generalized metric, ${}^\star\mathcal{F}_{AB} = (\mathcal{T}\mathcal{F}\mathcal{T})_{AB}$, and T-dual currents, ${}^\star\bar{\mathcal{J}}_{+A} = \mathcal{T}_A{}^B\bar{\mathcal{J}}_{+B}$ and ${}^\star\mathcal{J}_{-A} = \mathcal{T}_A{}^B\mathcal{J}_{-B}$, in terms of the initial ones. These expressions produce the expression for fermionic T-dual NS–R fields and R–R field strength. Expressions for fermionic T-dual metric and Kalb–Ramond field are obtained separately under some assumptions. We conclude that only symmetric part of R–R field strength, $F_s^{\alpha\beta} = \frac{1}{2}(F^{\alpha\beta} + F^{\beta\alpha})$, and symmetric part of its fermionic T-dual, ${}^\star F_{\alpha\beta}^s = \frac{1}{2}({}^\star F_{\alpha\beta} + {}^\star F_{\beta\alpha})$, give contribution to the dilaton field transformation under fermionic T-duality. We also investigate the dilaton field in double space.

2. Type II superstring and fermionic T-duality

In this section we will introduce the action of type II superstring theory in pure spinor formulation and perform fermionic T-duality [16,17,20] using fermionic analogue of Buscher rules [3].

2.1. Action and supergravity constraints

In this manuscript we use the action of type II superstring theory in pure spinor formulation [24] up to the quadratic terms with constant background fields. Here we will derive the final form of the action which will be exploited in the further analysis. It corresponds to the actions used in Refs. [25–28].

The sigma model action for type II superstring of Ref. [29] is of the form

$$S = S_0 + V_{SG}, \tag{2.1}$$

where S_0 is the action in the flat background

$$S_0 = \int\limits_{\Sigma} d^2\xi \left(\frac{\kappa}{2} \eta^{mn} \eta_{\mu\nu} \partial_m x^\mu \partial_n x^\nu - \pi_\alpha \partial_- \theta^\alpha + \partial_+ \bar{\theta}^\alpha \bar{\pi}_\alpha \right) + S_\lambda + S_{\bar{\lambda}} , \tag{2.2}$$

and it is deformed by integrated form of the massless type II supergravity vertex operator

$$V_{SG} = \int\limits_{\Sigma} d^2\xi (X^T)^M A_{MN} \bar{X}^N . \tag{2.3}$$

The vectors X^M and $\bar{X}^N$ are defined as

$$X^M = \begin{pmatrix} \partial_+ \theta^\alpha \\ \Pi_+^\mu \\ d_\alpha \\ \frac{1}{2} N_+^{\mu\nu} \end{pmatrix} , \quad \bar{X}^N = \begin{pmatrix} \partial_- \bar{\theta}^\alpha \\ \Pi_-^\mu \\ \bar{d}_\alpha \\ \frac{1}{2} \bar{N}_-^{\mu\nu} \end{pmatrix} , \tag{2.4}$$

and supermatrix A_{MN} is of the form

$$A_{MN} = \begin{pmatrix} A_{\alpha\beta} & A_{\alpha\nu} & E_\alpha{}^\beta & \Omega_{\alpha,\mu\nu} \\ A_{\mu\beta} & A_{\mu\nu} & \bar{E}_\mu{}^\beta & \Omega_{\mu,\nu\rho} \\ E^\alpha{}_\beta & E^\alpha_\nu & P^{\alpha\beta} & C^\alpha{}_{\mu\nu} \\ \Omega_{\mu\nu,\beta} & \Omega_{\mu\nu,\rho} & \bar{C}_{\mu\nu}{}^\beta & S_{\mu\nu,\rho\sigma} \end{pmatrix} , \tag{2.5}$$

where notation and definitions are taken from Ref. [29]. The actions for pure spinors, S_λ and $S_{\bar{\lambda}}$, are free field actions and fully decoupled from the rest of action S_0. The world sheet Σ is parameterized by $\xi^m = (\xi^0 = \tau, \xi^1 = \sigma)$ and $\partial_\pm = \partial_\tau \pm \partial_\sigma$. Bosonic part of superspace is spanned by coordinates x^μ ($\mu = 0, 1, 2, \ldots, 9$), while the fermionic one is spanned by θ^α and $\bar{\theta}^\alpha$ ($\alpha = 1, 2, \ldots, 16$). The variables π_α and $\bar{\pi}_\alpha$ are canonically conjugated momenta to θ^α and $\bar{\theta}^\alpha$, respectively. All spinors are Majorana–Weyl ones, which means that each of them has 16 independent real components. Matrix with superfields generally depends on x^μ, θ^α and $\bar{\theta}^\alpha$.

The superfields $A_{\mu\nu}$, $\bar{E}_\mu{}^\alpha$, $E^\alpha{}_\mu$ and $P^{\alpha\beta}$ are known as physical superfields, while the fields given in the first column and first row are auxiliary superfields because they can be expressed in terms of the physical ones [29]. The rest ones, $\Omega_{\mu,\nu\rho}(\Omega_{\mu\nu,\rho})$, $C^\alpha{}_{\mu\nu}(\bar{C}_{\mu\nu}{}^\alpha)$ and $S_{\mu\nu,\rho\sigma}$, are curvatures (field strengths) for physical superfields.

The expanded form of the vertex operator (2.3) is [29]

$$\begin{aligned} V_{SG} = \int d^2\xi \Big[&\partial_+ \theta^\alpha A_{\alpha\beta} \partial_- \bar{\theta}^\beta + \partial_+ \theta^\alpha A_{\alpha\mu} \Pi_-^\mu + \Pi_+^\mu A_{\mu\alpha} \partial_- \bar{\theta}^\alpha + \Pi_+^\mu A_{\mu\nu} \Pi_-^\nu \\ &+ d_\alpha E^\alpha{}_\beta \partial_- \bar{\theta}^\beta + d_\alpha E^\alpha{}_\mu \Pi_-^\mu + \partial_+ \theta^\alpha E_\alpha{}^\beta \bar{d}_\beta + \Pi_+^\mu E_\mu{}^\beta \bar{d}_\beta + d_\alpha P^{\alpha\beta} \bar{d}_\beta \\ &+ \frac{1}{2} N_+^{\mu\nu} \Omega_{\mu\nu,\beta} \partial_- \bar{\theta}^\beta + \frac{1}{2} N_+^{\mu\nu} \Omega_{\mu\nu,\rho} \Pi_-^\rho + \frac{1}{2} \partial_+ \theta^\alpha \Omega_{\alpha,\mu\nu} \bar{N}_-^{\mu\nu} + \frac{1}{2} \Pi_+^\mu \Omega_{\mu,\nu\rho} \bar{N}_-^{\nu\rho} \\ &+ \frac{1}{2} N_+^{\mu\nu} \bar{C}_{\mu\nu}{}^\beta \bar{d}_\beta + \frac{1}{2} d_\alpha C^\alpha{}_{\mu\nu} \bar{N}_-^{\mu\nu} + \frac{1}{4} N_+^{\mu\nu} S_{\mu\nu,\rho\sigma} \bar{N}_-^{\rho\sigma} \Big] . \end{aligned} \tag{2.6}$$

The supergravity constraints are the conditions obtained as a consequence of nilpotency and (anti)holomorphicity of BRST operators $Q = \int \lambda^\alpha d_\alpha$ and $\bar{Q} = \int \bar{\lambda}^\alpha \bar{d}_\alpha$, where λ^α and $\bar{\lambda}^\alpha$ are pure spinors and d_α and $\bar{d}_\alpha$ are independent variables. Let us discuss the choice of background fields satisfying superspace equations of motion in the context of supergravity constraints which are explained in details for pure spinor formalism in Refs. [32,29].

In order to implement T-duality many restrictions should be imposed. For example, in bosonic case one should assume the existence of Killing vectors, which in fact means background fields independence on corresponding suitably selected coordinates. The idea is to avoid dependence on the coordinate x^μ and allow only dependence on the σ and τ derivatives of the coordinates, $\dot{x}^\mu$ and x'^μ. The case with explicit dependence on the coordinate requires particular attention and has been considered in Ref. [30]. Similar simplifications must be imposed in consideration of the non-commutativity of the coordinates [31,30].

A similar situation occurs in the supersymmetric case. In order to perform fermionic T-duality we must avoid explicit dependence of background fields on the fermionic coordinates θ^α and $\bar{\theta}^\alpha$ (fermionic coordinates are Killing spinors) and allow only dependence on the σ and τ derivatives of these coordinates. Assumption of existence of Killing spinors produces that the auxiliary superfields should be taken to be zero what can be seen from Eq. (5.5) of Ref. [29].

The right-hand side of the equations of motion for background fields (see for example [33]) is energy-momentum tensor which is generally square of field strengths. In our case physical super-fields $G_{\mu\nu}$, $B_{\mu\nu}$, Φ, Ψ^α_μ and $\bar{\Psi}^\alpha_\mu$ are constant (do not depend on x^μ, θ^α, $\bar{\theta}^\alpha$) and corresponding field strengths, $\Omega_{\mu,\nu\rho}(\Omega_{\mu\nu,\rho})$, $C^\alpha{}_{\mu\nu}(\bar{C}_{\mu\nu}{}^\alpha)$ and $S_{\mu\nu,\rho\sigma}$, are zero. The only nontrivial contribution of the quadratic terms in equations of motion comes from constant field strength $P^{\alpha\beta}$. It can induce back-reaction to the background fields. In order to analyze this issue we will use relations from Eq. (3.6) of Ref. [29] labeled by $(\frac{1}{2}, \frac{3}{2}, \frac{3}{2})$

$$D_\alpha P^{\beta\gamma} - \frac{1}{4}(\Gamma^{\mu\nu})_\alpha{}^\beta \bar{C}_{\mu\nu}{}^\gamma = 0, \quad \bar{D}_\alpha P^{\beta\gamma} - \frac{1}{4}(\Gamma^{\mu\nu})_\alpha{}^\gamma C^\beta{}_{\mu\nu} = 0, \tag{2.7}$$

in which derivative of $P^{\alpha\beta}$ appears. Here

$$D_\alpha = \frac{\partial}{\partial\theta^\alpha} + \frac{1}{2}(\Gamma^\mu\theta)_\alpha \frac{\partial}{\partial x^\mu}, \quad \bar{D}_\alpha = \frac{\partial}{\partial\bar{\theta}^\alpha} + \frac{1}{2}(\Gamma^\mu\bar{\theta})_\alpha \frac{\partial}{\partial x^\mu}, \tag{2.8}$$

are superspace covariant derivatives and $C^\alpha{}_{\mu\nu}$ and $\bar{C}_{\mu\nu}{}^\alpha$ are field strengths for gravitino fields Ψ^α_μ and $\bar{\Psi}^\alpha_\mu$, respectively. In order to perform fermionic T-dualization along all fermionic directions, θ^α and $\bar{\theta}^\alpha$, we assume that they are Killing spinors which means

$$\frac{\partial P^{\beta\gamma}}{\partial\theta^\alpha} = \frac{\partial P^{\beta\gamma}}{\partial\bar{\theta}^\alpha} = 0. \tag{2.9}$$

Taking into account that gravitino fields, Ψ^α_μ and $\bar{\Psi}^\alpha_\mu$, are constant (corresponding field strengths are zero), from the equations (2.7) it follows

$$(\Gamma^\mu)_{\alpha\delta} \partial_\mu P^{\beta\gamma} = 0. \tag{2.10}$$

Note that this is more general case than equation of motion for R–R field strength,

$$(\Gamma^\mu)_{\alpha\beta} \partial_\mu P^{\beta\gamma} = 0,$$

given in Eq. (3.11) of Ref. [29] where there is summation over spinor indices. Our choice of constant $P^{\alpha\beta}$ is consistent with this condition. It is known fact that even constant R–R field strength produces back-reaction on background fields. In order to cancel non-quadratic terms originating from back-reaction, the constant R–R field strength must satisfy additional conditions – $AdS_5 \times S_5$ coset geometry or self-duality condition.

Taking into account these assumptions there exists solution

$$\Pi^\mu_\pm \to \partial_\pm x^\mu, \quad d_\alpha \to \pi_\alpha, \quad \bar{d}_\alpha \to \bar{\pi}_\alpha, \tag{2.11}$$

and only nontrivial superfields take the form

$$A_{\mu\nu} = \kappa(\frac{1}{2}g_{\mu\nu} + B_{\mu\nu}), \quad E_\nu^\alpha = -\Psi_\nu^\alpha, \quad \bar{E}_\mu^\alpha = \bar{\Psi}_\mu^\alpha, \quad P^{\alpha\beta} = \frac{2}{\kappa}P^{\alpha\beta} = \frac{2}{\kappa}e^{\frac{\Phi}{2}}F^{\alpha\beta}, \quad (2.12)$$

where $g_{\mu\nu}$ is symmetric and $B_{\mu\nu}$ is antisymmetric tensor.

The final form of the vertex operator under these assumptions is

$$V_{SG} = \int_\Sigma d^2\xi \left[\kappa(\frac{1}{2}g_{\mu\nu} + B_{\mu\nu})\partial_+ x^\mu \partial_- x^\nu - \pi_\alpha \Psi_\mu^\alpha \partial_- x^\mu + \partial_+ x^\mu \bar{\Psi}_\mu^\alpha \bar{\pi}_\alpha + \frac{2}{\kappa}\pi_\alpha P^{\alpha\beta} \pi_\beta \right].$$

$$(2.13)$$

Consequently, the action S is of the form

$$S = \kappa \int_\Sigma d^2\xi \left[\partial_+ x^\mu \Pi_{+\mu\nu} \partial_- x^\nu + \frac{1}{4\pi\kappa} \Phi R^{(2)} \right] \qquad (2.14)$$

$$+ \int_\Sigma d^2\xi \left[-\pi_\alpha \partial_- (\theta^\alpha + \Psi_\mu^\alpha x^\mu) + \partial_+ (\bar{\theta}^\alpha + \bar{\Psi}_\mu^\alpha x^\mu)\bar{\pi}_\alpha + \frac{2}{\kappa}\pi_\alpha P^{\alpha\beta} \bar{\pi}_\beta \right],$$

where $G_{\mu\nu} = \eta_{\mu\nu} + g_{\mu\nu}$ and

$$\Pi_{\pm\mu\nu} = B_{\mu\nu} \pm \frac{1}{2}G_{\mu\nu}. \qquad (2.15)$$

All terms containing pure spinors vanished because curvatures are zero under our assumption that physical superfields are constant. Actions S_λ and $S_{\bar{\lambda}}$ are fully decoupled from the rest action and can be neglected in the further analysis. The action, in its final form, is ghost independent.

Here we work both with type IIA and type IIB superstring theory. The difference is in the chirality of NS–R background fields and content of R–R sector. In NS–R sector there are two gravitino fields Ψ_μ^α and $\bar{\Psi}_\mu^\alpha$ which are Majorana–Weyl spinors of the opposite chirality in type IIA and same chirality in type IIB theory. The same feature stands for the pairs of spinors $(\theta^\alpha, \bar{\theta}^\alpha)$ and $(\pi_\alpha, \bar{\pi}_\alpha)$. The R–R field strength $F^{\alpha\beta}$ is expressed in terms of the antisymmetric tensors $F_{(k)} = F_{\mu_1\mu_2...\mu_k}$ [1]

$$F^{\alpha\beta} = \sum_{k=0}^{D} \frac{1}{k!} F_{(k)} (C\Gamma_{(k)})^{\alpha\beta}, \quad \left[\Gamma_{(k)}^{\alpha\beta} = (\Gamma^{[\mu_1...\mu_k]})^{\alpha\beta} \right] \qquad (2.16)$$

where

$$\Gamma^{[\mu_1\mu_2...\mu_k]} \equiv \Gamma^{[\mu_1}\Gamma^{\mu_2} \dots \Gamma^{\mu_k]}, \qquad (2.17)$$

is basis of completely antisymmetrized product of gamma matrices and C is charge conjugation operator. For more technical details regarding gamma matrices see the first reference in [1].

R–R field strength satisfies the chirality condition $\Gamma^{11} F = \pm F\Gamma^{11}$, where Γ^{11} is a product of gamma matrices in $D = 10$ dimensional space–time. The sign $+$ corresponds to type IIA while sign $-$ corresponds to type IIB superstring theory. Consequently, type IIA theory contains only even rank tensors $F_{(k)}$, while type IIB contains only odd rank tensors. For type IIA the independent tensors are $F_{(0)}$, $F_{(2)}$ and $F_{(4)}$, while independent tensors for type IIB are $F_{(1)}$, $F_{(3)}$ and self-dual part of $F_{(5)}$.

2.2. Fixing the chiral gauge invariance

The fermionic part of the action (2.14) has the form of the first order theory. We want to eliminate the fermionic momenta and work with the action expressed in terms of coordinates and their derivatives. So, on the equations of motion for fermionic momenta π_α and $\bar{\pi}_\alpha$,

$$\pi_\alpha = -\frac{\kappa}{2}\partial_+\left(\bar{\theta}^\beta + \bar{\Psi}^\beta_\mu x^\mu\right)(P^{-1})_{\beta\alpha}, \quad \bar{\pi}_\alpha = \frac{\kappa}{2}(P^{-1})_{\alpha\beta}\partial_-\left(\theta^\beta + \Psi^\beta_\mu x^\mu\right), \tag{2.18}$$

the action gets the form

$$S(\partial_\pm x, \partial_-\theta, \partial_+\bar{\theta}) = \kappa \int_\Sigma d^2\xi \, \partial_+x^\mu \Pi_{+\mu\nu}\partial_-x^\nu + \frac{1}{4\pi}\int_\Sigma d^2\xi \, \Phi R^{(2)}$$

$$+ \frac{\kappa}{2}\int_\Sigma d^2\xi \, \partial_+\left(\bar{\theta}^\alpha + \bar{\Psi}^\alpha_\mu x^\mu\right)(P^{-1})_{\alpha\beta}\partial_-\left(\theta^\beta + \Psi^\beta_\nu x^\nu\right)$$

$$= \kappa \int_\Sigma d^2\xi \, \partial_+x^\mu\left[\Pi_{+\mu\nu} + \frac{1}{2}\bar{\Psi}^\alpha_\mu(P^{-1})_{\alpha\beta}\Psi^\beta_\nu\right]\partial_-x^\nu + \frac{1}{4\pi}\int_\Sigma d^2\xi \, \Phi R^{(2)} \tag{2.19}$$

$$+ \frac{\kappa}{2}\int_\Sigma d^2\xi \left[\partial_+\bar{\theta}^\alpha(P^{-1})_{\alpha\beta}\partial_-\theta^\beta + \partial_+\bar{\theta}^\alpha(P^{-1}\Psi)_{\alpha\mu}\partial_-x^\mu + \partial_+x^\mu(\bar{\Psi}P^{-1})_{\mu\alpha}\partial_-\theta^\alpha\right].$$

In the above action θ^α appears only in the form $\partial_-\theta^\alpha$ and $\bar{\theta}^\alpha$ in the form $\partial_+\bar{\theta}^\alpha$. This means that the theory has a local symmetry

$$\delta\theta^\alpha = \varepsilon^\alpha(\sigma^+), \quad \delta\bar{\theta}^\alpha = \bar{\varepsilon}^\alpha(\sigma^-), \quad (\sigma^\pm = \tau \pm \sigma). \tag{2.20}$$

We will treat this symmetry within BRST formalism. The corresponding BRST transformations are

$$s\theta^\alpha = c^\alpha(\sigma^+), \quad s\bar{\theta}^\alpha = \bar{c}^\alpha(\sigma^-), \tag{2.21}$$

where for each gauge parameter $\varepsilon^\alpha(\sigma^+)$ and $\bar{\varepsilon}^\alpha(\sigma^-)$ we introduced the ghost fields $c^\alpha(\sigma^+)$ and $\bar{c}^\alpha(\sigma^-)$, respectively. Here s is BRST nilpotent operator.

To fix gauge freedom we introduce gauge fermion with ghost number -1

$$\Psi = \frac{\kappa}{2}\int d^2\xi \left[\bar{C}_\alpha\left(\partial_+\theta^\alpha + \frac{\alpha^{\alpha\beta}}{2}b_{+\beta}\right) + \left(\partial_-\bar{\theta}^\alpha + \frac{1}{2}\bar{b}_{-\beta}\alpha^{\beta\alpha}\right)C_\alpha\right], \tag{2.22}$$

where $\alpha^{\alpha\beta}$ is arbitrary non-singular matrix, $\bar{C}_\alpha$ and C_α are antighost fields, while $b_{+\alpha}$ and $\bar{b}_{-\alpha}$ are Nakanishi–Lautrup auxiliary fields which satisfy

$$sC_\alpha = b_{+\alpha}, \quad s\bar{C}_\alpha = \bar{b}_{-\alpha}, \quad sb_{+\alpha} = 0 \quad s\bar{b}_{-\alpha} = 0. \tag{2.23}$$

BRST transformation of gauge fermion Ψ produces the gauge fixed and Fadeev–Popov action

$$s\Psi = S_{gf} + S_{FP},$$

$$S_{gf} = \frac{\kappa}{2}\int d^2\xi \left[\bar{b}_{-\alpha}\partial_+\theta^\alpha + \partial_-\bar{\theta}^\alpha b_{+\alpha} + \bar{b}_{-\alpha}\alpha^{\alpha\beta}b_{+\beta}\right],$$

$$S_{FD} = \frac{\kappa}{2}\int d^2\xi \left[\bar{C}_\alpha\partial_+c^\alpha + (\partial_-\bar{c}^\alpha)C_\alpha\right]. \tag{2.24}$$

The Fadeev–Popov action is decoupled from the rest and, consequently, it can be omitted in further analysis. On the equations of motion for b-fields

$$b_{+\alpha} = -(\alpha^{-1})_{\alpha\beta}\partial_+\theta^\beta, \quad \bar{b}_{-\alpha} = -\partial_-\bar{\theta}^\beta(\alpha^{-1})_{\beta\alpha}, \tag{2.25}$$

we obtain the final form of the BRST gauge fixed action

$$S_{gf} = -\frac{\kappa}{2}\int d^2\xi\,\partial_-\bar{\theta}^\alpha(\alpha^{-1})_{\alpha\beta}\partial_+\theta^\beta. \tag{2.26}$$

2.3. Fermionic T-duality

We will perform fermionic T-duality using fermionic version of Buscher procedure similarly to Refs. [20] where we worked without chiral gauge fixing. After introducing S_{gf} the action still has a global shift symmetry in θ^α and $\bar{\theta}^\alpha$ directions. We introduce gauge fields $v_\pm^\alpha$ and $\bar{v}_\pm^\alpha$ and replace ordinary world-sheet derivatives with covariant ones

$$\partial_\pm\theta^\alpha \to D_\pm\theta^\alpha \equiv \partial_\pm\theta^\alpha + v_\pm^\alpha, \quad \partial_\pm\bar{\theta}^\alpha \to D_\pm\bar{\theta}^\alpha \equiv \partial_\pm\bar{\theta}^\alpha + \bar{v}_\pm^\alpha. \tag{2.27}$$

In order to make the fields $v_\pm^\alpha$ and $\bar{v}_\pm^\alpha$ to be unphysical we add the following terms in the action

$$\begin{aligned}
S_{gauge}(\vartheta, v_\pm, \bar{\vartheta}, \bar{v}_\pm) = \frac{1}{2}\kappa\int_\Sigma d^2\xi\,\bar{\vartheta}_\alpha(\partial_+v_-^\alpha - \partial_-v_+^\alpha) \\
+ \frac{1}{2}\kappa\int_\Sigma d^2\xi(\partial_+\bar{v}_-^\alpha - \partial_-\bar{v}_+^\alpha)\vartheta_\alpha,
\end{aligned} \tag{2.28}$$

where ϑ_α and $\bar{\vartheta}_\alpha$ are Lagrange multipliers. The full gauge invariant action is of the form

$$\begin{aligned}
S_{inv}(x, \theta, \bar{\theta}, \vartheta, \bar{\vartheta}, v_\pm, \bar{v}_\pm) = S(\partial_\pm x, D_-\theta, D_+\bar{\theta}) \\
+ S_{gf}(D_-\theta, D_+\bar{\theta}) + S_{gauge}(\vartheta, \bar{\vartheta}, v_\pm, \bar{v}_\pm).
\end{aligned} \tag{2.29}$$

Fixing θ^α and $\bar{\theta}^\alpha$ to zero we obtain the gauge fixed action

$$S_{fix} = \kappa\int_\Sigma d^2\xi\,\partial_+x^\mu\left[\Pi_{+\mu\nu} + \frac{1}{2}\bar{\Psi}_\mu^\alpha(P^{-1})_{\alpha\beta}\Psi_\nu^\beta\right]\partial_-x^\nu + \frac{1}{4\pi}\int_\Sigma d^2\xi\,\Phi R^{(2)} \tag{2.30}$$

$$+ \frac{\kappa}{2}\int_\Sigma\left[\bar{v}_+^\alpha(P^{-1})_{\alpha\beta}v_-^\beta + \bar{v}_+^\alpha(P^{-1})_{\alpha\beta}\Psi_\nu^\beta\partial_-x^\nu + \partial_+x^\mu\bar{\Psi}_\mu^\alpha(P^{-1})_{\alpha\beta}v_-^\beta - \bar{v}_-^\alpha(\alpha^{-1})_{\alpha\beta}v_+^\beta\right]$$

$$+ \frac{\kappa}{2}\int_\Sigma d^2\xi\,\bar{\vartheta}_\alpha(\partial_+v_-^\alpha - \partial_-v_+^\alpha) + \frac{\kappa}{2}\int_\Sigma d^2\xi(\partial_+\bar{v}_-^\alpha - \partial_-\bar{v}_+^\alpha)\vartheta_\alpha.$$

Varying the above action with respect to the Lagrange multipliers ϑ_α and $\bar{\vartheta}_\alpha$ we obtain the initial action (2.19) because

$$\partial_+v_-^\alpha - \partial_-v_+^\alpha = 0 \Longrightarrow v_\pm^\alpha = \partial_\pm\theta^\alpha, \quad \partial_+\bar{v}_-^\alpha - \partial_-\bar{v}_+^\alpha = 0 \Longrightarrow \bar{v}_\pm^\alpha = \partial_\pm\bar{\theta}^\alpha. \tag{2.31}$$

The equations of motion for $v_\pm^\alpha$ and $\bar{v}_\pm^\alpha$ give

$$\bar{v}_-^\alpha = \partial_-\bar{\vartheta}_\beta\alpha^{\beta\alpha}, \quad \bar{v}_+^\alpha = \partial_+\bar{\vartheta}_\beta P^{\beta\alpha} - \partial_+x^\mu\bar{\Psi}_\mu^\alpha, \tag{2.32}$$

$$v_+^\alpha = -\alpha^{\alpha\beta}\partial_+\vartheta_\beta\,, \quad v_-^\alpha = -P^{\alpha\beta}\partial_-\vartheta_\beta - \Psi_\mu^\alpha\partial_-x^\mu\,. \tag{2.33}$$

Substituting these expressions in the action S_{fix} we obtain the fermionic T-dual action

$$^\star S(\partial_\pm x, \partial_-\vartheta, \partial_+\bar\vartheta) = \kappa\int_\Sigma d^2\xi\,\partial_+x^\mu\Pi_{+\mu\nu}\partial_-x^\nu + \frac{1}{4\pi}\int_\Sigma d^2\xi\,{}^\star\Phi R^{(2)}\,, \tag{2.34}$$

$$+\frac{\kappa}{2}\int_\Sigma d^2\xi\left[\partial_+\bar\vartheta_\alpha P^{\alpha\beta}\partial_-\vartheta_\beta - \partial_+x^\mu\bar\Psi_\mu^\alpha\partial_-\vartheta_\alpha + \partial_+\bar\vartheta_\alpha\Psi_\mu^\alpha\partial_-x^\mu - \partial_-\bar\vartheta_\alpha\alpha^{\alpha\beta}\partial_+\vartheta_\beta\right]\,.$$

It should be in the form of the initial action (2.19)

$$^\star S = \kappa\int_\Sigma d^2\xi\,\partial_+x^\mu\left[{}^\star\Pi_{+\mu\nu} + \frac{1}{2}{}^\star\Psi^{\alpha\mu}({}^\star P^{-1})^{\alpha\beta}{}^\star\Psi_{\beta\nu}\right]\partial_-x^\nu + \frac{1}{4\pi}\int_\Sigma d^2\xi\,{}^\star\Phi R^{(2)} \tag{2.35}$$

$$+\frac{\kappa}{2}\int_\Sigma d^2\xi\left[\partial_+\bar\vartheta_\alpha({}^\star P^{-1})^{\alpha\beta}\partial_-\vartheta_\beta + \partial_+x^\mu({}^\star\bar\Psi{}^\star P^{-1})_\mu^\alpha\partial_-\vartheta^\alpha + \partial_+\bar\vartheta_\alpha({}^\star P^{-1\star}\Psi)_\mu^\alpha\partial_-x^\nu\right]$$

$$-\frac{\kappa}{2}\int d^2\xi\,\partial_-\bar\vartheta_\alpha({}^\star\alpha^{-1})^{\alpha\beta}\partial_+\vartheta_\beta\,, \tag{2.36}$$

and so we get

$$^\star\Psi_{\alpha\mu} = (P^{-1}\Psi)_{\alpha\mu}\,, \quad {}^\star\bar\Psi_{\mu\alpha} = -(\bar\Psi P^{-1})_{\mu\alpha}\,, \tag{2.37}$$

$$^\star P_{\alpha\beta} = (P^{-1})_{\alpha\beta}\,, \quad {}^\star\alpha_{\alpha\beta} = (\alpha^{-1})_{\alpha\beta}\,. \tag{2.38}$$

From the condition

$$^\star\Pi_{+\mu\nu} + \frac{1}{2}{}^\star\bar\Psi_{\alpha\mu}({}^\star P^{-1})^{\alpha\beta}{}^\star\Psi_{\beta\nu} = \Pi_{+\mu\nu}\,, \tag{2.39}$$

we read the fermionic T-dual metric and Kalb–Ramond field

$$^\star G_{\mu\nu} = G_{\mu\nu} + \frac{1}{2}\left[(\bar\Psi P^{-1}\Psi)_{\mu\nu} + (\bar\Psi P^{-1}\Psi)_{\nu\mu}\right]\,,$$

$$^\star B_{\mu\nu} = B_{\mu\nu} + \frac{1}{4}\left[(\bar\Psi P^{-1}\Psi)_{\mu\nu} - (\bar\Psi P^{-1}\Psi)_{\nu\mu}\right]\,. \tag{2.40}$$

Dilaton transformation under fermionic T-duality will be presented in the section 4. Let us note that two successive dualizations give the initial background fields.

The T-dual transformation laws are connection between initial and T-dual coordinates. We can obtain them combining the different solutions of equations of motion for $v_\pm^\alpha$ and $\bar v_\pm^\alpha$ (2.31) and (2.32)–(2.33)

$$\partial_-\theta^\alpha \cong -P^{\alpha\beta}\partial_-\vartheta_\beta - \Psi_\mu^\alpha\partial_-x^\mu\,, \quad \partial_+\bar\theta^\alpha \cong \partial_+\bar\vartheta_\beta P^{\beta\alpha} - \partial_+x^\mu\bar\Psi_\mu^\alpha\,, \tag{2.41}$$

$$\partial_+\theta^\alpha \cong -\alpha^{\alpha\beta}\partial_+\vartheta_\beta\,, \quad \partial_-\bar\theta^\alpha \cong \partial_-\bar\vartheta_\beta\alpha^{\beta\alpha}\,. \tag{2.42}$$

Here the symbol $\cong$ denotes the T-duality relation. From these relations we can obtain inverse transformation rules

$$\partial_-\vartheta_\alpha \cong -(P^{-1})_{\alpha\beta}\partial_-\theta^\beta - (P^{-1})_{\alpha\beta}\Psi_\mu^\beta\partial_-x^\mu\,,$$

$$\partial_+\bar\vartheta_\alpha \cong \partial_+\bar\theta^\beta(P^{-1})_{\beta\alpha} + \partial_+x^\mu\bar\Psi_\mu^\beta(P^{-1})_{\beta\alpha}\,, \tag{2.43}$$

$$\partial_+\vartheta_\alpha \cong -(\alpha^{-1})_{\alpha\beta}\partial_+\theta^\beta\,, \quad \partial_-\bar\vartheta_\alpha \cong \partial_-\bar\theta^\beta(\alpha^{-1})_{\beta\alpha}\,. \tag{2.44}$$

Note that without gauge fixing in subsection 2.2, instead expressions for $\bar{v}_-^\alpha$ and v_+^α (first relations of (2.32) and (2.33)), we would have only constraints on the T-dual variables, $\partial_- \bar{\vartheta}_\alpha = 0$ and $\partial_+ \vartheta_\alpha = 0$. Consequently, integration over $v_\pm^\alpha$ and $\bar{v}_\pm^\alpha$ would be singular and we would lose the part of T-dual transformations (2.42) and (2.44).

3. Fermionic T-dualization in fermionic double space

In this section we will extend the meaning of the double space. We will introduce double fermionic space adding to the fermionic coordinates, θ^α and $\bar{\theta}^\alpha$, the fermionic T-dual ones, ϑ_α and $\bar{\vartheta}_\alpha$. Then we will show that fermionic T-dualization can be represented as permutation of the appropriate fermionic coordinates and their T-dual partners.

3.1. Transformation laws in fermionic double space

In the same way as the double bosonic coordinates were introduced [4,14,15], we double both fermionic coordinate as

$$\Theta^A = \begin{pmatrix} \theta^\alpha \\ \vartheta_\alpha \end{pmatrix}, \quad \bar{\Theta}^A = \begin{pmatrix} \bar{\theta}^\alpha \\ \bar{\vartheta}_\alpha \end{pmatrix}. \tag{3.1}$$

Each double coordinate has 32 real components. In terms of the double fermionic coordinates the transformation laws, (2.41)–(2.44), can be rewritten in the form

$$\partial_- \Theta^A \cong -\Omega^{AB} \left[\mathcal{F}_{BC} \partial_- \Theta^C + \mathcal{J}_{-B} \right], \quad \partial_+ \bar{\Theta}^A \cong \left[\partial_+ \bar{\Theta}^C \mathcal{F}_{CB} + \bar{\mathcal{J}}_{+B} \right] \Omega^{BA}, \tag{3.2}$$

$$\partial_+ \Theta^A \cong -\Omega^{AB} \mathcal{A}_{BC} \partial_+ \Theta^C, \quad \partial_- \bar{\Theta}^A \cong \partial_- \bar{\Theta}^C \mathcal{A}_{CB} \Omega^{BA}, \tag{3.3}$$

where "fermionic generalized metric" $\mathcal{F}_{AB}$ has the form

$$\mathcal{F}_{AB} = \begin{pmatrix} (P^{-1})_{\alpha\beta} & 0 \\ 0 & P^{\gamma\delta} \end{pmatrix}, \tag{3.4}$$

and

$$\mathcal{A}_{AB} = \begin{pmatrix} (\alpha^{-1})_{\alpha\beta} & 0 \\ 0 & \alpha^{\gamma\delta} \end{pmatrix}. \tag{3.5}$$

$\mathcal{F}_{AB}$ is bosonic variable but we put the name fermionic because it appears in the case of fermionic T-duality.

The double currents, $\bar{\mathcal{J}}_{+A}$ and $\mathcal{J}_{-A}$, are fermionic variables of the form

$$\bar{\mathcal{J}}_{+A} = \begin{pmatrix} (\bar{\Psi} P^{-1})_{\mu\alpha} \partial_+ x^\mu \\ -\bar{\Psi}_\mu^\alpha \partial_+ x^\mu \end{pmatrix}, \quad \mathcal{J}_{-A} = \begin{pmatrix} (P^{-1} \Psi)_{\alpha\mu} \partial_- x^\mu \\ \Psi_\mu^\alpha \partial_- x^\mu \end{pmatrix}, \tag{3.6}$$

while the matrix Ω^{AB} is constant

$$\Omega^{AB} = \begin{pmatrix} 0 & 1 \\ 1 & 0 \end{pmatrix}, \tag{3.7}$$

where identity matrix is 16×16. By straightforward calculation we can prove the relations

$$\Omega^2 = 1, \quad \det \mathcal{F}_{AB} = 1. \tag{3.8}$$

Consistency of the transformation laws (3.2) produces

$$(\Omega \mathcal{F})^2 = 1, \quad \mathcal{J}_- = \mathcal{F} \Omega \mathcal{J}_-, \quad \bar{\mathcal{J}}_+ = -\bar{\mathcal{J}}_+ \Omega \mathcal{F}. \tag{3.9}$$

3.2. Double action

It is well known that equations of motion of initial theory are Bianchi identities in T-dual picture and vice versa. As a consequence of the identities

$$\partial_+\partial_-\Theta^A - \partial_-\partial_+\Theta^A = 0, \quad \partial_+\partial_-\bar{\Theta}^A - \partial_-\partial_+\bar{\Theta}^A = 0, \tag{3.10}$$

known as Bianchi identities, and relations (3.2) and (3.3), we obtain the consistency conditions

$$\partial_+(\mathcal{F}_{AB}\partial_-\Theta^B + J_{-A}) - \partial_-(\mathcal{A}_{AB}\partial_+\Theta^B) = 0, \tag{3.11}$$

$$\partial_-(\partial_+\bar{\Theta}^B\mathcal{F}_{BA} + \bar{J}_{+A}) - \partial_+(\partial_-\bar{\Theta}^B\mathcal{A}_{BA}) = 0. \tag{3.12}$$

The equations (3.11) and (3.12) are equations of motion of the following action

$$S_{double}(\Theta, \bar{\Theta}) = \tag{3.13}$$

$$= \frac{\kappa}{2}\int d^2\xi\left[\partial_+\bar{\Theta}^A\mathcal{F}_{AB}\partial_-\Theta^B + \bar{J}_{+A}\partial_-\Theta^A + \partial_+\bar{\Theta}^A J_{-A} - \partial_-\bar{\Theta}^A\mathcal{A}_{AB}\partial_+\Theta^B + L(x)\right],$$

where $L(x)$ is arbitrary functional of the bosonic coordinates.

3.3. Fermionic T-dualization of type II superstring theory as permutation of fermionic coordinates in double space

In order to exchange θ^α with ϑ_α and $\bar{\theta}$ with $\bar{\vartheta}_\alpha$, let us introduce the permutation matrix

$$\mathcal{T}^A{}_B = \begin{pmatrix} 0 & 1 \\ 1 & 0 \end{pmatrix}, \tag{3.14}$$

so that double T-dual coordinates are

$$^\star\Theta^A = \mathcal{T}^A{}_B\Theta^B, \quad ^\star\bar{\Theta}^A = \mathcal{T}^A{}_B\bar{\Theta}^B. \tag{3.15}$$

We demand that T-dual transformation laws for double T-dual coordinates $^\star\Theta^A$ and $^\star\bar{\Theta}^A$ have the same form as for initial ones Θ^A and $\bar{\Theta}^A$ (3.2) and (3.3)

$$\partial_-{}^\star\Theta^A \cong -\Omega^{AB}\left[{}^\star\mathcal{F}_{BC}\partial_-{}^\star\Theta^C + {}^\star J_{-B}\right], \quad \partial_+{}^\star\bar{\Theta}^A \cong \left[\partial_+{}^\star\bar{\Theta}^{C\star}\mathcal{F}_{CB} + {}^\star\bar{J}_{+B}\right]\Omega^{BA}, \tag{3.16}$$

$$\partial_+{}^\star\Theta^A \cong -\Omega^{AB\star}\mathcal{A}_{BC}\partial_+{}^\star\Theta^C, \quad \partial_-{}^\star\bar{\Theta}^A \cong \partial_-{}^\star\bar{\Theta}^{C\star}\mathcal{A}_{CB}\Omega^{BA}. \tag{3.17}$$

Then the fermionic T-dual "generalized metric" $^\star\mathcal{F}_{AB}$ and T-dual currents, $^\star\bar{J}_{+A}$ and $^\star J_{-A}$, with the help of (3.15) and (3.2), can be expressed in terms of initial ones

$$^\star\mathcal{F}_{AB} = \mathcal{T}_A{}^C\mathcal{F}_{CD}\mathcal{T}^D{}_B, \quad ^\star\bar{J}_{+A} = \mathcal{T}_A{}^B\bar{J}_{+B}, \quad ^\star J_{-A} = \mathcal{T}_A{}^B J_{-B}. \tag{3.18}$$

The matrix $\mathcal{A}_{AB}$ transforms as

$$^\star\mathcal{A}_{AB} = \mathcal{T}_A{}^C\mathcal{A}_{CD}\mathcal{T}^D{}_B = (\mathcal{A}^{-1})_{AB}. \tag{3.19}$$

Note that, as well as bosonic case, double space action (3.13) has global symmetry under transformations (3.15) if the conditions (3.18) are satisfied.

From the first relation in (3.18) we obtain the form of the fermionic T-dual R–R background field

$$^\star P_{\alpha\beta} = (P^{-1})_{\alpha\beta}\,, \tag{3.20}$$

while from the second and third equation we obtain the form of the fermionic T-dual NS–R background fields

$$^\star\Psi_{\alpha\mu} = (P^{-1})_{\alpha\beta}\Psi_\mu^\beta\,, \quad ^\star\bar\Psi_{\alpha\mu} = -\bar\Psi_\mu^\beta(P^{-1})_{\beta\alpha}\,. \tag{3.21}$$

The non-singular matrix $\alpha^{\alpha\beta}$ transforms as

$$(^\star\alpha)_{\alpha\beta} = (\alpha^{-1})_{\alpha\beta}\,. \tag{3.22}$$

The expressions (3.20)–(3.22) are in full agreement with the relations (2.37) and (2.38) obtained by the standard fermionic Buscher procedure. Consequently, we showed that permutation of fermionic coordinates defined in (3.14) and (3.15) completely reproduces fermionic T-dual R–R and NS–R background fields.

3.4. Fermionic T-dual metric $^\star G_{\mu\nu}$ and Kalb–Ramond field $^\star B_{\mu\nu}$

The expression $\Pi_{+\mu\nu} + \frac{1}{2}\Psi_\mu^\alpha(P^{-1})_{\alpha\beta}\Psi_\nu^\beta$ appears in the action (2.19) coupled with $\partial_\pm x^\mu$, along which we do not T-dualize. It is an analogue of ij-term of Refs. [9,10] where x^i coordinates are not T-dualized, and $\alpha\beta$-term in [22] where fermionic directions are undualized.

Taking into account the form of the doubled action (3.13) we suppose that term $L(x)$ has the form

$$L(x) = 2\partial_+ x^\mu \left(\Pi_{+\mu\nu} + {}^\star\Pi_{+\mu\nu}\right)\partial_- x^\nu \equiv \mathcal{L} + {}^\star\mathcal{L}\,, \tag{3.23}$$

where $\Pi_{+\mu\nu}$ is defined in (2.15) and $^\star\Pi_{+\mu\nu}$ is fermionic T-dual which we are going to find. This term should be invariant under T-dual transformation

$$^\star\mathcal{L} = \mathcal{L} + \Delta\mathcal{L}\,. \tag{3.24}$$

Using the fact that two successive T-dualizations are identity transformation, we obtain

$$\mathcal{L} = {}^\star\mathcal{L} + {}^\star\Delta\mathcal{L}\,. \tag{3.25}$$

Combining last two relations we get

$$^\star\Delta\mathcal{L} = -\Delta\mathcal{L}\,. \tag{3.26}$$

If $\Delta\mathcal{L} = 2\partial_+ x^\mu \Delta_{\mu\nu}\partial_- x^\nu$, we obtain the condition for $\Delta_{\mu\nu}$

$$^\star\Delta_{\mu\nu} = -\Delta_{\mu\nu}\,. \tag{3.27}$$

Using the relations (2.37) and (2.38) we realize that, up to multiplication constant, combination

$$\Delta_{\mu\nu} = \bar\Psi_\mu^\alpha(P^{-1})_{\alpha\beta}\Psi_\nu^\beta\,, \tag{3.28}$$

satisfies the condition (3.27). So, we conclude that

$$^\star\Pi_{+\mu\nu} = \Pi_{+\mu\nu} + c\bar\Psi_\mu^\alpha(P^{-1})_{\alpha\beta}\Psi_\nu^\beta\,, \tag{3.29}$$

where c is an arbitrary constant. For $c = \frac{1}{2}$ we obtain the relations (2.40). So, in double space formulation the fermionic T-dual NS–NS background fields can be obtained up to an arbitrary constant under assumption that two successive T-dualizations produce initial action.

4. Dilaton field in double fermionic space

Dilaton field transformation under fermionic T-duality is considered [16]. Here we will discuss some new features of dilaton transformation under fermionic T-duality as well as the dilaton field in fermionic double space.

Because the dilaton transformation has quantum origin we start with the path integral for the gauge fixed action given in Eq. (2.30)

$$Z = \int d\bar{v}_{+}^{\alpha} d\bar{v}_{-}^{\alpha} dv_{+}^{\alpha} dv_{-}^{\alpha} d\bar{\vartheta}_{\alpha} d\vartheta_{\alpha} e^{i \, S_{fix}(v_{\pm}, \bar{v}_{\pm}, \partial_{\pm}\vartheta, \partial_{\pm}\bar{\vartheta})} . \tag{4.1}$$

For constant background case, after integration over the fermionic gauge fields $\bar{v}_{\pm}^{\alpha}$ and $v_{\pm}^{\alpha}$, we obtain the generating functional Z in the form

$$Z = \int d\bar{\vartheta}_{\alpha} d\vartheta_{\alpha} \det\left[(P^{-1}\alpha^{-1})_{\alpha\beta} \right] e^{i \, {}^{\star}S(\vartheta, \bar{\vartheta})} , \tag{4.2}$$

where ${}^{\star}S(\vartheta, \bar{\vartheta})$ is T-dual action given in Eq. (2.34). We are able to perform such integration thank to the facts that we fixed the gauge in subsection 2.2.

Note that here we multiplied with determinants of P^{-1} and α^{-1} because we integrate over Grassman fields $v_{\pm}^{\alpha}$ and $\bar{v}_{\pm}^{\alpha}$. We can choose that $\det \alpha = 1$, and the generating functional gets the form

$$Z = \int d\bar{\vartheta}_{\alpha} d\vartheta_{\alpha} \det\left[(P^{-1})_{\alpha\beta} \right] e^{i \, {}^{\star}S(\vartheta, \bar{\vartheta})} . \tag{4.3}$$

This produces the fermionic T-dual transformation of dilaton field

$$^{\star}\Phi = \Phi + \ln \det\left[(P^{-1})_{\alpha\beta} \right] = \Phi - \ln \det P^{\alpha\beta} . \tag{4.4}$$

Let us calculate $\det P^{\alpha\beta}$ using the expression

$$(P P_{s}^{-1} P^{T})^{\alpha\beta} = P_{s}^{\alpha\beta} - P_{a}^{\alpha\gamma} (P_{s}^{-1})_{\gamma\delta} P_{a}^{\delta\beta} , \tag{4.5}$$

where we introduce the symmetric and antisymmetric parts for initial background fields

$$P_{s}^{\alpha\beta} = \frac{1}{2}\left(P^{\alpha\beta} + P^{\beta\alpha} \right) , \quad P_{a}^{\alpha\beta} = \frac{1}{2}\left(P^{\alpha\beta} - P^{\beta\alpha} \right) , \tag{4.6}$$

and similar expressions for T-dual background fields, ${}^{\star}P_{\alpha\beta}^{s}$ and ${}^{\star}P_{\alpha\beta}^{a}$. Taking into account that

$$(P \cdot {}^{\star}P)^{\alpha}{}_{\beta} = \delta^{\alpha}{}_{\beta} , \tag{4.7}$$

we obtain

$$P_{s}^{\alpha\gamma} \, {}^{\star}P_{\gamma\beta}^{s} + P_{a}^{\alpha\gamma} \, {}^{\star}P_{\gamma\beta}^{a} = \delta^{\alpha}{}_{\beta} , \quad P_{s}^{\alpha\gamma} {}^{\star}P_{\gamma\beta}^{a} + P_{a}^{\alpha\gamma} {}^{\star}P_{\gamma\beta}^{s} = 0 . \tag{4.8}$$

From these two equations we obtain

$$^{\star}P_{\alpha\beta}^{s} = \left[(P_{s} - P_{a} P_{s}^{-1} P_{a})^{-1} \right]_{\alpha\beta} , \tag{4.9}$$

and, consequently, we have

$$(P P_{s}^{-1} P^{T})^{\alpha\beta} = \left[({}^{\star}P_{s})^{-1} \right]^{\alpha\beta} . \tag{4.10}$$

Taking determinant of the left and right-hand side of the above equation we get

$$\det P^{\alpha\beta} = \sqrt{\frac{\det P_s}{\det {}^{\star}P_s}},$$

(4.11)

which produces

$$^{\star}\Phi = \Phi - \ln\sqrt{\frac{\det P_s}{\det {}^{\star}P_s}}.$$

(4.12)

Using the fact that $P^{\alpha\beta} = e^{\frac{\Phi}{2}}F^{\alpha\beta}$ and $^{\star}P^{\alpha\beta} = e^{\frac{{}^{\star}\Phi}{2}}{}^{\star}F^{\alpha\beta}$, fermionic T-dual transformation law for dilaton takes the form

$$^{\star}\Phi = \Phi - \ln\sqrt{e^{8(\Phi - {}^{\star}\Phi)}\frac{\det F_s}{\det {}^{\star}F_s}},$$

(4.13)

and finally we have

$$^{\star}\Phi = \Phi + \frac{1}{6}\ln\frac{\det F_s}{\det {}^{\star}F_s}.$$

(4.14)

It is obvious that two successive T-dualizations act as identity transformation

$$^{\star\star}\Phi = \Phi.$$

(4.15)

We can conclude that only symmetric parts of the R–R field strengths give contribution to the transformation of dilaton field under fermionic T-duality. In type IIA superstring theory R–R field strength $F^{\alpha\beta}$ contains tensors F_0^A, $F_{\mu\nu}^A$ and $F_{\mu\nu\rho\lambda}^A$, while in type IIB $F^{\alpha\beta}$ contains F_{μ}^B, $F_{\mu\nu\rho}^B$ and self dual part of $F_{\mu\nu\rho\lambda\omega}^B$. Using the conventions for gamma matrices from the appendix of the first reference in [1] (see Appendix A), we conclude that symmetric part of $F^{\alpha\beta}$ in type IIA contains scalar F_0^A and 2-rank tensor $F_{\mu\nu}^A$, while in type IIB superstring theory it contains 1-rank F_{μ}^B and self dual part of 5-rank tensor $F_{\mu\nu\rho\lambda\omega}^B$.

Let us write the path integral for double action (3.13)

$$Z_{double} = \int d\Theta^A d\bar{\Theta}^A e^{i S_{double}(\Theta, \bar{\Theta})}.$$

(4.16)

Because $\det \mathcal{F} = 1$ and $\det \mathcal{A} = 1$ we obtain that dilaton field in double space is invariant under fermionic T-duality. Consequently, a new dilaton should be introduced (see [14,15]), invariant under T-duality transformations. Because of the relation (4.15) we define the T-duality invariant dilaton as

$$\Phi_{inv} = \frac{1}{2}\left(^{\star}\Phi + \Phi\right) = \Phi + \frac{1}{12}\ln\frac{\det F_s}{\det {}^{\star}F_s}, \qquad ^{\star}\Phi_{inv} = \Phi_{inv}.$$

(4.17)

5. Concluding remarks

In this article we considered the fermionic T-duality of the type II superstring theory using the double space approach. We used the action of the type II superstring theory in pure spinor formulation neglecting ghost terms and keeping all terms up to the quadratic ones which means that all background fields are constant.

Using equations of motion with respect to the fermionic momenta π_α and $\bar{\pi}_\alpha$ we eliminated them from the action. We obtained the action expressed in terms of the derivatives $\partial_\pm x^\mu$, $\partial_- \theta^\alpha$ and $\partial_+ \bar{\theta}^\alpha$, where θ^α and $\bar{\theta}^\alpha$ are fermionic coordinates. Because θ^α appears in the action in the form $\partial_- \theta^\alpha$ and $\bar{\theta}^\alpha$ in the form $\partial_+ \bar{\theta}^\alpha$, there is a local chiral gauge symmetry with parameters depending on $\sigma^\pm = \tau \pm \sigma$. We fixed this gauge invariance using BRST approach.

Using the Buscher approach we performed fermionic T-duality procedure and obtained the form of the fermionic T-dual background fields. It is obvious that two successive fermionic T-dualizations produce initial theory i.e. they are equivalent to the identity transformation.

In the central point of the article we generalize the idea of double space and show that fermionic T-duality can be represented as permutation in fermionic double space. In the bosonic case double space spanned by coordinates $Z^M = (x^\mu, y_\mu)$ can be obtained adding T-dual coordinates y_μ to the initial ones x^μ. Using analogy with the bosonic case we introduced double fermionic space doubling the initial coordinates θ^α and $\bar{\theta}^\alpha$ with their fermionic T-duals, ϑ_α and $\bar{\vartheta}_\alpha$. Double fermionic space is spanned by the coordinates $\Theta^A = (\theta^\alpha, \vartheta_\alpha)$ and $\bar{\Theta}^A = (\bar{\theta}^\alpha, \bar{\vartheta}_\alpha)$.

T-dual transformation laws and their inverse ones are rewritten in fermionic double space by single relation introducing the fermionic generalized metric $\mathcal{F}_{AB}$ and currents $\mathcal{J}_{-A}$ and $\bar{\mathcal{J}}_{+A}$. Demanding that transformation laws for fermionic T-dual double coordinates, $^\star\Theta^A = \mathcal{T}^A{}_B \Theta^B$ and $^\star\bar{\Theta}^A = \mathcal{T}^A{}_B \bar{\Theta}^B$, are of the same form as those for Θ^A and $\bar{\Theta}^A$, we obtained fermionic T-dual generalized metric $^\star\mathcal{F}_{AB}$ and currents $^\star\mathcal{J}_{-A}$ and $^\star\bar{\mathcal{J}}_{+A}$. These transformations act as symmetry transformations of the double action (3.13). They produce the form of the fermionic T-dual NS–R and R–R background fields which are in full accordance with the results obtained by standard Buscher procedure.

The expressions for T-dual metric $^\star G_{\mu\nu}$ and Kalb–Ramond field $^\star B_{\mu\nu}$ cannot be found from double space formalism because they do not appear in the T-dual transformation laws. These expressions, up to arbitrary constant, are obtained assuming that two successive T-dualizations act as identity transformation.

We considered transformation of dilaton field under fermionic T-duality. We derived the transformation law for dilaton field and concluded that just symmetric parts of R–R field strengths, $F_s^{\alpha\beta}$ and $^\star F_{\alpha\beta}^s$, affected the dilaton transformation law. This means that in the case of type IIA scalar and 2-rank tensor have influence on the dilaton transformation, while in the case of type IIB 1-rank tensor and self-dual part of 5-rank tensor take that role.

Therefore, we extended T-dualization in double space to the fermionic case. We proved that permutation of fermionic coordinates with corresponding T-dual ones in double space is equivalent to the fermionic T-duality along initial coordinates θ^α and $\bar{\theta}^\alpha$.

Appendix A. Gamma matrices

In the appendix of the first reference of [1] one specific representation of gamma matrices is given. Here we will calculate the transpositions of basis matrices $(C\Gamma_{(k)})^{\alpha\beta}$ for $k = 1, 2, 3, 4, 5$, where C is charge conjugation operator.

The charge conjugation operator is antisymmetric matrix, $C^T = -C$, and it acts on gamma matrices as

$$C\Gamma^\mu C^{-1} = -(\Gamma^\mu)^T. \tag{A.1}$$

Now we have

$$(C\Gamma^\mu)^T = (\Gamma^\mu)^T C^T = -(\Gamma^\mu)^T C = C\Gamma^\mu C^{-1} C = C\Gamma^\mu, \tag{A.2}$$

$$(C\Gamma^\mu\Gamma^\nu)^T = C\Gamma^\mu\Gamma^\nu \Longrightarrow (C\Gamma^{[\mu\nu]})^T = C\Gamma^{[\mu\nu]}\,, \tag{A.3}$$

$$(C\Gamma^\mu\Gamma^\nu\Gamma^\rho)^T = -C\Gamma^\mu\Gamma^\nu\Gamma^\rho \Longrightarrow (C\Gamma^{[\mu\nu\rho]})^T = -C\Gamma^{[\mu\nu\rho]}\,, \tag{A.4}$$

$$(C\Gamma^\mu\Gamma^\nu\Gamma^\rho\Gamma^\lambda)^T = -C\Gamma^\mu\Gamma^\nu\Gamma^\rho\Gamma^\lambda \Longrightarrow (C\Gamma^{[\mu\nu\rho\lambda]})^T = -C\Gamma^{[\mu\nu\rho\lambda]}\,, \tag{A.5}$$

$$(C\Gamma^\mu\Gamma^\nu\Gamma^\rho\Gamma^\lambda\Gamma^\omega)^T = C\Gamma^\mu\Gamma^\nu\Gamma^\rho\Gamma^\lambda\Gamma^\omega \Longrightarrow (C\Gamma^{[\mu\nu\rho\lambda\omega]})^T = C\Gamma^{[\mu\nu\rho\lambda\omega]}\,. \tag{A.6}$$

References

[1] J. Polchinski, String Theory – Volume II, Cambridge University Press, 1998;
K. Becker, M. Becker, J.H. Schwarz, String Theory and M-Theory – A Modern Introduction, Cambridge University Press, 2007.

[2] E. Alvarez, L. Alvarez-Gaume, Y. Lozano, An introduction to T-duality in string theory, arXiv:hep-th/9410237;
A. Giveon, M. Porrati, E. Rabinovici, Target space duality in string theory, Phys. Rep. 244 (1994) 77–202, arXiv:hep-th/9401139;
I. Bandos, B. Julia, J. High Energy Phys. 08 (2003) 032;
D. Luest, J. High Energy Phys. 12 (2010) 084.

[3] T.H. Buscher, Phys. Lett. B 194 (1987) 59;
T.H. Buscher, Phys. Lett. B 201 (1988) 466.

[4] M. Duff, Nucl. Phys. B 335 (1990) 610.

[5] A.A. Tseytlin, Phys. Lett. B 242 (1990) 163.

[6] A.A. Tseytlin, Nucl. Phys. B 350 (1991) 395.

[7] W. Siegel, Phys. Rev. D 48 (1993) 2826.

[8] W. Siegel, Phys. Rev. D 47 (1993) 5453.

[9] C.M. Hull, J. High Energy Phys. 10 (2005) 065.

[10] C.M. Hull, J. High Energy Phys. 10 (2007) 057;
C.M. Hull, J. High Energy Phys. 07 (2007) 080.

[11] D.S. Berman, M. Cederwall, M.J. Perry, J. High Energy Phys. 09 (2014) 066;
D.S. Berman, C.D.A. Blair, E. Malek, M.J. Perry, Int. J. Mod. Phys. A 29 (15) (2014) 1450080;
C.D.A. Blair, E. Malek, A.J. Routh, Class. Quantum Gravity 31 (20) (2014) 205011.

[12] C.M. Hull, R.A. Reid-Edwards, J. High Energy Phys. 09 (2009) 014.

[13] O. Hohm, B. Zwiebach, J. High Energy Phys. 11 (2014) 075.

[14] B. Sazdović, T-duality as coordinates permutation in double space, arXiv:1501.01024.

[15] B. Sazdović, J. High Energy Phys. 08 (2015) 055.

[16] N. Berkovits, J. Maldacena, J. High Energy Phys. 09 (2008) 062;
I. Bakhmatov, D.S. Berman, Nucl. Phys. B 832 (2010) 89–108;
K. Sfetsos, K. Siampos, D.C. Thompson, QMUL-PH-10-08, arXiv:1007.5142;
I. Bakhmatov, Fermionic T-duality and U-duality in type II supergravity, arXiv:1112.1983.

[17] N. Beisert, R. Ricci, A.A. Tseytlin, M. Wolf, Phys. Rev. D 78 (2008) 126004;
R. Ricci, A.A. Tseytlin, M. Wolf, J. High Energy Phys. 12 (2007) 082.

[18] M. Hatsuda, K. Kamimura, W. Siegel, J. High Energy Phys. 06 (2014) 039.

[19] W. Siegel, Phys. Rev. D 50 (1994) 2799–2805.

[20] B. Nikolić, B. Sazdović, Phys. Rev. D 84 (2011) 065012;
B. Nikolić, B. Sazdović, J. High Energy Phys. 06 (2012) 101.

[21] R. Benichou, G. Policastro, J. Troost, Phys. Lett. B 661 (2008) 192–195.

[22] B. Nikolić, B. Sazdović, arXiv:1505.06044.

[23] C.M. Hull, J. High Energy Phys. 07 (1998) 021.

[24] N. Berkovits, arXiv:hep-th/0209059;
P.A. Grassi, G. Policastro, P. van Nieuwenhuizen, J. High Energy Phys. 10 (2002) 054;
P.A. Grassi, G. Policastro, P. van Nieuwenhuizen, J. High Energy Phys. 11 (2002) 004;
P.A. Grassi, G. Policastro, P. van Nieuwenhuizen, Adv. Theor. Math. Phys. 7 (2003) 499;
P.A. Grassi, G. Policastro, P. van Nieuwenhuizen, Phys. Lett. B 553 (2003) 96.

[25] J. de Boer, P.A. Grassi, P. van Nieuwenhuizen, Phys. Lett. B 574 (2003) 98.

[26] B. Nikolić, B. Sazdović, Phys. Lett. B 666 (2008) 400.

[27] B. Nikolić, B. Sazdović, J. High Energy Phys. 08 (2010) 037.

[28] B. Nikolić, B. Sazdović, Nucl. Phys. B 836 (2010) 100–126.

[29] P.A. Grassi, L. Tamassia, J. High Energy Phys. 07 (2004) 071.

[30] Lj. Davidović, B. Sazdović, Eur. Phys. J. C 74 (2014) 2683.

[31] Lj. Davidović, B. Nikolić, B. Sazdović, Eur. Phys. J. C 74 (2014) 2734.

[32] N. Berkovits, P. Howe, Nucl. Phys. B 635 (2002) 75–105.

[33] M.J. Duff, arXiv:hep-th/9912164v2.

5

Resolving the Schwarzschild singularity in both classic and quantum gravity

Ding-fang Zeng [a,b,*]

[a] *Theoretical Physics Division, College of Applied Sciences, Beijing University of Technology, China*
[b] *State Key Laboratory of Theoretical Physics, Institute of Theoretical Physics, Chinese Academy of Sciences, Beijing, 100124, China*

Editor: Stephan Stieberger

Abstract

The Schwarzschild singularity's resolution has key values in cracking the key mysteries related with black holes, the origin of their horizon entropy and the information missing puzzle involved in their evaporations. We provide in this work the general dynamic inner metric of collapsing stars with horizons and with non-trivial radial mass distributions. We find that static central singularities are not the final state of the system. Instead, the final state of the system is a periodically zero-cross breathing ball. Through 3+1 decomposed general relativity and its quantum formulation, we establish a functional Schrödinger equation controlling the micro-state of this breathing ball and show that, the system configuration with all the matter concentrating on the central point is not the unique eigen-energy-density solution. Using a Bohr–Sommerfield like "orbital" quantisation assumption, we show that for each black hole of horizon radius r_h, there are about $e^{r_h^2/\ell_{\rm pl}^2}$ allowable eigen-energy-density profiles. This naturally leads to physic interpretations for the micro-origin of horizon entropy, as well as solutions to the information missing puzzle involved in Hawking radiations.

---

[*] Correspondence to: Theoretical Physics Division, College of Applied Sciences, Beijing University of Technology, China.

E-mail address: dfzeng@bjut.edu.cn.

1. Motivation and logic

Although string theory and loop gravity [1–4] both give interpretations for the microscopic origin of some — loop claims any — black holes' entropy [5], partly due to lacks of a common semi-classic picture, none of them is considered the final answer [6]. Related with the physic of micro-states, is the black hole's information missing puzzle. That is, when a black hole evaporates, where does the information it carries go away [7–11]? In principles, any interpretation for the micro-states of a black hole should also tell us how they change when it evaporates. In practices, almost all existing resolutions [12–18] to this puzzle are regardless of the quantum theories' micro-state interpretation. Very recently, Hawking, Perry and Strominger [19] propose to solve this question+puzzle in a unifying framework of infinite number of hidden symmetries. Their proposal is still in completion but seems very hard to be dis/verified observationally. The purpose of this work is to provide a simple but dis/verifiable semi-classic picture, as well as quantisation method for the micro-state of black holes and the corresponding resolutions to the information missing puzzle involved in Hawking radiations. The core of the work is the Schwarzschild singularity's resolution.

Our logic is, if[1] central singularities are not the final fate of collapsing stars, then all question+puzzles related with the micro-state of black holes must be understandable from inner structures of the collapsing star leading to its formation. Obviously, for a Schwarzschild black hole, the most natural micro-structure inheritable from its parental star is the radial mass distribution $m(0, r)$ and evolving speed $\dot{m}(0, r)$ at some initial epochs. Non-radial local random motions inside an externally-looking spherical symmetric star are although possible, due to the fact that $n \propto m_{\text{total}} \propto r_h$, i.e. the particle number linearly depends on the mass thus on the horizon size of the black hole, they contribute to the entropy of the system only of $\mathcal{O}[r_h]$, obviously negligible relative to the horizon entropy $\mathcal{O}[r_h^{n-1}]$ in $n + 1$ dimensional space–times. We will show that in the quantum formulation of 3+1 decomposed general relativities, the micro-states of the collapsing star are defined by eigen-energy-density solutions of a functional Schrödinger equation. For very large this kind of star, through a Bohr–Sommerfield like "orbital" quantisation assumption, we show that the degeneracy of eigen-solutions is about $e^{r_h^2/r_{\text{pl}}^2}$. Since each of these degenerating stars has its own characteristic de-horizon/expansion speed determined by its radial mass distribution and could be measured as its identifying accordance, no information will be missed during a black hole's evaporation.

The content of this work is organised as follows. The next section will focus on classic metric exploration of collapsing stars with general radial mass distributions. While the next section provides quantum descriptions for the physic pictures uncovered in section 2. We then cost two sections discussing the micro-states' number counting of black holes and the resolution of information missing puzzle involved in Hawking radiations. The last section is our conclusion and prospects for future works.

2. Inner structure of black holes, classic picture

Historically, Oppenhemer and Snyder (OS) [20,21] are the earliest physicists to consider the inner structure of Schwarschild black holes. But they assume that matter contents inside the

---

[1] We will provide exact classic solution examples displaying that the final state is indeed a zero-cross breathing ball, thus no contradictions with Penrose and Hawking's singularity theorem occurs here.

horizon are uniformly distributed thus excludes the possibility of non-unique micro-states. Yo-dzis, Seifert and Müller (YSM) [22,23] considered layering matter contents inside horizons in constructing counter examples to the cosmic census hypothesis. But they noted nothing about this layering structure with the micro-state of black holes. In both OS and YSM's works, inner metrics of the black holes were written in co-moving spatial coordinates, which due to shell-crossing phenomenas will become invalid before central singularities formation thus of no use in quantum resolutions of the singularity. As comparisons, our metrics in this work use only Schwarzschild-like spatial coordinates [24,25]. They are thus valid during the whole process of the central singularity's formation.

We find that full geometries of a collapsing star with general radial mass distributions could be written as

$$ds^2 = -h^{-1}A(\tau,r)d\tau^2 + h^{-1}dr^2 + r^2 d\Omega_2^2 \tag{1}$$

$$h = 1 - \frac{2m(\tau,r)}{r}, \ r < r_0$$

$$A = \frac{\dot{m}^2}{m'^2} + h$$

where r_0 is the initial radius of the dust star and $m(\tau,r)$, the mass of all contents inside the sphere of radius r at time τ, with τ being the proper time of freely collapsing matter contents. To connect with the Schwarzschild metric on the boundary of the star, it is required that

$$A(0,r_0) = 1, \ d\tau = hdt \tag{2}$$

On there, τ happens to be the proper time of freely falling observers in the Schwarzschild back-ground, whose equations of motion just read $h\dot{t} = 1$, $\dot{r}^2 = 1 - h$. Inside the collapsing star, those observers will co-move with the matter contents are thus controlled by $u^0 = 1$, $u^1 = -\frac{\dot{m}}{m'}$ and Einstein equations $R_{\mu\nu} - \frac{1}{2}g_{\mu\nu}R = \rho u_\mu u_\nu$ in the zero-pressure dust star case

$$\frac{m''}{m'} = \frac{\dot{m}'}{\dot{m}} - \frac{m}{r^2}\frac{m'^2}{\dot{m}^2} - \frac{2m}{r(r-2m)} \tag{3}$$

$$\frac{\ddot{m}}{\dot{m}}\frac{m'}{\dot{m}} = \frac{\dot{m}'}{\dot{m}} + \frac{2m}{r^2}\frac{m'^2}{\dot{m}^2} + \frac{2m}{r(r-2m)} \tag{4}$$

It should be noted that the metric ansatz (1) is valid regardless matter contents consisting the col-lapsing star has pressures or not. However, zero-pressure condition enters equations (3) and (4). They are thus valid only for zero-pressure dust stars.

Equations (3)–(4) being valid simultaneously imply a redundancy, i.e. two equations con-trolling one variable $m(\tau,r)$'s evolution. We only need $m(0,r)$ instead of $\{m(0,r), \dot{m}(0,r)\}$ as a whole to specify initial status of the system. This is a general feature of Einstein equation. Similar things also occur in cosmologies. There the evolution of the homogeneous and isotropic universe is controlled by two Friedmann equations

$$\frac{\dot{a}^2}{a^2} + \frac{k}{a^2} = \frac{1}{3}(\rho + \Lambda) \tag{5}$$

$$2\frac{\ddot{a}}{a} + \frac{\dot{a}^2}{a^2} + \frac{k}{a^2} = -(\rho - \Lambda) \tag{6}$$

The first is time-first-order while the second is time-second-order. Obviously we need only know the initial value of $a(0)$ to predict its future evolutions.

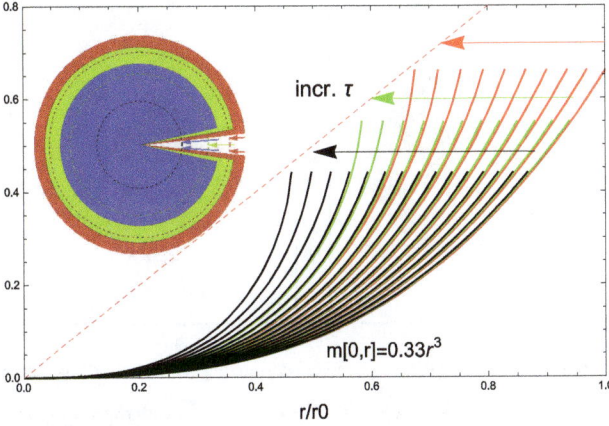

Fig. 1. Red lines display the variation of mass distributions inside the whole star $r < 1.00r_0$; green lines, the variation of mass distributions inside the sphere $r < 0.94r_0$; while the blue ones, that inside the sphere of $r < 0.87r_0$. Along the red dashed line, $1 - \frac{2m}{r} = 0$. The initial distribution is assumed $m(0, r) = c \cdot \min\{r, r_0\}^3$. Changing parameters c, q and r_0 will not change this picture qualitatively. But more general initials like $m(0, r) = \sum_i c_i \min\{r, r_i\}^{q_i}$ could lead to collapsing stars with multi-horizons among which some may be finished forming earlier than the outmost one. (For interpretation of the references to colour in this figure legend, the reader is referred to the web version of this article.)

Due to redundancies in the equation of motion, evolutions of a collapsing star are completely determined by its initial mass distribution $m(0, r)$. For example, for the following non-singular, no-horizon initial distributions,

$$m(0, r) = c \cdot r^q, \; 0 < 1 - \frac{2c \cdot r^q}{r}, \; 0 < r < r_0 \tag{7}$$

$$m(0, r) = cr_0^q, \; r_0 < r,$$

the corresponding $\dot{m}(0, r)$ is non-freely settable, it is determined by the constraint (3) ($\pm$ correspond collapsing/expanding respectively)

$$\dot{m}(0, r) = \pm r^{q-1} \left[\frac{b - \frac{c^2 q^2}{q+1}(1 - 2cr^{q-1})^{\frac{q+1}{q-1}}}{(1 - 2cr^{q-1})^{\frac{2}{q-1}}} \right]^{\frac{1}{2}} \tag{8}$$

$$\dot{m}(0, r) = 0, \; r_0 < r$$

In these formulas, $m(0, r)$ is the initial mass distribution, r_0, cr_0^q and $q(>1)$ are the initial star radius, total mass and pattern parameter of distributions respectively. Obviously, more general initials could be implemented by superpositions of the form $m(0, r) = \sum_i c_i \min\{r, r_i\}^{q_i}$. With initial conditions (7)+(8) as a concrete example, second order forward Runge–Kutta algorithm could be used to integrate equations (4) and (3) simultaneously. We displayed the results in Fig. 1.

From Fig. 1, we firstly note that no matter how the initial distribution is, near the outmost horizon entrance point, the mass function has linear-inversely divergent first order derivative $m'(\tau, r \to r_h) \propto (r - r_h)^{-1}$, so that

$$m(\tau, r) \xrightarrow{2m \to r_h} a \log(1 - r/r_h) + b \tag{9}$$

This will play key roles in our derivations of the black hole entropy's area law and can be seen from the limit analysis of equations (3) and (4) directly, in which $\frac{m'}{\dot{m}} = (\frac{dr}{d\tau})^{-1}$ is finite, but $\frac{m''}{m'}$,

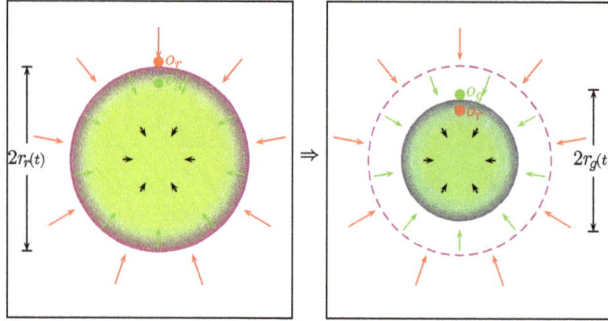

Fig. 2. $r_r(t)$ and $r_g(t)$ are radial coordinates of two observers co-moving with a collapsing star, r_{r0}, r_{g0} are their initial values. Newton mechanics tell us that $\ddot{r}_i = -\frac{G\rho_{i0}r_{i0}^3}{r_i^2} \Rightarrow r_i = r_{i0}(1 - \frac{t}{t_{i0}})^{\frac{2}{3}}$, with $t_{i0}^2 = \frac{1}{G\rho_{i0}}$ denoting the time observer i falling to the central point. Obviously, if the average density ρ_{r0} of masses inside the sphere r_{r0} is larger than that inside r_{g0}, then the observer r will fall earlier than g to the centre of the star, during which shell crossing happens somewhere inside sphere r_{g0}.

$\frac{\ddot{m}}{m}$, $\frac{\dot{m}'}{\dot{m}}$ and $\frac{2m}{r-2m}$ are all linear-inversely divergent. This forms a technique firewall prohibiting us from evolving the differential system beyond the horizon formation epoch. However, if we calculate the physical mass/energy density

$$\rho = -T_0^0 - T_1^1 = \frac{2(r - 2m)m'^3}{r^2[(r - 2m)m'^2 + r\dot{m}^2]} \tag{10}$$

we will find that the result is everywhere regular at this epoch. So this firewall is not a wall of infinite physical energy density. We guess they may be related the AMPS [8,9] firewalls technically.

The second point we can see from Fig. 1 is that, although the collapsing star as a whole will quickly contract into its horizon surface, its inner sub-star will not do so! The more inner sub-star needs more longer time to contract into their own horizon surface. The most inner sub-star almost needs infinite length of time to fall onto the central point. Combining this fact with Penrose and Hawking's singularity theorem [26–28] which says that any of this collapsing stars will collapse to the central point in finite proper times, we infer that during the outmost mass-shell's collapsing to the central point, it must shell-cross all mass-shells initially more close to the central point, see Fig. 2 for pictures. This shell-crossing phenomena were firstly mentioned by YSM in Ref. [22, 23] as the origin of naked singularity thus counter examples to the cosmic consensus hypothesis. For this reason, its physic value is long-termly ignored, or even negatively viewed.

In fact, the most important shell-crossing occurs on the central point. The crossing events there are unavoidable in both general relativity and Newton mechanics and have no dependence on the stars' having a high density outer skin or not. They are results of momentum-energy conservation laws. Consider an observer co-moving with the collapsing star, when it arrives near the central point

$$r = r_0\left(1 - \frac{t}{t_0}\right)^{\frac{2}{3}}, \ \dot{r} = -\frac{2r_0}{3t_0}\left(1 - \frac{t}{t_0}\right)^{-\frac{1}{3}}, \ t \to t_0 - \epsilon \tag{11}$$

Its radial speed is divergent. So it cannot stop there immediately and has to shell-cross to the anti-direction

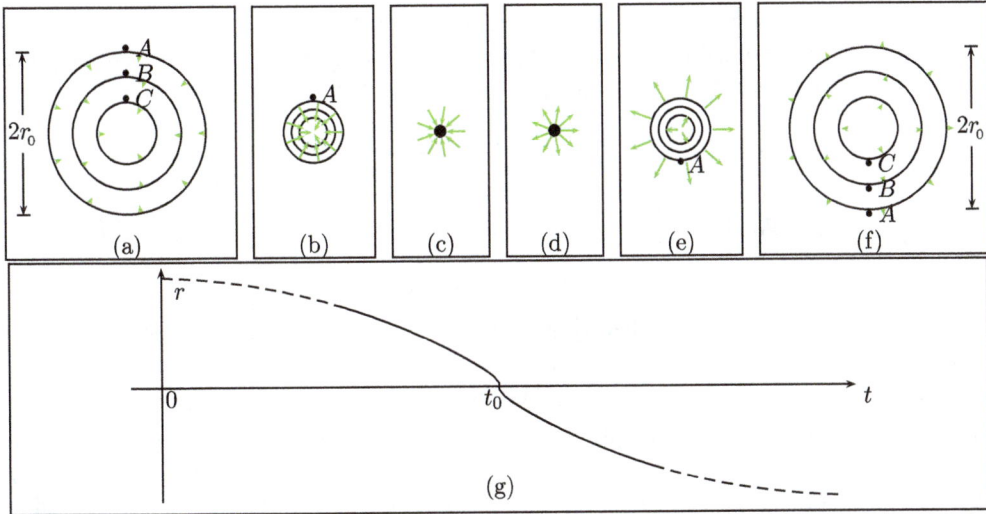

Fig. 3. A complete evolution cycle of a collapsing star is from (a) to (b), then to (c), then to (d), then to (e), then to (f) and finally to (a) again. Depending on the initial conditions, evolutions from (a) to (b) and (e) to (f) could contain shell-crossing events. But the evolution from (c) to (d) contains shell-crossing events no matter how the initial conditions are. Subfigure (g) displays the radius evolution of the collapsing star, the fact that $t(< t_0) \to t_0$, $r = r_0(1 - t/t_0)^{\frac{2}{3}}$ and as $t(> t_0) \to t_0$, $r = -r_0(t/t_0 - 1)^{\frac{2}{3}}$ follows from Newton mechanics $\ddot{r} = \frac{GM_{\text{tot}}}{r^2}$ directly. General relativity would not change this fact qualitatively.

$$r = -r_0\Big(1 - \frac{t}{t_0}\Big)^{\frac{2}{3}}, \ t \to t_0 + \epsilon \tag{12}$$

This means that static central singular point is not the final state of a collapsing star. The proper final state should be a periodically zero-cross breathing ball, see Fig. 3 for pictures. In real collapsing star consisting of fermion particles, shell-crossing phenomenas are also unavoidable. Although near such crossing point, infinite pressures could appear due to Pauli-exclusion principles and lead to bounces of the crossing-wish shells, the bouncing itself could be thought as a shell-crossing when the identity principle is considered. Obviously, this zero-cross breathing ball provides us a very smart way to resolve the central singularity of Schwarzshild black hole but successfully avoids contradictions [29] with Penrose and Hawking's singularity theorem. That is, central singularities indeed happen in finite times after the collapsing begins. However, if we continuously take photos inside the horizon of the system, what we get will be mostly of regular stars with various radius mass distributions instead a single point carries all the mass of system exclusively. We will show in the following that, this radial mass distribution provides just the physic basis for the micro-state of black holes, thus origins for the horizon entropy.

3. Inner structure of black holes, quantum description

Obviously, if we can provide a quantum description for above classic pictures, our declaration that radial mass distributions inside the horizon of a collapsing star is just the micro-state of the equal mass black holes will be more believable to peoples. This is directive in the quantum theory of gravitations originally proposed by B. S. DeWitt [30] and developed latter mainly in quantum cosmologies [31]. It is also applied to black holes exploration in references [32–37].

However, none of these works tries to understand the micro-state of black holes by this method, although it is so natural and directive. To implement such descriptions, we firstly consider the 3+1 decomposed dynamics of matter and geometries inside the collapsing star

$$ds^2 = -N^2 dt^2 + h^{-1} dr^2 + r^2 d\Omega_2^2, \quad h = 1 - \frac{2m(t,r)}{r} \tag{13}$$

$$\frac{S_L}{4\pi} = \int dt \, dr \, N h^{-\frac{1}{2}} r^2 \left[\frac{2m'}{r} - \frac{\rho}{2} (\dot{x} \cdot \dot{x} + 1) - \frac{p}{2} (\dot{x} \cdot \dot{x} - 1) \right] \tag{14}$$

$$+ \text{ local total derivative terms}$$

where ρ, p and $\dot{x}^\mu$ are the energy density, pressure and four velocity of fluid elements inside the collapsing star respectively. In the final equation of motion, normalisation $\dot{x} \cdot \dot{x} = -N^2 \dot{x}^0 \cdot \dot{x}^0 + h^{-1} \dot{x}^r \cdot \dot{x}^r = -1$ should be set everywhere. And because $\dot{x}^r = -\frac{\dot{m}}{m'}$, $x^0 = t$, $N^2 = \frac{\dot{m}^2}{m'^2} h^{-1} + 1$ follows from the 4-velocity's normalisation naturally. N^2 here being not independent variable has also counter sayings in cosmologies, where it is usually set as $N = 1$ for the co-moving observers. So, in this 3+1 decomposed system (13)–(14), only $m(t,r)$ and ρ, p are possible dynamic variables. Turning to the Hamiltonian language

$$P_m \equiv \frac{\delta S_L}{\delta \dot{m}(t,r)}, \quad S_H = \int dr \, \dot{m}(t,r) P_m - S_L \tag{15}$$

$$\frac{S_H}{4\pi} = \int dt \, dr \, N h^{-\frac{1}{2}} r^2 \left[-\frac{2m'}{r} - \frac{\rho}{2} (N^2 + h^{-1} \frac{\dot{m}^2}{m'^2} - 1) \right. $$
$$\left. - \frac{p}{2} \cdots \right] - \text{local total derivative terms} \tag{16}$$

In the case $p = 0$, Hamiltonian constraint following from this action $\delta S_H / \delta N = 0$ and the 4-velocity's normalisation will bring us expressions for ρ completely the same as (10)

$$\mathcal{H}(m, P_m) = h^{-\frac{1}{2}} r^2 \left[-\frac{2m'}{r^2} - \rho \left(h^{-1} \frac{\dot{m}^2}{m'^2} + 1 \right) \right] = 0 \tag{17}$$

On the other hand, from the Hamilton–Jaccobi equation following from this action and the conservation law following from the vanishing of local total derivative terms, equations (3) and (4) could also be derived out routinely. This justifies the correctness of equations of motion written in the previous sections from the aspect of action principles.

Now, following ideas completely the same as quantum cosmologies [30,31], we consider $m(r)$ as a general coordinate and introduce a wave function $\Psi[m(r)]$ to denote the probability amplitude of the system with mass distributions $m(r)$. $\Psi[m(r)]$ satisfies the operator version of constraint (17), with $\dot{m}$ replaced by functions of m and P_m, the latter by functional derivatives $\frac{-i\hbar\delta}{\delta m(r)}$

$$\left[8 h^{-\frac{3}{2}} r^2 m'^{-1} \rho - \frac{\hbar^2 \delta^2}{\delta m(r)^2} + 4 h^{-\frac{3}{2}} r^4 m'^{-2} \rho^2 \right] \Psi[m(r)] = 0 \tag{18}$$

This functional differential equation together with the following boundary condition [we use $H(x)$ denoting the usual Heaviside step function, so $H(x) = \begin{cases} 0, & x < 0 \\ 1, & 0 < x \end{cases}$]

$$\Psi[m_h H(r - 0)] \neq \infty, \quad \Psi[m(r = r_h) < m_h] = 0 \tag{19}$$

define a functional eigenvalue problem for $\Psi[m(r)]$. Similar to the usual eigenvalue problems in quantum mechanics, it can be imagined that only some special eigen-energy-densities $\{\rho_i(r), i = 0, 1, 2 \cdots\}$ could lead to normalisable wave-functional $\Psi_i[m(r)]$. Besides (18) and (19), the eigen-energy-density should also satisfy constraints

$$\int_0^{r_h} \rho_i(x)4\pi x^2 dx \approx m_h \tag{20}$$

the approximation symbol here indicates our neglecting of the curved space fact in its written down. With this final constraints, it's natural to conjecture that the index i of eigenvalue/states has upper bound and the wave-functions $\{\Psi_i[m(r)], i = 0, 1, 2 \cdots, i_{\max}\}$ have one-to-one correspondence with the micro-state of the black holes in consideration.

To understand the fact that equations (18)–(20) indeed define the quantum micro-state of black holes, let us try to solve them by the following strategies, i) constraining $m(r)$ to the form r^ξ so that $\frac{\delta}{\delta m} = (m \ln r)^{-1} \frac{\partial}{\partial \xi}$; ii) writing functionals $\Psi[m(r)]$ to usual functions $\Psi(\xi)$, thus changing the functional equation into a differential array

$$\forall r \in [0, r_h], \; \left[\left(\frac{8mr}{\xi}r^2\rho + \frac{4r^2}{\xi^2}r^4\rho^2\right)\left(1 - \frac{2m}{r}\right)^{-\frac{3}{2}}(\ln r)^2 \right. \tag{21}$$

$$\left. + \hbar^2\left(\ln r \, \partial_\xi - \partial_\xi^2\right)\right]\Psi(\xi) = 0, \; m = \frac{r_h}{2}\left(\frac{r}{r_h}\right)^\xi$$

$$\Psi(0) \neq \infty, \; \Psi(\infty) = 0, \; \int_0^{r_h} \rho(x)4\pi x^2 dx \approx m_h \tag{22}$$

This array contains infinite components, because its master equation need be satisfied as r varies in the continuous range $[0, r_h]$. Operationally we can choose to let it be satisfied only on some discrete values of r. For example, in the $1\ell_{pl}$-sized black hole, we can choose such discrete points as $r = \frac{1}{6}, \frac{3}{6}, \frac{5}{6}\ell_{pl}$ and specify the $r^2\rho(r)$ function by its values on three equal-width interval $(0, \frac{1}{3}), (\frac{1}{3}, \frac{2}{3}), (\frac{2}{3}, 1)$. Assuming that mass/energy densities on each of these intervals be uniform, considering the total mass constraints (22), all possible $r^2\rho$ profiles could be listed as follows

$r^2\rho$ \ r	$0\sim\frac{1}{3}$ ($\frac{1}{6}$)	$\frac{1}{3}\sim\frac{2}{3}$ ($\frac{3}{6}$)	$\frac{2}{3}\sim1$ ($\frac{5}{6}$)	m/e.dist.	ei.soln?
1	3	0	0		★
2	0	3	0		○
3	0	0	3		○
4	2	1	0		★
5	1	2	0		○
6	0	2	1		○
7	0	1	2		○
8	2	0	1		★
9	1	0	2		○
0	1	1	1		★

$$\tag{23}$$

For each of these $r^2\rho$ profiles we solve equations (21) [all solutions are normalised to $\Psi(\xi = 1) = 1$ and $\Psi(\xi = \infty) = 0$] on the interval centrals $r = \frac{1}{6}, \frac{3}{6}, \frac{5}{6}$. The results are displayed in

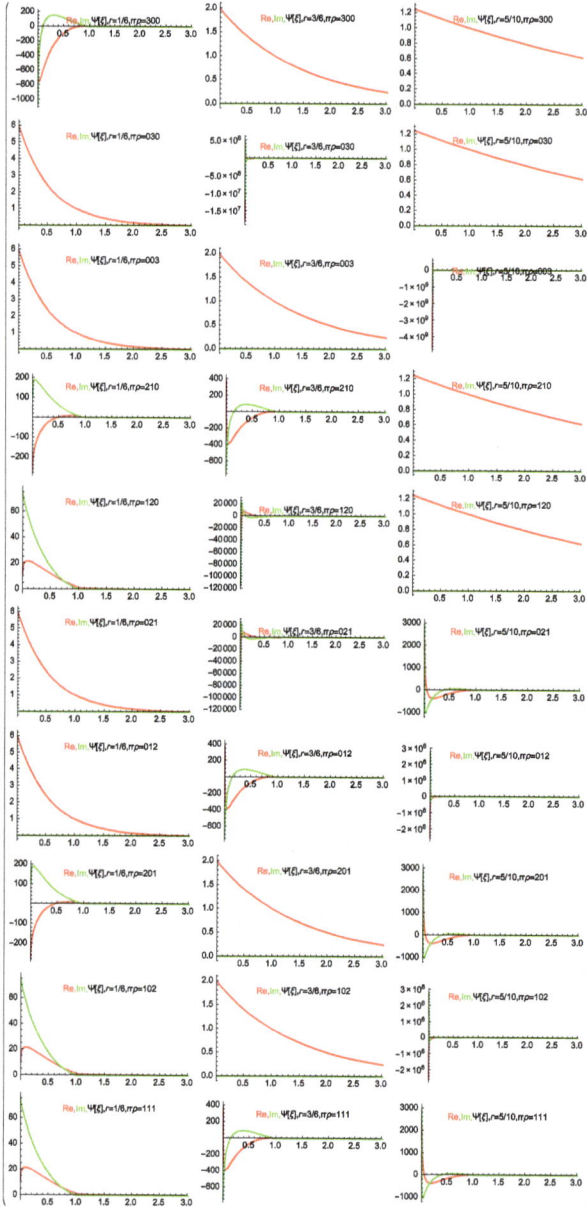

Fig. 4. Numerical solutions to the differential equation (21) with conditions $\Psi(\xi = 1) = 1$, $\Psi(\xi = \infty) = 0$ for 10 possible $r^2\rho$ profiles listed in (23) on positions $r = \frac{1}{6}, \frac{3}{6}, \frac{5}{6}\ell_{\rm pl}$. Good eigen-wave-function should be such ones that i) $\Psi(0) \neq \infty$, ii) normalisable and iii) r-independent as possible as can be. (For interpretation of the references to colour in this figure, the reader is referred to the web version of this article.)

Fig. 4, from which it can be easily see that, if we insist a good eigen-energy-density need satisfy i) $\Psi(0) \neq \infty$, ii) normalisable and iii) being r-independent exactly, then none of the ten mass/energy distributions listed in (23) is a good one. However, if we discretised function $r^2\rho(r)$ on more finer grids, we can obtain eigen-energy-density profiles more close to these judgements. On the

3-interval discretising level, the 1st, 4th, 8th and 10th profile in (23) could be looked as good ones. The key point here is that, the system configuration with all mass/energy concentrating on the central point — the 1st one — is not the unique good eigen-energy-distribution. Instead, the good eigen-distribution is $4 \sim e^{1^2}$-times degenerate.

Further, if we consider the $2\ell_{\mathrm{pl}}$-sized black hole, we will find that if the same precision as $1\ell_{\mathrm{pl}}$-sized black hole is wished, then the discretising of function $r^2\rho(r)$ should be on 6 equal-length interval, the number of all possible profiles adds up to 462, some of them could listed explicitly as follows

$r^2\rho$ \\ r	$0\sim\frac{1}{3}$ $\frac{1}{6}$	$\frac{1}{3}\sim\frac{2}{3}$ $\frac{3}{6}$	$\frac{2}{3}\sim\frac{3}{3}$ $\frac{5}{6}$	$\frac{3}{3}\sim\frac{4}{3}$ $\frac{7}{6}$	$\frac{4}{3}\sim\frac{5}{3}$ $\frac{9}{6}$	$\frac{5}{3}\sim\frac{6}{3}$ $\frac{11}{6}$
1	6	0	0	0	0	0
2	0	6	0	0	0	0
⋮	⋮	⋮	⋮	⋮	⋮	⋮
6	0	0	0	0	0	6
7	5	1	0	0	0	0
8	0	5	1	0	0	0
⋮	⋮	⋮	⋮	⋮	⋮	⋮
11	0	0	0	0	5	1
12	1	5	0	0	0	0
13	0	1	5	0	0	0
⋮	⋮	⋮	⋮	⋮	⋮	⋮
462	1	1	1	1	1	1

$$(24)$$

Similar to $1\ell_{\mathrm{pl}}$-sized black holes, we find that not all these mass/energy profiles are equally good eigen-energy-densities that make the quantum wave-function i) $\Psi(0) \neq \infty$, ii) normalisable and iii) r-independent to the highest degree. We find that, the good eigen-energy-distribution scheme is approximately $55 \sim e^{2^2}$-times degenerate. Now, if we want to use this same idea as in 1 and $2\ell_{\mathrm{pl}}$-sized black holes to more larger ones, we will need to numerically solve exponentially-many Schrödinger equation to find the good eigen-energy-density solutions, which is obviously impossible operationally. However, explorations in the small black hole examples indeed provide us supporting evidence that eigen-energy-densities defined by eqs. (18)–(20) have one-to-one correspondence with the micro-state of black holes. We introduce in the following an approximate method for the number counting of eigen-states of large black holes by the so called correspondence principles [38].

4. The micro-states' number counting and horizon entropies

As is well know, in collapsing stars corresponding to very large black holes, the average density of the system is very small $\rho_{\mathrm{av}} \approx M/(2GM)^3$. According to Newtonian mechanics, the collapsing speed of these large stars is correspondingly very small due to the fact that the collapsing time square $t^2 \propto 1/G\rho_{\mathrm{av}}$. According to equation (10), the local energy density of the system is thus approximately $\rho \approx \frac{m'}{4\pi r^2}$, which is just the density definition is conventional Newton mechanics. This means that for large black holes, the number counting of proper eigen-energy-density profiles $\rho(r)$ could be replaced by the number counting of mass function $m(r)$ directly. On the other hand, our numeric examples in the second section of work also tell us that for the

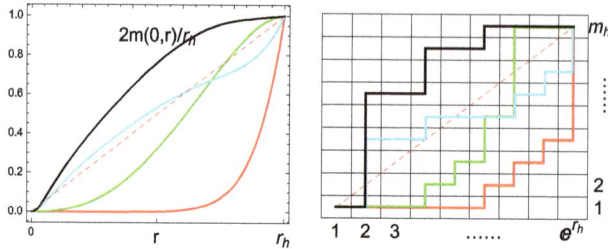

Fig. 5. The left hand side displays four mass distribution ways inside a collapsing star an infinitesimal time before/after it collapses into the horizon. Along the diagonal line $2m(0, r) = r$. The right hand side is a discrete representation of the left. Each continuous distribution way corresponds to a regular Young diagram of m_h/m_{pl} row and $1 \sim e^{r_h/\ell_{pl}}$ column.

initially non-singular collapsing stars, the horizon always forms earlier than central singularities. So matter distributions an infinitesimal time before or after the horizon forms could be looked as ideal proxies of the system's quantum states. Obviously, the idea here is very similar to the correspondence principle firstly introduced by N. Bohr in early quantum mechanics in establishing relations between the quantum wave function and classic orbits of electrons in atoms [38]. The key question here is, how to make the continuous mass function $m(r)$ become discrete object thus count them one by one.

Our idea is, introduce an "distribution quantisation" assumption so that any two collapsing stars with equal total mass but different radial distributions by 1-m_{pl} mass shells of radius $r \leqslant r_h$ and concentric with the parental star could be identified as distinguishable quantum states of the corresponding black hole. Similar to Bohr–Sommerfield's orbital quantisation condition leading to discretised orbit for electrons in atoms, our distribution quantisation condition will lead to discretised radial mass distributing ways. Referring to Fig. 5, our quantisation condition requires the vertical line of the distribution function space be discretised by m_{pl}, while to distinguish those distributions differ by only 1-m_{pl} mass shells near the horizon edges, the horizontal line must be discretised exponentially, which arises from the logarithmic divergence of mass functions there, see equation (9).

Obviously, in the discretised function space, each radial mass distribution corresponds to a regular Young diagram of m_h/m_{pl} row and $1 \sim e^{r_h/r_{pl}}$ column. So the total number of such distributions is $W = (e^{\frac{r_h}{r_{pl}}})^{\frac{m_h}{m_{pl}}}$. Since $r_h m_h \propto r_h^{n-1} \propto$ horizon area in $n+1$ dimensional space–time, this implies that, associating with every Schwarzschild black hole of horizon area A, is a micro-canonical ensemble of $e^{A/A_{pl}}$ collapsing stars, all with the same mass and surface area but each with different inner mass distributions. According to definitions, the entropy of such black holes reads

$$S = k_B \log W = \frac{A}{A_{pl}} \tag{25}$$

This is nothing but the Bekenstein–Hawking formula [5] up to a numeric factor of order 1. It is worth to emphasise that, our micro-states counting involves only initial distributions $m(0, r)$ instead of $m(0, r)$ and $\dot{m}(0, r)$ simultaneously. This is because, Einstein equation gives two superficially redundant components controlling the evolution $m(\tau, r)$. Given initial distributions, component (3) will fix the speed $\dot{m}(0, r)$ while (4) will yield dynamical evolutions $\ddot{m}(0, r)$ to the next epoch. This is remarkably different from other dynamical systems which are controlled by only one differential equation and is the key reason for area laws.

It should be emphasised that the micro-state we counted here is not local motion modes of particles inside the black hole. Such local degrees of freedom contribute to the micro-state of the system only of order $e^{n(\text{i.e. particle number})} \sim e^{m_h} \sim e^{r_h}$, which is obviously negligible relative to the radial mass distribution modes $e^{r_h^{n-1}}$ in $n+1$ dimensions. The micro-state we counted here is non-local collective motion of matter contents inside the black hole, they are essentially geometrical degrees of freedom because their form uniquely determines inner geometries of the system. In the series of works [39–42], Stojkovic et al. provide many concrete evidences that, non-locality plays key roles in both the central singularities resolution and the Hawking radiations's unitarity recovering.

5. Solutions to the information missing puzzle

The above pictures for the micro-state of black holes imply a direct method to resolve the information missing puzzle. To see this more explicitly, we rewrite equation (3) in first order forms, but in this time understand the mass function $m(\tau, r)$ as the inner mass distribution of the black holes in special states,

$$\dot{m}^2(\tau, r) = m'^2(\tau, r) \cdot \exp\Big[\int_0^r \frac{4m(\tau, x)dx}{x(x - 2m)} \Big] \times \tag{26}$$

$$\int_0^r 2m(\tau, x)x^{-2} \exp\Big[-\int_0^x \frac{4m(\tau, y)dy}{y(y - 2m)} \Big] dx$$

Obviously, each micro-state of the black hole has its own characteristic inner mass distribution, thus characteristic speed of total mass variation $\dot{m}(\tau, r = r_{\text{edge}})$ when they evaporate/accrete. By recording this speed of mass/size variation, we could reproduce all the information related with its inner mass distributions. So there are no information missing puzzles related with the Hawking radiation! This almost classic general relativistic resolution of information missing puzzles is possible because, in both Hawking's original calculation [7] and the latter advanced version of F. Wilcek [43], the background black holes are assumed to have fixed horizon sizes, thus imposes no effects on the evaporation speed. These calculations could provide dynamic mechanisms by which particles escape from the horizon. But they have no chances to catch kinematics of the background black hole's size variation. It is just this kinematics that carries away the missed information.

The area law and micro-interpretation for the black holes' entropy lie on centres in string theory and loop gravity's achievements. However, none of them, including the recent interpretations of Hawking, Perry and Strominger, is verifiable experimentally. As comparisons, our interpretations in this work are dis/verifiable observationally. Since the information of black holes in our interpretation is identified with radial mass distributions of the corresponding collapsing stars, it could be extracted or released through certain classic process. For example, in astrophysical events such as binary black hole's mergering [44,45], signals other than gravitational waves such as gamma ray bursts could be produced when matters going from one hole to the other. Reference [46–48] may have given us such evidences already. Even when gamma rays are non-available, the form of gravitational waves would have different shapes when produced from binaries with different inner distributions. With the development of gravitational wave and gamma ray astronomies, this verification may already be at our technique abilities.

6. Conclusion

We provide in this work the most general dynamic inner metric of collapsing stars with horizon and non-trivial radial mass distributions. We find that near the central singular point, shell-crossing phenomena are unavoidable and static central singularities are not the final state of all such collapsing stars. Instead, their final state is something we called zero-cross breathing balls. This naturally resolves the central singularity of Schwarzschild black holes but avoids contradiction with Penrose and Hawking's singularity theorem. If we take photos for these breathing balls in their horizon continuously, then what we get will be mostly of collapsing star with various regular radial mass distribution instead of singular points concentrating all masses of the system exclusively. The radial mass distribution here is nothing but micro-states that lead to horizon entropies for the black holes with equal masses. Non-radial local random motions of particles inside an externally-looking spherical symmetric collapsing star are although possible, due to the fact that the particle number inside the horizon linearly depends on the mass thus on the horizon size of the black hole in consideration, they contribute to the entropy of the system only of $\mathcal{O}[r_h]$, obviously negligible relative to the horizon entropy $\mathcal{O}[r_h^{n-1}]$ in $n+1$ dimensional space–times.

We then enhance the above classic picture in quantum formulations of the 3+1 decomposed general relativity further. We find that the micro-state of the zero-cross breathing ball is defined by the eigen-energy-density solution of a functional Schrödinger equation. For 1 and $2\ell_{pl}$ sized this kind of ball, we provide numeric evidence that the eigen-solution is about e^{1^2} and e^{2^2} times degenerate. While for large this kind of ball, by assuming that any two distributions with equal total mass but different radial profiles by any $1m_{pl}$-weighted mass-shells correspond to distinguishable quantum states, we show that the degeneracy is approximately of $e^{r_h^2/r_{pl}^2}$ order. Since each of these degenerating balls has special de-horizon/expansion speed determined by its radial mass distribution and could be measured as its identifying accordance, no information will be missed during a black hole's evaporation. We thus provide not only a microscopic interpretation for the horizon entropy of black holes, but also a concrete resolution to the information missing puzzle involved in their Hawking radiations.

Obviously, it is a progress to translate the question of micro-state definition and number-counting related with the black hole entropy into solution's searching of a functional eigen-value-problem. However, since we find no exact solutions, we still have distances to the final solutions to these questions exactly. So, as the first suggestion for future works, we think that, finding highly-effective numeric algorithm or systematic approximation scheme to solve equations (18)–(20) maybe the most important work to do. Our second suggestion is, since our physic picture implies that black holes are nothing but micro-ensemble of collapsing stars with the same mass but different radial mass distributions, it is very interesting to quantitatively investigate differences between the shape of gravitational waves produced in the mergering of binary black holes with different inner mass distribution. Such investigations are still absent on the market [49–51] and will be very useful for the future observational dis/verification of our pictures. Thirdly, considering the non-local essence of the micro-state corresponding to the horizon entropy, it is very important to investigate the relation-ship between our definitions through functional Schrodinger equation and that through quantum entanglements [52,53]. Other prospects such as generalising our discussions to some more general black holes or de Sitter space–time itself may also be possible and interesting.

Acknowledgements

The author thanks very much to J.-b. Wu, Y. H. Gao, G. Horowitz, D. Gross and E. Witten (during the 2016 international string theory conference) for their valuable comments and suggestions in the preparation and revisions of this work. This work is supported by Beijing Municipal Natural Science Foundation, Grant No. Z2006015201001 and partly by the Open Project Program of State Key Laboratory of Theoretical Physics, Institute of Theoretical Physics, Chinese Academy of Sciences, China.

References

[1] A. Strominger, C. Vafa, Microscopic origin of the Bekenstein–Hawking entropy, Phys. Lett. B 379 (1996) 99, arXiv:hep-th/9601029.

[2] S. Mathur, The quantum structure of black holes, Class. Quantum Gravity 23 (2006) R115, arXiv:hep-th/0510180.

[3] C. Rovelli, Black hole entropy from loop quantum gravity, Phys. Rev. Lett. 77 (1996) 3288, arXiv:gr-qc/9603063.

[4] A. Ashtekar, J. Baez, A. Corichi, K. Krasnov, Quantum geometry and black hole entropy, Phys. Rev. Lett. 80 (1998) 904, arXiv:gr-qc/9710007.

[5] R. Wald, The thermodynamics of black holes, Living Rev. Relativ. 4 (2001) 6.

[6] J. Polchinski, The black hole information problem, Lecture Notes in 2014–15 Jerusalem Winter School and the 2015 TASI.

[7] S. Hawking, Breakdown of predictability in gravitational collapse, Phys. Rev. D 14 (1976) 2460.

[8] A. Almheiri, D. Marolf, J. Polchinski, et al., Black holes: complementarity or firewalls?, JHEP 1302 (2013) 062, arXiv:1207.3123.

[9] A. Almheiri, D. Marolf, J. Polchinski, et al., An apologia for firewalls, JHEP 1309 (2013) 018, arXiv:1304.6483.

[10] P. Chen, Y. Ong, D. Page, et al., Naked black hole firewalls, Phys. Rev. Lett. 116 (2016) 161304, arXiv:1511.05695.

[11] S. Hawking, The information paradox for black holes, arXiv:1509.01147.

[12] J. Preskill, Do black holes destroy information?, in: International Symposium on Black Holes, Membranes, Wormholes, and Superstrings, 1992, arXiv:hep-th/9209058.

[13] S. Giddings, The black hole information paradox, in: Particles, Strings and Cosmology. Johns Hopkins Workshop on Current Problems in Particle Theory 19 and the PASCOS Interdisciplinary Symposium 5, 1995, arXiv:hep-th/9508151.

[14] J. Hartle, Generalized quantum theory in evaporating black hole spacetimes, in: Black Holes and Relativistic Stars, 1997, p. 195, arXiv:gr-qc/9705022.

[15] S. Giddings, Comments on information loss and remnants, Phys. Rev. D 49 (1993) 4078, arXiv:hep-th/9310101.

[16] H. Nikolic, Resolving the black-hole information paradox by treating time on an equal footing with space, Phys. Lett. B 678 (2009) 218–221, arXiv:0905.0538.

[17] S. Hawking, Information preservation and weather forecasting for black holes, arXiv:1401.5761.

[18] H. Nikolic, Gravitational crystal inside the black hole, Mod. Phys. Lett. A 30 (2015) 1550201, arXiv:1505.04088.

[19] S. Hawking, M. Perry, A. Strominger, Soft hair on black holes, Phys. Rev. Lett. 116 (2016) 231301, arXiv:1601.00921.

[20] J.R. Oppenheimer, H. Snyder, On continued gravitational contraction, Phys. Rev. 56 (1939) 455.

[21] C. Misner, K. Thorne, J. Wheeler, Gravitation, in: Gravitation, W. H. Freeman and Company Version, 1973, §32.3, 32.4, 32.6.

[22] P. Yodzis, H.J. Seifert, H. Müller zum Hagen, On the occurrence of naked singularities in general relativity, Commun. Math. Phys. 34 (1973) 135.

[23] P. Yodzis, H.J. Seifert, H. Müller zum Hagen, On the occurrence of naked singularities in general relativity. II, Commun. Math. Phys. 37 (1974) 29.

[24] P.C. Vaidya, Proc. Indian Acad. Sci. A 33 (1951) 264, Phys. Rev. 83 (1951) 10, Gen. Relativ. Gravit. 31 (1999) 119.

[25] W. Bonnor, P. Vaidya, Spherically symmetric radiation of charge in Einstein–Maxwell theory, Gen. Relativ. Gravit. 1 (1970) 127.

[26] R. Penrose, Gravitational collapse and space–time singularities, Phys. Rev. Lett. 14 (3) (1965) 57; Gravitational collapse: the role of general relativity, Riv. Nuovo Cimento 1 (1969) 252–276, Gen. Relativ. Gravit. 34 (2002) 1141.

[27] R. Geroch, G. Horowitz, Global structure of spacetimes, in: General Relativity: An Einstein Centenary Survey, 1979, pp. 212–293.

[28] S. Hawking, G.F.R. Ellis, The Large Scale Structure of Space–Time, Cambridge University Press, ISBN 0-521-09906-4, 1973.

[29] N. Engelhardt, G. Horowitz, New insights into quantum gravity from gauge/gravity duality, arXiv:1605.04335.

[30] B.S. DeWitt, Quantum theory of gravity. I. The canonical theory, Phys. Rev. 160 (1967) 1113.

[31] J.J. Halliwell, Introductory lectures on quantum cosmology, in: S. Coleman, J.B. Hartle, T. Piran, S. Weiberg (Eds.), Proceedings of the Jerusalem Winter School on Quantum Cosmology and Baby Universes, World Scientific, Singapore, 1991.

[32] M. Allen, Canonical quantization of a spherically symmetric, massless scalar field interacting with gravity in $2+1$ dimensions, Class. Quantum Gravity 4 (1987) 149.

[33] L.Z. Fang, M. Li, Formation of black holes in quantum cosmology, Phys. Lett. B 169 (1986) 28.

[34] R. Laflamme, Wavefunction of a black hole interior, in: J. Demaret (Ed.), Origin and Early History of the Universe: Proceedings of the 26th Liege International Astrophysical Colloquium, 1987.

[35] Y. Nambu, M. Sasaki, The wavefunction of a collapsing dust sphere inside the black hole horizon, Prog. Theor. Phys. 79 (1988) 96.

[36] H. Nagai, Wavefunction of the de Sitter–Schwarzschild universe, Prog. Theor. Phys. 82 (1989) 322.

[37] L. Rodrigues, I. Soares, J. Zanelli, Black hole decay and topological stability in quantum gravity, Phys. Rev. Lett. 62 (1989) 989.

[38] S. Weinberg, Lecture Notes on Quantum Mechanics, Cambridge University Press, 2013.

[39] E. Greenwood, D. Stojkovic, Quantum gravitational collapse: non-singularity and non-locality, JHEP 0806 (2008) 042, arXiv:0802.4087.

[40] J. Wang, E. Greenwood, D. Stojkovic, Schrodinger formalism, black hole horizons and singularity behavior, Phys. Rev. D 80 (2009) 124027, arXiv:0906.3250.

[41] A. Saini, D. Stojkovic, Nonlocal (but also nonsingular) physics at the last stages of gravitational collapse, Phys. Rev. D 89 (2014) 044003, arXiv:1401.6182.

[42] A. Saini, D. Stojkovic, Radiation from a collapsing object is manifestly unitary, Phys. Rev. Lett. 114 (2015) 111301, arXiv:1503.01487.

[43] S. Iso, H. Umetsu, F. Wilczek, Hawking radiation from charged black holes via gauge and gravitational anomalies, Phys. Rev. Lett. 96 (2006) 151302, arXiv:hep-th/0602146.

[44] The LIGO Scientific Collaboration, the Virgo Collaboration, Observation of gravitational waves from a binary black hole merger, Phys. Rev. Lett. 116 (2016) 061102, arXiv:1602.03837.

[45] LIGO Scientific and Virgo Collaborations, GW151226: observation of gravitational waves from a 22-solar-mass binary black hole coalescence, Phys. Rev. Lett. 116 (2016) 241103, arXiv:1606.04855.

[46] V. Connaughton, E. Burns, A. Goldstein, et al., Fermi GBM observations of LIGO gravitational wave event GW150914, Astrophys. J. Lett. 826 (1) (2016) L6, http://dx.doi.org/10.3847/2041-8205/826/1/L6, arXiv:1602.03920.

[47] A. Janiuk, M. Bejger, S. Charzynski, P. Sukova, On the gamma-ray burst – gravitational wave association in GW150914, arXiv:1604.07132.

[48] V. Cardoso, E. Franzin, P. Pani, Is the gravitational-wave ringdown a probe of the event horizon?, Phys. Rev. Lett. 116 (2006) 171101, arXiv:1602.07309.

[49] F. Pretorius, Evolution of binary black-hole spacetimes, Phys. Rev. Lett. 95 (2005) 121101, arXiv:gr-qc/0507014.

[50] M. Campanelli, C.O. Lousto, P. Marronetti, et al., Accurate evolutions of orbiting black-hole binaries without excision, Phys. Rev. Lett. 96 (2006) 111101, arXiv:gr-qc/0511048.

[51] J.G. Baker, J. Centrella, D.-I. Choi, et al., Gravitational-wave extraction from an inspiraling configuration of merging black holes, Phys. Rev. Lett. 96 (2006) 111102, arXiv:gr-qc/0511103.

[52] M. Srednicki, Entropy and area, Phys. Rev. Lett. 71 (1993) 666, arXiv:hep-th/9303048.

[53] S. Ryu, T. Takayanagi, Holographic derivation of entanglement entropy from the anti-de Sitter/conformal field theory correspondence, Phys. Rev. Lett. 96 (2006) 181602, arXiv:hep-th/0603001.

6

Revisiting van der Waals like behavior of $f(R)$ AdS black holes via the two point correlation function

Jie-Xiong Mo [a,b], Gu-Qiang Li [a,b], Ze-Tao Lin [b], Xiao-Xiong Zeng [c,d,*]

[a] *Institute of Theoretical Physics, Lingnan Normal University, Zhanjiang, 524048, Guangdong, China*
[b] *Department of Physics, Lingnan Normal University, Zhanjiang, 524048, Guangdong, China*
[c] *School of Material Science and Engineering, Chongqing Jiaotong University, Chongqing, 400074, China*
[d] *Institute of Theoretical Physics, Chinese Academy of Sciences, Beijing 100190, China*

Editor: Stephan Stieberger

Abstract

Van der Waals like behavior of $f(R)$ AdS black holes is revisited via two point correlation function, which is dual to the geodesic length in the bulk. The equation of motion constrained by the boundary condition is solved numerically and both the effect of boundary region size and $f(R)$ gravity are probed. Moreover, an analogous specific heat related to δL is introduced. It is shown that the $T - \delta L$ graphs of $f(R)$ AdS black holes exhibit reverse van der Waals like behavior just as the $T - S$ graphs do. Free energy analysis is carried out to determine the first order phase transition temperature T_* and the unstable branch in $T - \delta L$ curve is removed by a bar $T = T_*$. It is shown that the first order phase transition temperature is the same at least to the order of 10^{-10} for different choices of the parameter b although the values of free energy vary with b. Our result further supports the former finding that charged $f(R)$ AdS black holes behave much like RN-AdS black holes. We also check the analogous equal area law numerically and find that the relative errors for both the cases $\theta_0 = 0.1$ and $\theta_0 = 0.2$ are small enough. The fitting functions between $\log | T - T_c |$ and $\log | \delta L - \delta L_c |$ for both cases are also obtained. It is shown that the slope is around 3, implying that the critical exponent is about $2/3$. This result is in accordance with those in former literatures of specific heat related to the thermal entropy or entanglement entropy.

[*] Corresponding author.
E-mail addresses: mojiexiong@gmail.com (J.-X. Mo), zsgqli@hotmail.com (G.-Q. Li), Albertlyn@163.com (Z.-T. Lin), xxzeng@itp.ac.cn (X.-X. Zeng).

1. Introduction

Van der Waals like behavior of black holes has long been an interesting topic in the black hole physics research for it discloses the close relation between black hole thermodynamics and ordinary thermodynamic systems. In the famous paper, Chamblin et al. [1,2] found that Reissner–Nordström-AdS (RN-AdS) black holes undergoes first order phase transition, which is analogous to the van der Waals liquid-gas phase transition. Carlip and Vaidya investigated the thermodynamics of a four-dimensional charged black hole in a finite cavity in asymptotically flat and asymptotically de Sitter space and discovered similar phase transition [3]. Lu et al. [4] studied the phase structure of asymptotically flat nondilatonic as well as dilatonic black branes in a cavity in arbitrary dimensions. It was shown that the phase diagram has a line of first-order phase transition in a certain range of temperatures which ends up at a second order phase transition point when the charge is below a critical value [4]. In this sense, van der Waals like behavior is such a universal phenomenon that it exists not only in AdS black holes, but also in asymptotically flat and asymptotically de Sitter black holes and black branes. Treating the cosmological constant as thermodynamic pressure, Kubizňák and Mann [5] investigated the $P - V$ criticality of RN-AdS black holes in the extended phase space and further enhanced the relation between charged AdS black holes and van der Waals liquid-gas systems. It has been shown that black holes are in general quite analogous to van der Waal fluids and exhibit the diverse behavior of different substances in everyday life. See the nice reviews [6–9] and references therein.

Here, we would like to focus on the van der Waals like behavior of charged AdS black holes in the $R + f(R)$ gravity with constant curvature [10], whose entropy, heat capacity and Helmholtz free energy was obtained in Ref. [10]. Ref. [11] studied their $P - V$ criticality in the extended phase space and showed that van der Waals like behavior exists in the $P - V$ graph of $f(R)$ AdS black holes. Recently, we investigated their phase transition in the canonical ensemble and further showed that $T - S$ graphs of $f(R)$ AdS black holes exhibit reverse van der Waals like behavior [12]. In this paper, we would like to revisit the van der Waals like behavior from a totally different perspective. Namely, the two point correlation function. Studying the properties of $f(R)$ AdS black holes [10–32] is of interest, because $f(R)$ gravity is one of modified gravity theories which successfully mimics the history of universe, especially the cosmic acceleration.

On the other hand, investigating the properties of the two point correlation function is also intriguing itself. The two point correlation function is dual to the geodesic length according to the famous AdS/CFT correspondence [33–35]. Interesting but intractable phenomena in strongly coupled system can be traced elegantly via the nonlocal observables such as two point correlation function, Wilson loop and entanglement entropy. Examples can be found in the researches of superconducting phase transition [36–43], the holographic thermalization [44–57] and cosmological singularity [58,59]. Recently, the isocharges in the entanglement entropy-temperature plane was investigated in Ref. [60], where both the critical temperature and critical exponent were proved to be exactly the same as the case of entropy-temperature plane. This finding is really intriguing and is attracting more and more attention [61–70]. Especially, it was proved in Ref. [62] that the entanglement entropy-temperature plane obeys the equal area law just as $T - S$ curve does [71]. In this paper, we would like to generalize these research to see whether the two point correlation function of $f(R)$ AdS black holes exhibits similar behavior. If it does, it would be a totally different perspective to observe the van der Waals like behavior of $f(R)$ AdS black holes other than $T - S$ graph and $P - V$ graph. To the best of our knowledge, this issue has not been covered in literature yet. As described above, the former literatures mainly focus on the entanglement entropy, including its van der Waals like behavior [60] and equal area law [62].

Our paper mainly focus on the two point correlation function (not the entanglement entropy) of $f(R)$ AdS black holes. So the results presented in this paper is independent and will certainly contribute to the knowledge of both $f(R)$ gravity and AdS black holes.

The metrics of $f(R)$ AdS black holes and RN-AdS black holes look similar via rescaling of parameters. Then one may expect that the results presented in this paper can be compared with those of RN-AdS black holes by performing that rescaling. However, this is not the whole story. Similarities and differences coexist. So it is worth probing from two different perspectives. On the one hand, the more similarities they have, the more importance they gain. The amazing similarities imply that this black hole solution may serve as a bridge across the Einstein gravity and $f(R)$ gravity. The similarities will also call for further investigation which may shed light on some deeper physics which has not been disclosed yet. On the other hand, one will expect to search for the possible unique features that are different from the features in Einstein gravity. These differences are certainly of interest since the metrics look similar. In this paper, we will show that although the free energy changes with b, the first order phase transition temperature is the same for different b. If we interpret $q/\sqrt{b}$ as the rescaled charge that can be compared with that of RN-AdS black holes, then our result implies that the first order phase transition temperature does not vary with the rescaled charge Q (at least it is true for the cases when b varies). And it is quite different from the RN-AdS black holes whose first order phase transition temperature depends on Q.

The organization of this paper is as follows. In Sec. 2 we will have a brief review of critical phenomena of $f(R)$ AdS black hole. Two point correlation function of $f(R)$ AdS black holes will be investigated numerically in Sec. 3. Maxwell equal area law will be numerically checked in Sec. 4. Conclusions will be drawn in Sec. 5.

2. A brief review of critical phenomena of charged AdS black holes in $f(R)$ gravity

Ref. [10] obtained in the $R + f(R)$ gravity with constant curvature scalar $R = R_0$ a charged AdS black hole solution, whose metric reads

$$ds^2 = -N(r)dt^2 + \frac{dr^2}{N(r)} + r^2(d\theta^2 + \sin^2\theta d\phi^2), \tag{1}$$

where

$$N(r) = 1 - \frac{2m}{r} + \frac{q^2}{br^2} - \frac{R_0}{12}r^2, \tag{2}$$

$$b = 1 + f'(R_0). \tag{3}$$

Note that $b > 0$, $R_0 < 0$.

This black hole solution is asymptotically AdS when one identify the curvature scalar as [10]

$$R_0 = -\frac{12}{l^2} = 4\Lambda. \tag{4}$$

The black hole ADM mass M and the electric charge Q are related to the parameters m and q respectively [10]

$$M = mb, \quad Q = \frac{q}{\sqrt{b}}. \tag{5}$$

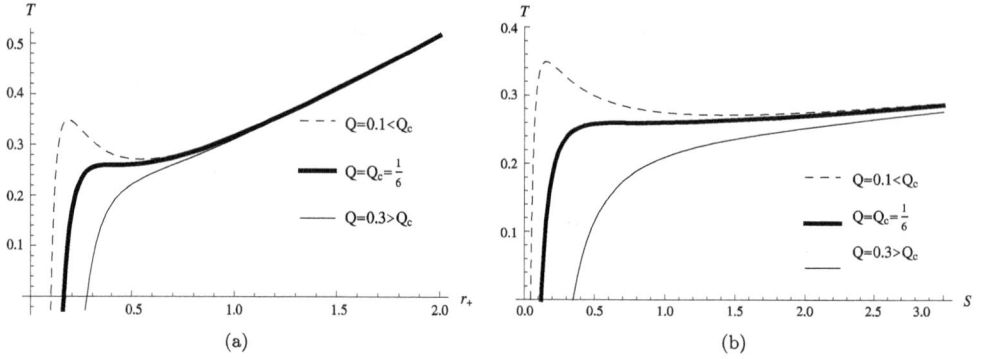

Fig. 1. (a) T vs. r_+ for $b = 1.5$, $R_0 = -12$ (b) T vs. S for $b = 1.5$, $R_0 = -12$ [15].

Ref. [11] reviewed its Hawking temperature, entropy and electric potential as

$$T = \frac{N'(r_+)}{4\pi} = \frac{1}{4\pi r_+}(1 - \frac{q^2}{br_+^2} - \frac{R_0 r_+^2}{4}). \tag{6}$$

$$S = \pi r_+^2 b. \tag{7}$$

$$\Phi = \frac{\sqrt{b}q}{r_+}. \tag{8}$$

In Ref. [12], we investigated in detail the critical phenomena in the canonical ensemble. As shown in Fig. 1, both $T - S$ curve and $T - r_+$ curve show reverse van der Waals behavior (reverse denotes that at small r_+ or small S, $T \to 0$ rather than $T \to \infty$) when $Q < Q_c$. The relevant critical quantities were derived as [12]

$$Q_c = \sqrt{\frac{-1}{3R_0}}, \quad r_c = \sqrt{\frac{-2}{R_0}}, \quad S_c = \frac{-2b\pi}{R_0}. \tag{9}$$

Numerical check of Maxwell equal area law for the cases $Q = 0.2Q_c, 0.4Q_c, 0.6Q_c, 0.8Q_c$ was also carried out in $T - S$ graph. It was shown that the relative errors are amazingly small and the Maxwell equal area law holds for $T - S$ curve of $f(R)$ black holes [12]. The analytic expression of free energy was obtained as

$$F = \frac{R_0 S^2 + 12\pi b(3b\pi Q^2 + S)}{48\pi^{3/2}\sqrt{bS}}. \tag{10}$$

It was shown that the classical swallow tails characteristic of first order phase transition appears in the case $Q < Q_c$ [12].

3. Two point correlation function of $f(R)$ AdS black holes and its van der Waals like behavior

The equal time two point correlation function in the large Δ limit reads [72]

$$\langle \mathcal{O}(t_0, x_i)\mathcal{O}(t_0, x_j) \rangle \approx e^{-\Delta L}, \tag{11}$$

where L is the length of the bulk geodesic between the points (t_0, x_i) and (t_0, x_j) on the AdS boundary while Δ denotes the conformal dimension of scalar operator $\mathcal{O}$ in the dual field theory.

Here, we choose two boundary points as $(\phi = \frac{\pi}{2}, \theta = 0)$ and $(\phi = \frac{\pi}{2}, \theta = \theta_0)$ for simplicity. Parameterizing the trajectory with θ, the proper length can be obtained as

$$L = \int_0^{\theta_0} \mathcal{L}(r(\theta), \theta) d\theta, \quad \mathcal{L} = \sqrt{\frac{\dot{r}^2}{N(r)} + r^2}, \tag{12}$$

where $\dot{r} = dr/d\theta$.

Utilizing the Euler–Lagrange equation $\frac{\partial L}{\partial r} = \frac{d}{d\theta}\left(\frac{\partial L}{\partial r'}\right)$, one can derive the equation of motion for $r(\theta)$ as

$$2r(\theta)N(r)\ddot{r}(\theta) - [r(\theta)N'(r) + 4N(r)]\dot{r}(\theta)^2 - 2N(r)^2 r(\theta)^2 = 0. \tag{13}$$

The boundary conditions can be fixed as

$$r(0) = r_0, \dot{r}(0) = 0. \tag{14}$$

Solving the equation of motion (13) constrained by the boundary condition (14), one can obtain $r(\theta)$. However, the analytic expression is difficult to derive and we have to appeal for numerical methods.

It is worth mentioning that the geodesic length should be regularized by subtracting off the geodesic length in pure AdS with the same boundary region to avoid the divergence. We use δL to denote the regularized geodesic length, which can be calculated through the definition $\delta L \equiv L - L_0$. Note that $r_{AdS}(\theta)$ corresponding to entanglement entropy in pure AdS L_0 has been obtained analytically as $r_{AdS}(\theta) = l[(\frac{\cos\theta}{\cos\theta_0})^2 - 1]^{-1/2}$ [73,74]. Note that this formula can be applied to the case of $f(R)$ AdS black holes by introducing an effective cosmological constant and rescaling the AdS length. And the outcome is in accord with the numerical treatment.

To probe the effect of boundary region size on the phase structure, we choose $\theta_0 = 0.1, 0.2$ as two specific examples and we set the cutoff $\theta_c = 0.099, 0.199$ respectively. On the other hand, we consider the case $b = 0.5, 1, 1.5$ to investigate the effect of $f(R)$ gravity on the phase structure. For convenience, we set the AdS radius l to be 1, which is equivalent to $R_0 = -12$. According to Eq. (9), $Q_c = 1/6$. As we are interested in the possible reverse van der Waals behaviors, we would like to pay more attention to the case $Q < Q_c$. So we set the charge Q to be 0.1 in most of the cases in this paper. The case $Q = 0$ is also probed for the purpose of comparison. As shown in Fig. 2 and Fig. 3, the $T - \delta L$ graphs for $Q = 0$ have only one inflection point while the graphs for $Q < Q_c$ have two inflection points. From the $T - \delta L$ graphs for $Q < Q_c$, one can see clearly the reverse van der Waals like behavior. It is also shown that the effect of b is so small that the $T - \delta L$ graphs seem the same at the first glance, although the specific numeric data points are different. On the other hand, the effect of boundary region size is quite obvious as reflected in the range of δL axis.

4. Numerical check of equal area law in $T - \delta L$ graph

To investigate the stability of black holes reflected in the $T - \delta L$ graph, one can introduce an analogous definition of specific heat as

$$C = T \frac{\partial \delta L}{\partial T}. \tag{15}$$

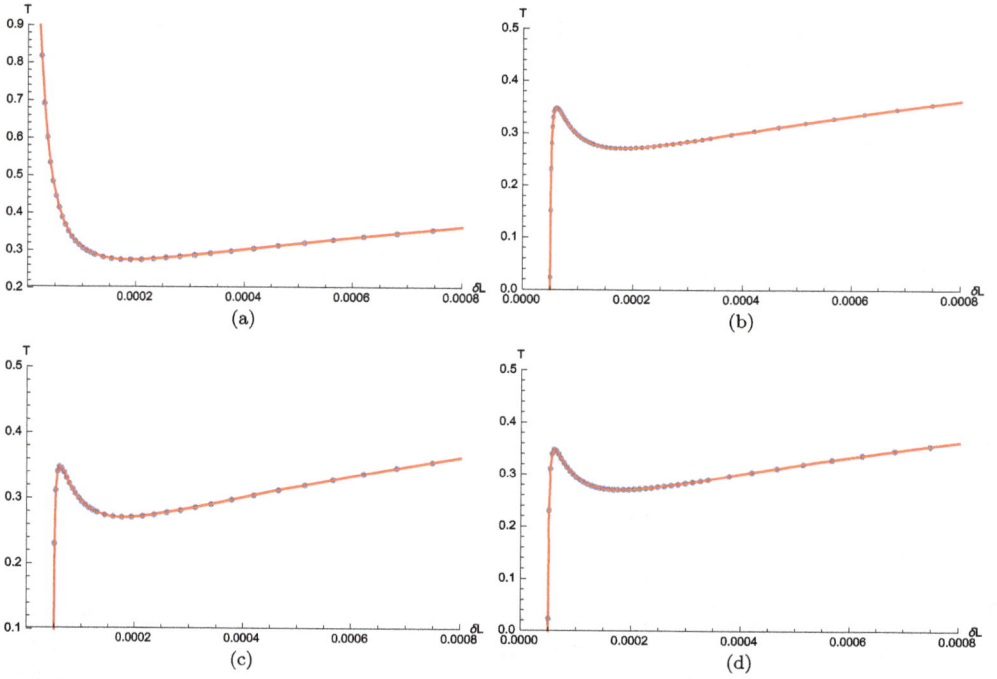

Fig. 2. T vs. δL for $\theta_0 = 0.1$ (a) $b = 1$, $Q = 0$ (b) $b = 0.5$, $Q = 0.1$ (c) $b = 1$, $Q = 0.1$ (d) $b = 1.5$, $Q = 0.1$.

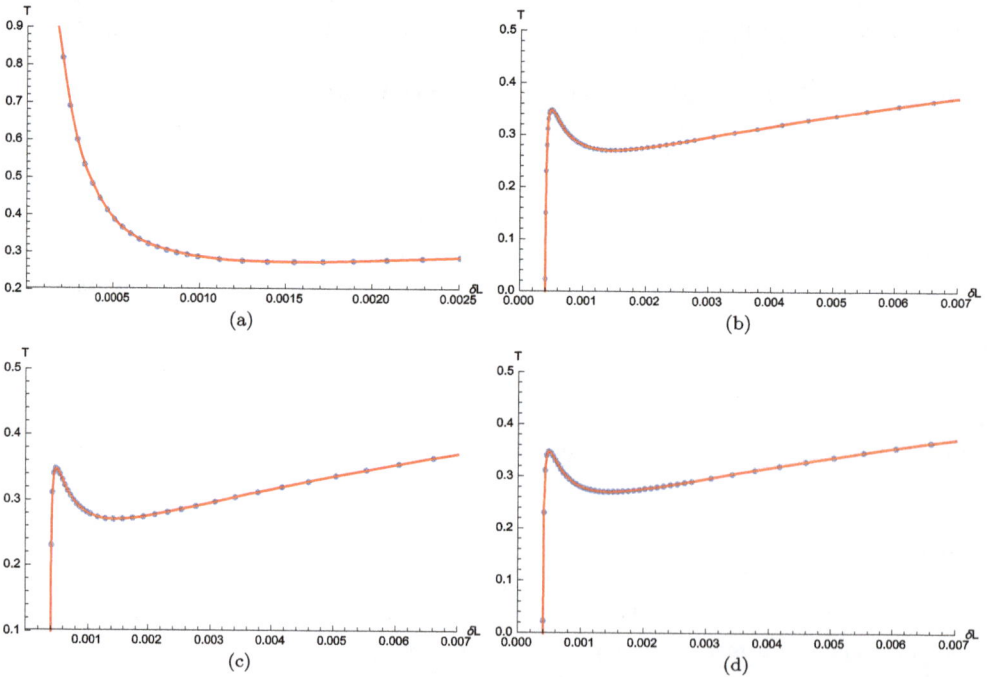

Fig. 3. T vs. δL for $\theta_0 = 0.2$ (a) $b = 1$, $Q = 0$ (b) $b = 0.5$, $Q = 0.1$ (c) $b = 1$, $Q = 0.1$ (d) $b = 1.5$, $Q = 0.1$.

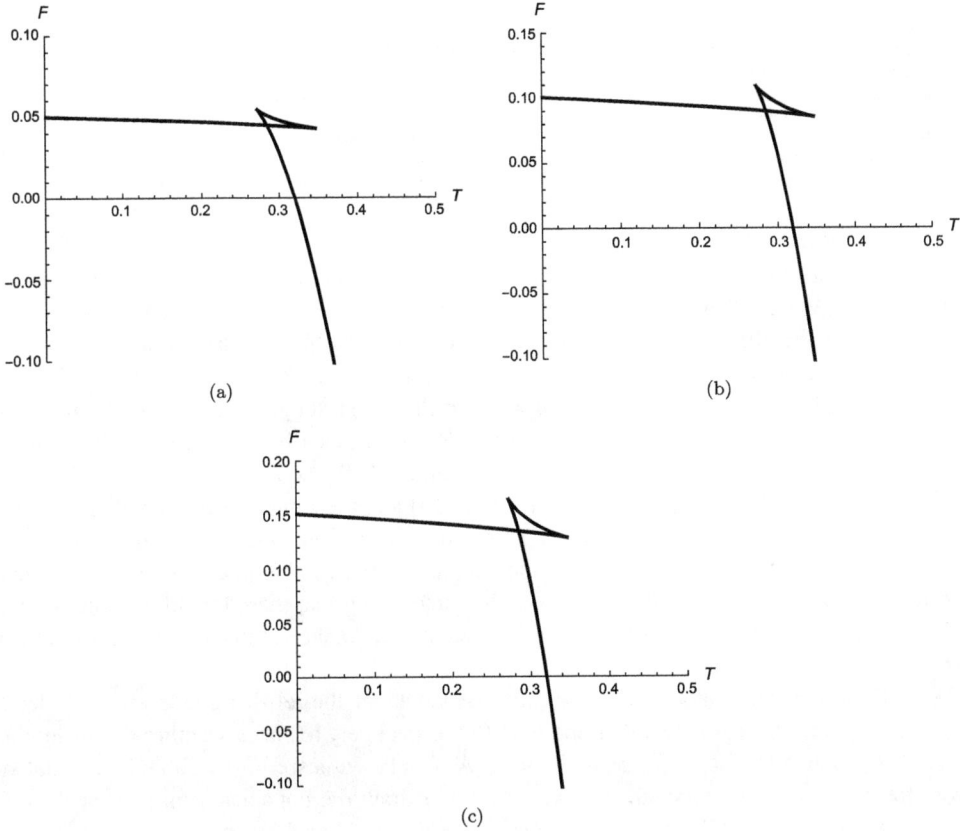

Fig. 4. F vs. T for $Q = 0.1$, $R_0 = -12$ (a) $b = 0.5$ (b) $b = 1$ (c) $b = 1.5$.

Similar to the $T - S$ graph, one can find in the $T - \delta L$ graph that when $Q < Q_c$, both the large radius branch and the small radius branch are stable with positive specific heat while the medium radius branch is unstable with negative specific heat. Following a similar approach as that for $T - S$ graph [71], one can remove the unstable branch in $T - \delta L$ curve with a bar $T = T_*$ vertical to the temperature axis. Physically, T_* denotes the Hawking temperature corresponding to the first order phase transition. The analogous Maxwell equal area law in the $T - \delta L$ graph may be written as

$$T_* \times (\delta L_3 - \delta L_1) = \int_{\delta L_1}^{\delta L_3} T d\delta L, \tag{16}$$

where δL_1, δL_2, δL_3 denote the three values of δL at $T = T_*$ with the assumption that $\delta L_1 < \delta L_2 < \delta L_3$.

T_* can be determined utilizing the free energy analysis. One can find T_* from the intersection point of two branches in the $F - T$ graph. The free energy graphs for different choice of b is plotted in Fig. 4. From Fig. 4, one can see clearly that the values of free energy are influenced by the parameter b showing the impact of $f(R)$ gravity. To one's surprise, the effect of b on T_* is very little at the first glance at Fig. 4. We further obtain numerical result of T_* to be

Table 1
Numerical check of Maxwell equal area law in $T - \delta L$ graph for $R_0 = -12$, $Q = 0.1$, $b = 1.5$.

θ_0	δL_1	δL_2	δL_3	$T_*(\delta L_3 - \delta L_1)$	$\int_{\delta L_1}^{\delta L_3} T d\delta L$	relative error
0.1	0.0000516933	0.0001149030	0.0002999516	0.00007068	0.00007099	0.4367%
0.2	0.0004189403	0.0009334939	0.0024361593	0.00057431	0.00057684	0.4386%

0.2847050173 for the three choices of b, namely, 0.5, 1, 1.5, implying that the first order phase transition temperature T_* is the same at least to the order of 10^{-10} for different choices of b. Considering the fact that the $f(R)$ AdS black holes reduce to RN-AdS black holes when $b = 1$, the result we obtain here is very interesting for it implies that $f(R)$ gravity does not influence the first order phase transition temperature. Our result further supports the former finding that charged $f(R)$ AdS black holes behave much like RN-AdS black holes. In former literature, it was reported in the reduced parameter space that charged $f(R)$ AdS black holes share the same equation of state [11], the same coexistence curve [13] and the same molecule number density difference [13] with RN-AdS black holes. This can be attributed to the observation that these two black hole metrics are identical when the charge is rescaled and an effective cosmological constant is defined for $f(R)$ AdS black holes. It is interesting that these two black holes in two different gravity theories ($f(R)$ gravity and Einstein gravity) share so much similarities in their metrics and other behaviors.

With T_* at hand, we calculate numerically the values of the left-hand side and right-hand side of Eq. (16) for the cases $\theta_0 = 0.1$ and $\theta_0 = 0.2$ respectively to check whether the analogous Maxwell equal area law holds for $T - \delta L$ curve. As can be witnessed from Table 1, the relative errors for both cases are so small that we can safely draw the conclusion that the analogous Maxwell equal area law holds for $T - \delta L$ curve of $f(R)$ AdS black holes.

Utilizing Eqs. (6) and (9), one can easily derive the critical temperature as

$$T_c = \frac{\sqrt{-R_0}}{3\sqrt{2\pi}}. \tag{17}$$

Substituting $R_0 = -12$ into the above equation, one can obtain $T_c = 0.2598989337$. Note that it is independent of θ_0. Then one can adopt the interpolating functions obtained from the numeric result to obtain δL_c. For the case $\theta_0 = 0.1$, $\delta L_c = 0.0001337812$ while for the case $\theta_0 = 0.2$, $\delta L_c = 0.0010728619$. The relation between $\log |T - T_c|$ and $\log |\delta L - \delta L_c|$ is plotted for the cases $\theta_0 = 0.1$ and $\theta_0 = 0.2$ in Fig. 5(a) and 5(b) respectively. Note that the points are chosen neighboring the critical point from the $T - \delta L$ graph corresponding to $Q = Q_c = 1/6$. And the fitting function for these two cases are obtained as

$$\log |T - T_c| = \begin{cases} 25.7008 + 3.08202 \log |\delta L - \delta L_c|, & \text{for } \theta_0 = 0.1, \\ 18.9784 + 3.06772 \log |\delta L - \delta L_c|, & \text{for } \theta_0 = 0.2. \end{cases} \tag{18}$$

From above, one can see clearly that the slope is around 3, implying that the critical exponent (defined through $C \sim |T - T_c|^{-\alpha}$) is about 2/3. Our result of critical exponent for the analogous specific heat related to the δL is in accordance with those in former literatures of specific heat related to the thermal entropy [2] or entanglement entropy [60].

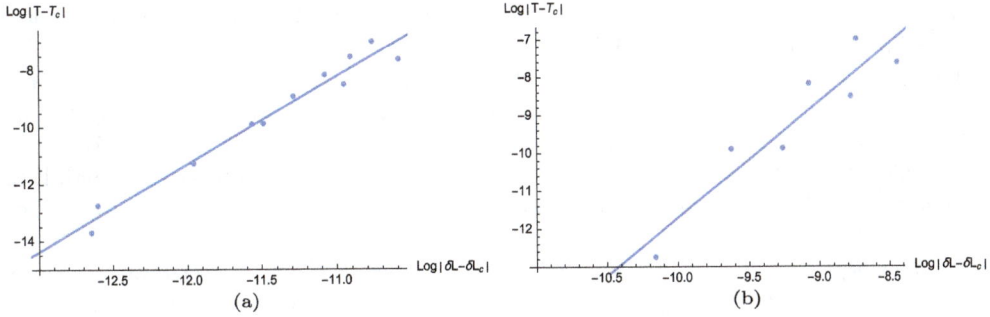

Fig. 5. $\log | T - T_c |$ vs. $\log | \delta L - \delta L_c |$ for $R_0 = -12$, $b = 1.5$, $Q = Q_c = 1/6$ (a) $\theta_0 = 0.1$ (b) $\theta_0 = 0.2$.

5. Conclusions

In this paper, we focus on the two point correlation function of $f(R)$ AdS black hole. First, we choose two boundary points as $(\phi = \frac{\pi}{2}, \theta = 0)$ and $(\phi = \frac{\pi}{2}, \theta = \theta_0)$ and obtain the proper length by parameterizing the trajectory with θ. Then utilizing the Euler–Lagrange equation, we derive the equation of motion for $r(\theta)$. Solving the equation of motion constrained by the boundary condition, we obtain $r(\theta)$ numerically. Moreover, we regularize the geodesic length by subtracting off the geodesic length in pure AdS with the same boundary region.

Second, to probe the effect of boundary region size on the phase structure, we choose $\theta_0 = 0.1, 0.2$ as two specific examples and we set the cutoff $\theta_c = 0.099, 0.199$ respectively. On the other hand, we consider the case $b = 0.5, 1, 1.5$ to investigate the effect of $f(R)$ gravity on the phase structure. The case $Q < Q_c$ are plotted for different cases. From the $T - \delta L$ graphs, one can see clearly the reverse van der Waals like behavior. It is also shown that the effect of b is so small that the $T - \delta L$ graphs seem the same at the first glance, although the specific numeric data points are different. On the other hand, the effect of boundary region size is quite obvious as reflected in the range of δL axis.

Third, we introduce an analogous definition of specific heat to investigate the stability of black holes reflected in the $T - \delta L$ graph. Similar to the $T - S$ graph, one can find in the $T - \delta L$ graph that when $Q < Q_c$, both the large radius branch and the small radius branch are stable with positive specific heat while the medium radius branch is unstable with negative specific heat. We remove the unstable branch in $T - \delta L$ curve with a bar $T = T_*$ vertical to the temperature axis, where T_* denotes the Hawking temperature corresponding to the first order phase transition. We carry out free energy analysis to determine T_*. It is shown that the values of free energy are influenced by the parameter b showing the impact of $f(R)$ gravity. However, to one's surprise, the numerical result of T_* to be 0.2847050173 for all the three choices of b, namely, 0.5, 1, 1.5, implying that the first order phase transition temperature T_* is the same at least to the order of 10^{-10} for different choices of b. Considering the fact that the $f(R)$ AdS black holes reduce to RN-AdS black holes when $b = 1$, the result we obtain here is very interesting for it implies that $f(R)$ gravity does not influence the first order phase transition temperature. Our result further supports the finding that charged $f(R)$ AdS black holes behave much like RN-AdS black holes reported in former literature, where charged $f(R)$ AdS black holes were shown to share the same equation of state [11], the same coexistence curve [13] and the same molecule number density difference [13] with RN-AdS black holes in the reduced parameter space. We also check the analogous equal area law numerically. It is shown that the relative errors for both cases are so small that we can safely draw the conclusion that the analogous Maxwell equal area law holds for $T - \delta L$ curve of $f(R)$ AdS black holes.

Last but not the least, we plot the relation between $\log | T - T_c |$ and $\log | \delta L - \delta L_c |$ for the cases $\theta_0 = 0.1$ and $\theta_0 = 0.2$ and obtain the fitting functions for these two cases. It is shown that the slope is around 3, implying that the critical exponent (defined through $C \sim | T - T_c |^{-\alpha}$) is about $2/3$. Our result of critical exponent for the analogous specific heat related to the δL is in accordance with those in former literatures of specific heat related to the thermal entropy [2] or entanglement entropy [60].

To summary, the $T - \delta L$ graph of $f(R)$ AdS black holes exhibits the reverse van der Waals like behavior just as the $T - S$ graphs do. And the analogous Maxwell equal area law holds for $T - \delta L$ graph. Moreover, the critical exponent for the analogous specific heat related to the δL is shown to be the same as those of specific heat related to the thermal entropy or entanglement entropy. So the two point correlation function may serves as an alternative perspective to observe van der Waals like behavior of $f(R)$ AdS black holes.

Acknowledgements

The authors want to express their sincere gratitude to both the editor and the referee for their joint effort to improve the quality of this paper significantly. This research is supported by National Natural Science Foundation of China (Grant No. 11605082), and in part supported by Natural Science Foundation of Guangdong Province, China (Grant Nos. 2016A030310363, 2016A030307051, 2015A030313789). Xiao-Xiong Zeng is supported by the National Natural Science Foundation of China (Grant No. 11405016) and Natural Science Foundation of Education Committee of Chongqing (Grant No. KJ1500530).

References

[1] A. Chamblin, R. Emparan, C.V. Johnson, R.C. Myers, Charged AdS black holes and catastrophic holography, Phys. Rev. D 60 (1999) 064018.

[2] A. Chamblin, R. Emparan, C.V. Johnson, R.C. Myers, Holography, thermodynamics and fluctuations of charged AdS black holes, Phys. Rev. D 60 (1999) 104026.

[3] S. Carlip, S. Vaidya, Phase transitions and critical behavior for charged black holes, Class. Quantum Gravity 20 (2003) 3827–3838.

[4] J.X. Lu, S. Roy, Z. Xiao, Phase transitions and critical behavior of black branes in canonical ensemble, J. High Energy Phys. 1101 (2011) 133.

[5] D. Kubizňák, R.B. Mann, $P - V$ criticality of charged AdS black holes, J. High Energy Phys. 07 (2012) 033.

[6] D. Kubizňák, R.B. Mann, M. Teo, Black hole chemistry: thermodynamics with lambda, arXiv:1608.06147.

[7] N. Altamirano, D. Kubizňák, R.B. Mann, Z. Sherkatghanad, Thermodynamics of rotating black holes and black rings: phase transitions and thermodynamic volume, Galaxies 2 (2014) 89–159.

[8] B.P. Dolan, Black holes and Boyle's law-the thermodynamics of the cosmological constant, Mod. Phys. Lett. A 30 (2015) 1540002.

[9] D. Kubizňák, R.B. Mann, Black hole chemistry, Can. J. Phys. 93 (2015) 999–1002.

[10] T. Moon, Y.S. Myung, E.J. Son, $f(R)$ black holes, Gen. Relativ. Gravit. 43 (2011) 3079–3098.

[11] S. Chen, X. Liu, C. Liu, J. Jing, $P - V$ criticality of AdS black hole in $f(R)$ gravity, Chin. Phys. Lett. 30 (2013) 060401.

[12] J.X. Mo, G.Q. Li, Y.C. Wu, A consistent and unified picture for critical phenomena of $f(R)$ AdS black holes, J. Cosmol. Astropart. Phys. 04 (2016) 045.

[13] J.X. Mo, G.Q. Li, Coexistence curves and molecule number densities of AdS black holes in the reduced parameter space, Phys. Rev. D 92 (2015) 024055.

[14] G.Q. Li, J.X. Mo, Phase transition and thermodynamic geometry of $f(R)$ AdS black holes in the grand canonical ensemble, Phys. Rev. D 93 (2016) 124021.

[15] A. De Felice, S. Tsujikawa, $f(R)$ theories, Living Rev. Relativ. 13 (2010) 3.

[16] S. Capozziello, M. De Laurentis, Extended theories of gravity, Phys. Rep. 509 (2011) 167–321.

[17] A. de la C. Dombriz, A. Dobado, A.L. Maroto, Black holes in $f(R)$ theories, Phys. Rev. D 80 (2009) 124011.

[18] A. Larranaga, A rotating charged black hole solution in $f(R)$ gravity, Pramana 78 (2012) 697–703.

[19] J.A.R. Cembranos, A. de la C. Dombriz, P.J. Romero, Kerr–Newman black holes in $f(R)$ theories, Int. J. Geom. Methods Mod. Phys. 11 (2014) 1450001.

[20] A. Sheykhi, Higher-dimensional charged $f(R)$ black holes, Phys. Rev. D 86 (2012) 024013.

[21] L. Sebastiani, S. Zerbini, Static spherically symmetric solutions in $F(R)$ gravity, Eur. Phys. J. C 71 (2011) 1591.

[22] S.H. Hendi, The relation between $F(R)$ gravity and Einstein-conformally invariant Maxwell source, Phys. Lett. B 690 (2010) 220–223.

[23] S.H. Hendi, D. Momeni, Black hole solutions in $F(R)$ gravity with conformal anomaly, Eur. Phys. J. C 71 (2011) 1823.

[24] G.J. Olmo, D.R. Garcia, Palatini $f(R)$ black holes in nonlinear electrodynamics, Phys. Rev. D 84 (2011) 124059.

[25] S.H. Mazharimousavi, M. Halilsoy, Black hole solutions in $f(R)$ gravity coupled with non-linear Yang–Mills field, Phys. Rev. D 84 (2011) 064032.

[26] Y.S. Myung, T. Moon, E.J. Son, Stability of $f(R)$ black holes, Phys. Rev. D 83 (2011) 124009.

[27] T. Moon, Y.S. Myung, E.J. Son, Stability analysis of $f(R)$-AdS black holes, Eur. Phys. J. C 71 (2011) 1777.

[28] Y.S. Myung, Instability of rotating black hole in a limited form of $f(R)$ gravity, Phys. Rev. D 84 (2011) 024048.

[29] Y.S. Myung, Instability of a Kerr black hole in $f(R)$ gravity, Phys. Rev. D 88 (2013) 104017.

[30] S. Nojiri, S.D. Odintsov, Instabilities and anti-evaporation of Reissner–Nordström black holes in modified $F(R)$ gravity, Phys. Lett. B 735 (2014) 376–382.

[31] S. Nojiri, S.D. Odintsov, Anti-evaporation of Schwarzschild–de Sitter black holes in $F(R)$ gravity, Class. Quantum Gravity 30 (2013) 125003.

[32] S. Nojiri, S.D. Odintsov, Unified cosmic history in modified gravity: from $F(R)$ theory to Lorentz non-invariant models, Phys. Rep. 505 (2011) 59–144.

[33] J.M. Maldacena, Large N limit of superconformal field theories and supergravity, Int. J. Theor. Phys. 38 (1999) 1113.

[34] E. Witten, Anti-de Sitter space and holography, Adv. Theor. Math. Phys. 2 (1998) 253.

[35] S.S. Gubser, I.R. Klebanov, A.M. Polyakov, Gauge theory correlators from noncritical string theory, Phys. Lett. B 428 (1998) 105.

[36] T. Albash, C.V. Johnson, Holographic studies of entanglement entropy in superconductors, J. High Energy Phys. 1205 (2012) 079.

[37] R.G. Cai, S. He, L. Li, Y.L. Zhang, Holographic entanglement entropy in insulator/superconductor transition, J. High Energy Phys. 1207 (2012) 088.

[38] R.G. Cai, L. Li, L.F. Li, R.K. Su, Entanglement entropy in holographic P-wave superconductor/insulator model, J. High Energy Phys. 1306 (2013) 063.

[39] L.F. Li, R.G. Cai, L. Li, C. Shen, Entanglement entropy in a holographic p-wave superconductor model, Nucl. Phys. B 894 (2015) 15–28.

[40] R.G. Cai, S. He, L. Li, Y.L. Zhang, Holographic entanglement entropy in insulator/superconductor transition, J. High Energy Phys. 1207 (2012) 088.

[41] X. Bai, B.H. Lee, L. Li, J.R. Sun, H.Q. Zhang, Time evolution of entanglement entropy in quenched holographic superconductors, J. High Energy Phys. 040 (2015) 66.

[42] R.G. Cai, L. Li, L.F. Li, R.Q. Yang, Introduction to holographic superconductor models, Sci. China, Phys. Mech. Astron. 58 (2015) 060401.

[43] Y. Ling, P. Liu, C. Niu, J.P. Wu, Z.Y. Xian, Holographic entanglement entropy close to quantum phase transitions, J. High Energy Phys. 1604 (2016) 114.

[44] V. Balasubramanian, et al., Thermalization of strongly coupled field theories, Phys. Rev. Lett. 106 (2011) 191601.

[45] V. Balasubramanian, et al., Holographic thermalization, Phys. Rev. D 84 (2011) 026010.

[46] D. Galante, M. Schvellinger, Thermalization with a chemical potential from AdS spaces, J. High Energy Phys. 1207 (2012) 096.

[47] E. Caceres, A. Kundu, Holographic thermalization with chemical potential, J. High Energy Phys. 1209 (2012) 055.

[48] X.X. Zeng, B.W. Liu, Holographic thermalization in Gauss–Bonnet gravity, Phys. Lett. B 726 (2013) 481.

[49] X.X. Zeng, X.M. Liu, B.W. Liu, Holographic thermalization with a chemical potential in Gauss–Bonnet gravity, J. High Energy Phys. 03 (2014) 031.

[50] X.X. Zeng, D.Y. Chen, L.F. Li, Holographic thermalization and gravitational collapse in the spacetime dominated by quintessence dark energy, Phys. Rev. D 91 (2015) 046005.

[51] X.X. Zeng, X.M. Liu, B.W. Liu, Holographic thermalization in noncommutative geometry, Phys. Lett. B 744 (2015) 48–54.

[52] X.X. Zeng, X.Y. Hu, L.F. Li, Effect of phantom dark energy on the holographic thermalization, Chin. Phys. Lett. 34 (2017) 010401.

[53] Y.P. Hu, X.X. Zeng, H.Q. Zhang, Holographic thermalization and generalized Vaidya-AdS solutions in massive gravity, Phys. Lett. B 765 (2017) 120.

[54] H. Liu, S.J. Suh, Entanglement tsunami: universal scaling in holographic thermalization, Phys. Rev. Lett. 112 (2014) 011601.

[55] S.J. Zhang, E. Abdalla, Holographic thermalization in charged dilaton anti-de Sitter spacetime, Nucl. Phys. B 896 (2015) 569.

[56] A. Buchel, R.C. Myers, A.v. Niekerk, Nonlocal probes of thermalization in holographic quenches with spectral methods, J. High Energy Phys. 02 (2015) 017.

[57] B. Craps, et al., Gravitational collapse and thermalization in the hard wall model, J. High Energy Phys. 02 (2014) 120.

[58] N. Engelhardt, T. Hertog, G.T. Horowitz, Holographic signatures of cosmological singularities, Phys. Rev. Lett. 113 (2014) 121602.

[59] N. Engelhardt, T. Hertog, G.T. Horowitz, Further holographic investigations of Big Bang singularities, J. High Energy Phys. 1507 (2015) 044.

[60] C.V. Johnson, Large N phase transitions, finite volume, and entanglement entropy, J. High Energy Phys. 1403 (2014) 047.

[61] E. Caceres, P.H. Nguyen, J.F. Pedraza, Holographic entanglement entropy and the extended phase structure of STU black holes, J. High Energy Phys. 1509 (2015) 184.

[62] P.H. Nguyen, An equal area law for holographic entanglement entropy of the AdS-RN black hole, J. High Energy Phys. 1512 (2015) 139.

[63] X.X. Zeng, H. Zhang, L.F. Li, Phase transition of holographic entanglement entropy in massive gravity, Phys. Lett. B 756 (2016) 170–179.

[64] X.X. Zeng, L.F. Li, Holographic phase transition probed by non-local observables, Adv. High Energy Phys. 2016 (2016) 6153435.

[65] S. He, L.F. Li, X.X. Zeng, Holographic van der Waals-like phase transition in the Gauss–Bonnet gravity, Nucl. Phys. B 915 (2017) 243.

[66] A. Dey, S. Mahapatra, T. Sarkar, Thermodynamics and entanglement entropy with Weyl corrections, arXiv:1512.07117.

[67] X.X. Zeng, L.F. Li, Van der Waals phase transition in the framework of holography, arXiv:1512.08855.

[68] X.X. Zeng, X.M. Liu, L.F. Li, Phase structure of the born-infeld-anti-de Sitter black holes probed by non-local observables, arXiv:1601.01160.

[69] S. Kundu, J.F. Pedraza, Aspects of holographic entanglement at finite temperature and chemical potential, arXiv:1602.07353.

[70] D. Momeni, K. Myrzakulov, R. Myrzakulov, Fidelity susceptibility as holographic $P - V$ criticality, arXiv:1604.06909.

[71] E. Spallucci, A. Smailagic, Maxwell's equal area law for charged anti-de Sitter black holes, Phys. Lett. B 723 (2013) 436–441.

[72] V. Balasubramanian, S.F. Ross, Holographic particle detection, Phys. Rev. D 61 (2000) 044007.

[73] D.D. Blanco, H. Casini, L.-Y. Hung, R.C. Myers, Entropy and holography, J. High Energy Phys. 08 (2013) 060.

[74] H. Casini, M. Huerta, R.C. Myers, Towards a derivation of holographic entanglement entropy, J. High Energy Phys. 1105 (2011) 036.

7

The modified first laws of thermodynamics of anti-de Sitter and de Sitter space–times

Deyou Chen [a,*], Gan qingyu [b], Jun Tao [b]

[a] *Institute of Theoretical Physics, China West Normal University, Nanchong, 637009, China*
[b] *Center for Theoretical Physics, College of Physical Science and Technology, Sichuan University, Chengdu, 610064, China*

Editor: Professor Leonardo Rastelli

Abstract

We modify the first laws of thermodynamics of a Reissner–Nordstrom anti-de Sitter black hole and a pure de Sitter space–time by the surface tensions. The corresponding Smarr relations are obeyed. The cosmological constants are first treated as fixed constants, and then as variables associated to the pressures. For the black hole, the law is written as $\delta E = T\delta S - \sigma \delta A$ when the cosmological constant is fixed, where E is the Misner–Sharp mass and σ is the surface tension. Adopting the varied constant, we modify the law as $\delta E_0 = T\delta S - \sigma_{eff}\delta A + V\delta P$, where $E_0 = M - \frac{Q^2}{2r_+}$ is the enthalpy. The thermodynamical properties are investigated. For the de Sitter space–time, the expressions of the modified laws are different from these of the black hole. The differential way to derive the law is discussed.

1. Introduction

Thermodynamical properties of anti-de Sitter (AdS) and de Sitter (dS) space–times attract considerable attentions. The research on thermodynamical properties of AdS spaces originated in the seminal work of Hawking and Page. In their work, the existence of the phase transition

* Corresponding author.
E-mail addresses: dchen@cwnu.edu.cn (D. Chen), qingyugan@hotmail.com (G. qingyu), taojun@scu.edu.cn (J. Tao).

in the Schwarzschild AdS black hole was first found [1]. The phase transitions and the thermo-dynamical stabilities in the various complicated spacetimes were studied in the subsequent work [2,3].

The cosmological constants were usually seen as the fixed constants. However, they were recently treated as the variables [4–12]. The first reason is that the physical constants, such as Newtonian's constant, gauge coupling constants and the cosmological constant arise as vacuum expectation values and vary in more fundamental theories [7]. The second reason is reconciling the inconsistencies between the first law of black hole thermodynamics and the Smarr relation derived by the scaling method. Associating the cosmological constant with the pressure $P = -\frac{\Lambda}{8\pi}$, Kastor et al. first derived the modified first law of thermodynamics and the corresponding Smarr relation of the static AdS space–time by the geometric approach [4]. The key ingredient is the two-form potential for the Killing field. The new term VdP appeared in the law and the black hole mass was naturally interpreted as the enthalpy. In [6], Dolan obtained the equation of state for the AdS black hole and discussed the analogy with that of the Van der Waals system. This work got further development and some interesting results were found in [7]. Kubiznak and Mann revealed the identification of the Reissner–Nordstrom anti-de Sitter (RN–AdS) black hole with the liquid–gas system in the extended phase space by investigating the behavior of the Van der Waals system and that of the Gibbs free energy in the fixed charge ensemble. The modified first law in the extended phase space was gotten as

$$\delta M = T\delta S + \Phi\delta Q + V\delta P, \tag{1}$$

where Φ is the electromagnetic potential at the horizon and V is the black hole volume. The extensions to other complicated space–times are referred to [13–23]. In the AdS/CFT correspondence, the variation of the cosmological constant is associated to that of the degrees of freedom in the gauge theory. Treating the constant as the number of colors in the field, Dolan calculated the chemical potential conjugate to the number and studied the thermodynamics of the $AdS_5 \times S^5$ space–times in [24]. He found that the potential in the high temperature is negative and decreases as the increase of the temperature. When the temperature is lowered below the Hawking–Page temperature, the potential approaches to zero. This work was further discussed in [25,26]. It was found that the chemical potential becomes positive more easily due to the existence of the charge in RN–AdS black holes. The positive cosmology constant was first seen as a variable in [9]. Taking into account the variable cosmology constant, Wu and Sekiwa studied the thermodynamics of the dS space–times, respectively [9,10]. Subsequently, the constant was also treated as the pressure and the thermodynamic properties of the charged and rotating de Sitter black holes in the extended phase spaces were studied in [27–30].

On the other hand, Einstein's equations can be written as a thermodynamical identity [31–33]. It was first proved in the spherically symmetric space–time and the equations were written in the form of [31]

$$\delta E = T\delta S - P\delta V, \tag{2}$$

where δV is the change of space volume and P is the pressure provided by the source of Einstein's equations. This result could also be applicable for stationary space–times and other gravity theories [34–36]. Recently, this work was extended to the Kerr black hole [33]. The horizon energy E of the black hole was identified as its Misner–Sharp mass, and then the angular momentum was Ea. The horizon radius r_+ and the rotation parameter a were independent variables. These identifications are different from the traditional concept where a is a fixed constant. From the relation between the entropy and the horizon area, the surface tension σ and the work term

$\sigma \delta A$ were gotten. Thus, Einstein's equations were rewritten as the modified first law of black hole thermodynamics

$$\delta E = T\delta S + \Omega\delta J - \sigma\delta A. \tag{3}$$

In fact, the surface tension σ discussed in the first law was first appeared in the context of the thermodynamics of black holes in a cavity [37]. Its surface tension is conjugate to the area of a cavity enclosing the ensemble and so the cavity area and the black hole entropy can vary independently.

In this paper, we modify the first laws of thermodynamics of the RN–AdS black hole and the pure dS space–time by the surface tensions. We first treat the cosmological constant as a fixed constant and derive the modified first law of the RN–AdS black hole. The corresponding Smarr relation is obeyed. Then we revisit the law by associating the cosmological constant with the pressure. The effective surface tension and the modified law are gotten. The corresponding Smarr relation is also satisfied. Thermodynamics of dS space–times is more complex than that of AdS space–times. The reason is that the non-equilibrium state exists in this region between the black hole horizon and the cosmological horizon due to the different temperatures at the different horizons. Meanwhile, the asymptotic mass is difficult to be defined since the Killing vector is timelike outside the black hole horizon. To overcome these problems, some effective ways were put forward [12,38–41]. Here we modify the first law at the cosmological horizon of the pure dS space–time where only a cosmological horizon exists. In the dS space–time, the energy is measured as the negative value. We identify the negative Misner–Sharp mass as the horizon energy. The modified first laws and the corresponding Smarr relations are gotten when the cosmological constant is fixed and varied, respectively. The modified first laws at the black hole horizon and the cosmological horizon take on different forms.

The rest is outlined as follows. In the next section, using the expressions of the temperature and the radial Einstein's equations, we derive the surface tension and the modified first laws of thermodynamics of the RN–AdS black hole, where the cosmological constant is seen as a fixed constant and a variable, respectively. In section 3, the surface tension and the modified first laws at the cosmological horizon of the dS space–time are investigated. Section 4 is devoted to our discussion and conclusion.

2. Modifying the first law of thermodynamics at the black hole horizon

2.1. The first law with the fixed cosmology constant

The metric of the RN–AdS black hole is given by

$$ds^2 = -\Delta dt^2 + \frac{1}{\Delta}dr^2 + r^2(d\theta^2 + \sin^2\theta d\phi^2), \tag{4}$$

with the electromagnetic potential $A_\mu = \frac{Q}{r}dt$, where

$$\Delta = 1 - \frac{2M}{r} + \frac{Q^2}{r^2} + \frac{r^2}{l^2}, \tag{5}$$

$2M = r_+ + \frac{r_+^3}{l^2} + \frac{Q^2}{r_+}$. l is related to the cosmological constant Λ as $l^2 = -\frac{3}{\Lambda}$. The black hole horizon is located at the largest root r_+ obtained from $\Delta = 0$. M and Q are the physical mass and charge, respectively. There are several ways to define the mass. Here we adopt the definition

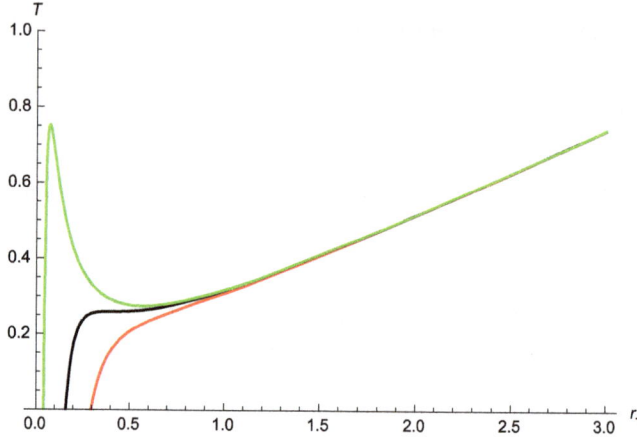

Fig. 1. The red, blue and green curves correspond to $Q = 1/3$, $Q = 1/6$ and $Q = 1/24$, respectively. (For interpretation of the references to color in this figure legend, the reader is referred to the web version of this article.)

of the Misner–Sharp mass to investigate the thermodynamical properties of the black hole [42]. The explicit form of the mass was presented in [43–45]. For the RN–AdS black hole, its mass is gotten as $E = M - \frac{Q^2}{2r_+} - \frac{r_+^3}{2l^2}$. Introducing the expression of the physical mass M yields a concise expression

$$E = \frac{r_+}{2}. \tag{6}$$

The black hole entropy and temperature are

$$S = \frac{A}{4} = \pi r_+^2,$$

$$T = \frac{\Delta'(r_+)}{4\pi} = \frac{1}{4\pi r_+}\left(1 + \frac{3r_+^2}{l^2} - \frac{Q^2}{r_+^2}\right), \tag{7}$$

respectively, where $\Delta'(r_+) = \frac{\partial \Delta}{\partial r}|_{r=r_+}$. The value of T is determined by r_+ and Q. The Fig. 1 describes the variation relation between T and r_+ at the fixed values Q and $l = 1$. When the horizon radius is small, the temperatures are different for the different Q. Finally, the temperatures approach to the same value at the large radius.

Einstein's equations can be written as a thermodynamic identity. To investigate the thermodynamics and the surface tension at the black hole horizon, we first calculate the radial Einstein equation at the horizon. The equation is

$$G_r^r|_{r_+} = 8\pi T_r^r|_{r_+} = \frac{r_+\Delta'(r_+) - 1}{r_+^2}. \tag{8}$$

Compared Eq. (7) and Eq. (8), the black hole temperature is rewritten as

$$T = \frac{8\pi r_+^2 T_r^r|_{r_+} + 1}{4\pi r_+}. \tag{9}$$

Carrying out differential on the entropy yields $\delta S = 2\pi r_+ \delta r_+$, where r_+ is seen as an independent variable. Multiplying by δS on the both sides of Eq. (9), we get

$$T\delta S = 2r_+ T_r^r \mid_{r_+} \delta S + \frac{\delta r_+}{2}. \tag{10}$$

The first term on the right hand side (rhs) in the above equation is relied on the matter, while the second term can be identified as the differential form of the Misner–Sharp mass, namely, $\delta E = \frac{1}{2}\delta r_+$. Thus, Eq. (10) is written as

$$\delta E = T\delta S - \sigma \delta A, \tag{11}$$

where $\sigma = \frac{r_+ T_r^r \mid_{r_+}}{2} = \frac{1}{16\pi}(-\frac{Q^2}{r_+^3} + \frac{3r_+}{l^2})$ describes the surface tension at the horizon and $A = 4S$ is the horizon area. Eq. (11) is the modified first law of thermodynamics at the black hole horizon. The value of the surface tension is dependent on both of the charge and the cosmological constant. When $3r_+^4 = Q^2 l^2$, the surface tension is zero. The positive surface tension is gotten for $3r_+^4 > Q^2 l^2$, while it is negative when $3r_+^4 < Q^2 l^2$. The corresponding Smarr relation

$$E = 2TS - 2\sigma A, \tag{12}$$

is satisfied by using the expressions of E, T, S, A and σ. Clearly, the first law in Eq. (11) is different from that gotten in [7,8]. The reason is that the Misner–Sharp mass was introduced as the energy and the temperature was expressed by the radial Einstein equation. When the ADM mass expresses the energy, Eq. (11) is reduced to Eq. (3.9) in [7].

The Gibbs free energy is

$$G = E - TS + \sigma A. \tag{13}$$

Using the expressions of E, T, S, σ and A, we get $G = \frac{r_+}{4}$. The above equation obeys the differential form

$$\delta G = -S\delta T + A\delta\sigma. \tag{14}$$

If we order the effective temperature be $T_{eff} = T - 4\sigma = \frac{1}{4\pi r_+}$, Eq. (11) is then written as $\delta E = T_{eff}\delta S$. The corresponding Gibbs free energy is $G_{eff} = E - T_{eff}S$ which obeys $\delta G_{eff} = -S\delta T_{eff}$. The $G - T$ and $G_{eff} - T_{eff}$ diagrams are displayed in the Figs. 2 and 3, respectively. In the Fig. 2, we find that the Gibbs free energies are different at the different surface tensions when the temperature is very low. Finally, the energies approach to the same value with the increase of the temperature.

2.2. The pressure

In this subsection, we discuss the pressure at the horizon. Using the relation between the area and the volume, $V = \frac{Ar_+}{3}$, yields $\sigma\delta A = P\delta V$, where $P = T_r^r \mid_{r_+} = \frac{1}{8\pi}(-\frac{Q^2}{r_+^4} + \frac{3}{l^2})$ is the pressure. Thus, the modified law becomes

$$\delta E = T\delta S - P\delta V. \tag{15}$$

The positive (zero and negative) pressures correspond to the positive (zero and negative) surface tensions. Using the expressions of the pressure and the temperature, we get

$$P = \frac{T}{2r_+} - \frac{1}{8\pi r_+^2}. \tag{16}$$

In the Van der Waals equation $P = \frac{kT}{v-b} - \frac{a}{v^2}$, v, T and k are the specific volume, the temperature and the Boltzmann constant, respectively. a is a positive constant described the attraction of

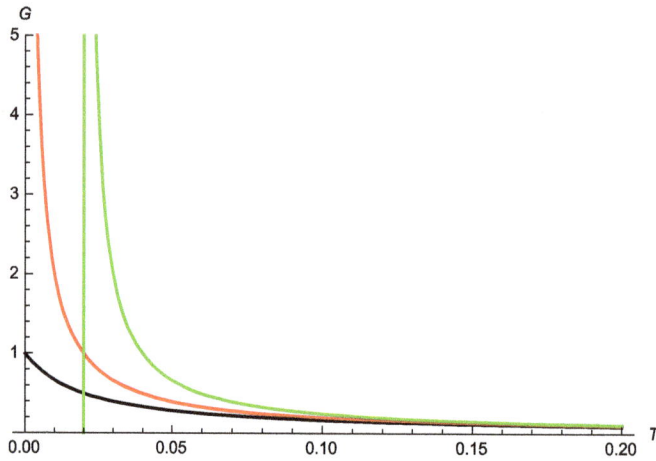

Fig. 2. The blue, red and green curves correspond to $\sigma = -0.005$, $\sigma = 0$ and $\sigma = 0.005$, respectively. (For interpretation of the references to color in this figure legend, the reader is referred to the web version of this article.)

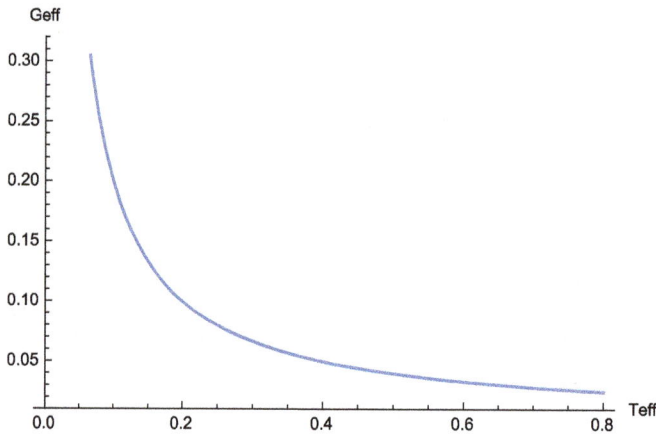

Fig. 3. The diagram shows that the effective Gibbs free energy decreases with the increase of the effective temperature.

molecules in the system, while b is the size of molecules. From Eq. (16), we get $a = \frac{1}{2\pi}$ and $b = 0$, which show that the molecules are ideal point particles and their interaction force is the attractive force. $v = 2r_+$ is identified as the specific volume. Then Eq. (17) is written as

$$P = \frac{T}{v} - \frac{1}{2\pi v^2}. \tag{17}$$

When $Q = 0$, the metric (4) is reduced to the Schwarzschild AdS black hole and the pressure gotten as $P = -\frac{\Lambda}{8\pi}$ is positive always. When $\Lambda = 0$, it describes the RN black hole and the negative pressure $P = -\frac{Q^2}{8\pi r_+^4}$ is naturally obtained. The pressure is zero when $Q = \Lambda = 0$, which is full inconsistence with that of the Schwarzschild black hole described the vacuum solution with no pressure.

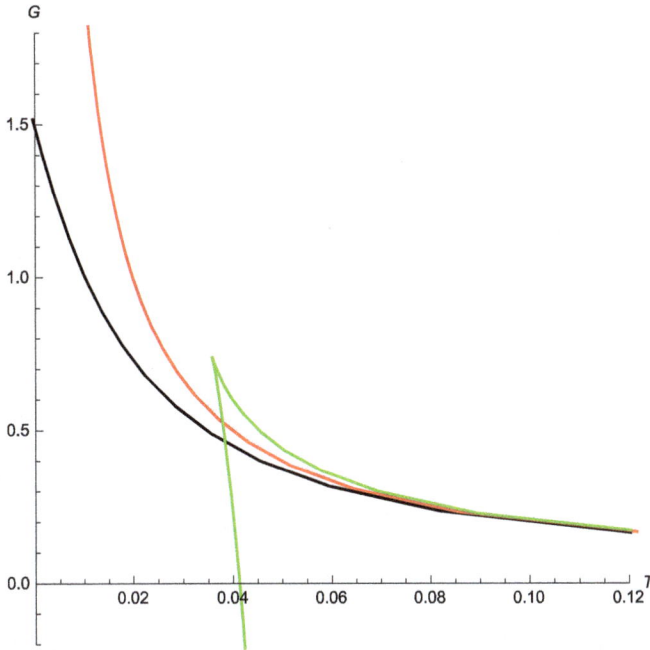

Fig. 4. The curves from left to right correspond to $P = -0.002$, $P = 0$ and $P = 0.002$, respectively.

The Gibbs free energy is $G = E - TS + PV$ and obeys

$$\delta G = -S\delta T + V\delta P. \tag{18}$$

In the recent work [46], Eq. (18) was seen as a new horizon first law, where the volume, pressure, and temperature were specified, and the horizon entropy and Gibbs free energy were derived concepts. Thus Eq. (15) is easily recovered by performing a degenerate Legendre transformation. This specification overcomes the ambiguity between the 'heat' and 'work' terms. The Fig. 4 depicts the Gibbs free energy varying with the temperature at the fixed pressures. It shows that a characteristic cusp exists when $P > 0$.

2.3. The first law with the varied cosmology constant

In the recent work, the cosmological constant was seen as an varied constant associated to the pressure. In this subsection, we adopt the varied cosmology constant and make further efforts to explore the thermodynamics of the RN–AdS black hole. Rewrite the first term on rhs of Eq. (10) as

$$2r_+ T_r^r \mid_{r_+} \delta S = -\frac{Q^2}{4\pi r_+^3}\delta S + \frac{3r_+}{4\pi l^2}\delta S = -\frac{Q^2}{16\pi r_+^3}\delta A + \frac{3}{8\pi l^2}\delta V. \tag{19}$$

The last equal sign in the above equation was gotten by using the relation among the entropy, horizon area and volume. The first term relies on the charge and the second term depends on the cosmological constant. We introduce the definition of the pressure $P = -\frac{\Lambda}{8\pi}$ and let $\sigma_{eff} = -\frac{Q^2}{16\pi r_+^3}$. Eq. (10) becomes

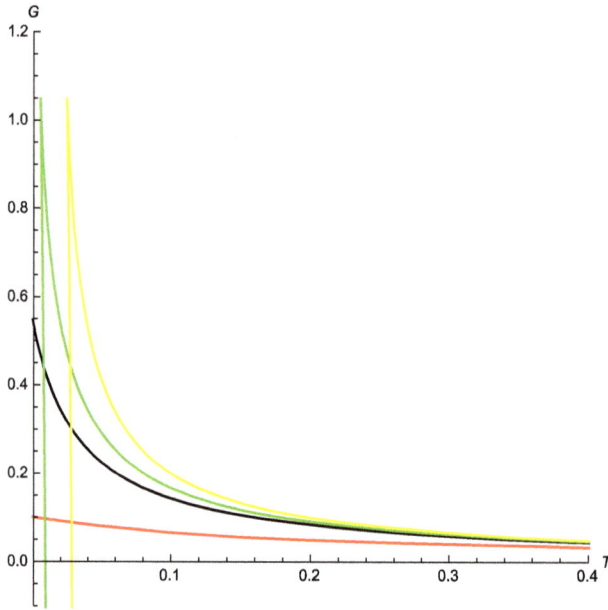

Fig. 5. The blue, red, green and yellow curves correspond to $\sigma = -0.01$, $\sigma = -0.05$, $\sigma = -0.005$ and $\sigma = -0.0001$, respectively. (For interpretation of the references to color in this figure legend, the reader is referred to the web version of this article.)

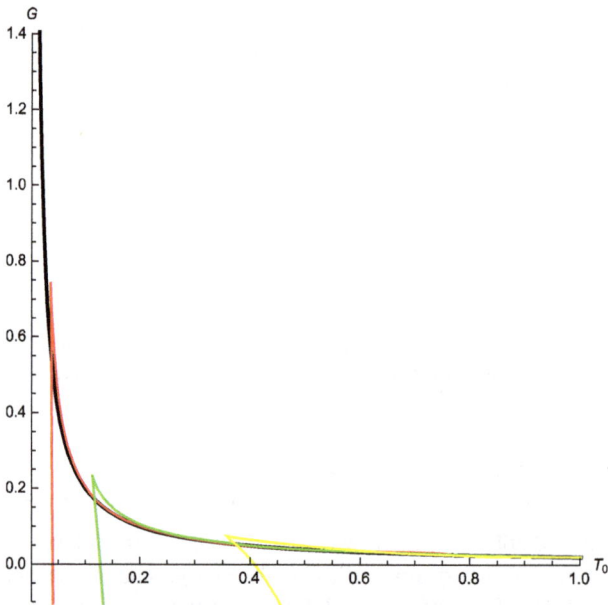

Fig. 6. The curves from left to right correspond to $P = 0$, $P = 0.002$, $P = 0.02$ and $P = 0.2$, respectively.

$$\delta E = T\delta S - \sigma_{eff}\delta A - P\delta V, \tag{20}$$

which can be rewritten as

$$\delta E_0 = T\delta S - \sigma_{eff}\delta A + V\delta P, \tag{21}$$

by using the transformation $P\delta V = \delta(PV) - V\delta P$, where $E_0 = E + PV = M - \frac{Q^2}{2r_+}$ is identified as the enthalpy. Further analysis shows E_0 is just the Misner–Sharp mass of the RN black hole. Taken into account the expressions of T, S, σ, A, P and V, the corresponding Smarr relation

$$E_0 = 2(TS - \sigma_{eff}A - PV), \tag{22}$$

is also satisfied. Therefore, Eq. (21) can also been seen as the modified first law of thermodynamics of the RN–AdS black hole. It is an important result in this paper. Due to the existence of the cosmological constant as the variable, there is an additional worked term $V\delta P$ in the law and the total energy is $M - \frac{Q^2}{2r_+}$. Thus, σ_{eff} is seen as the effective surface tension. Using the expressions of T and P, we get

$$P = \frac{T}{2r_+} - \frac{1}{8\pi r_+^2} + \frac{Q^2}{8\pi r_+^4}. \tag{23}$$

Compared with the Van der Waals equation, the properties of the RN–AdS black hole was discussed in detail in [7]. The difference is that the total energy is defined as $M - \frac{Q^2}{2r_+}$ in this subsection, while that is the physical mass M in [7]. The corresponding Gibbs free energy is $G = E - TS + \sigma_{eff}A + PV$. The $G - T$ diagram at the fixed σ and $P = 0.001$ is depicted in the Fig. 5. It shows that the characteristic cusps corresponded to the same Gibbs free energy when the surface tension is less than a certain value.

Eq. (21) can be furthermore written as

$$\delta E_0 = T_0\delta S + V\delta P, \tag{24}$$

which obeys the corresponding Smarr relation

$$E_0 = 2(T_0 S - PV), \tag{25}$$

where $T_0 = T - 4\sigma_{eff} = \frac{1}{4\pi r_+}\left(1 + \frac{3r_+^2}{l^2}\right)$. Therefore, Eq. (24) can also be regarded as the modified first law of the black hole thermodynamics. Now the Gibbs free energy is gotten as $G = E - T_0 S + PV$. The $G - T_0$ diagram is described in the Fig. 6. We find that the position of the characteristic cusp depends on the value of the pressure.

3. Modifying the first law of thermodynamics at the cosmological horizon

The research on thermodynamics of dS space–times is more complex than that of AdS space–times. In a dS space–time where a black hole horizon and a cosmological horizon exist at the same time, the temperatures at the different horizons are different. Therefore, there is a non-equilibrium state in this space–time. To discuss thermodynamical properties in the equilibrium state, one can treat the black hole horizon and the cosmological horizon as two independent thermodynamical systems [12,38–41]. Besides, one can adopt a global view to construct the globally effective temperature and other effective thermodynamic quantities.

In this section, we investigate the thermodynamics of the pure dS space–time by the surface tension. The modified first laws of thermodynamics at the cosmological horizon are gotten. The pure dS space–time only contains a cosmological horizon. The metric is given by

$$ds^2 = -F(r)dt^2 + \frac{1}{F(r)}dr^2 + r^2(d\theta^2 + sin^2\theta d\phi^2), \tag{26}$$

where $F(r) = 1 - \frac{r^2}{L^2}$, $L^2 = \frac{3}{\Lambda}$ and Λ is the cosmological constant. The cosmological horizon is located at $r_C = L$ derived from $F(r) = 0$. The entropy is $S = \frac{A}{4} = \pi r_C^2$. The temperature at the horizon is

$$T = \frac{|F'(r_C)|}{4\pi} = \frac{1}{2\pi L}. \tag{27}$$

In the above equation, $F'(r_C) = \frac{\partial F(r)}{\partial r}|_{r=r_C} = -\frac{2}{L}$. Since the temperature should be positive, we used the absolute value. Adopting the method in [44], we get the Misner–Sharp mass as

$$M = \frac{r_C}{2}. \tag{28}$$

The horizon thermodynamics are related to Einstein's equations. To discuss the surface tension and the thermodynamics, we first calculate the radial Einstein's equation and get

$$G_r^r|_{rc} = 8\pi T_r^r|_{rc} = \frac{r_C F'(r_C) - 1}{r_C^2}. \tag{29}$$

Solving the above equation yields

$$F'(r_C) = 8\pi r_C T_r^r|_{rc} + \frac{1}{r_C}. \tag{30}$$

Moving the first term on rhs in the above equation to the left hand side (lhs) and then multiplying by $-\frac{\delta S}{4\pi}$ on the both sides, we get

$$-\frac{F'(r_C)}{4\pi}\delta S + 2r_C T_r^r|_{rc} \delta S = -\frac{\delta r_C}{2}. \tag{31}$$

Here we have used the differential expression $\delta S = 2\pi r_C \delta r_C$ to derive the term on rhs of Eq. (31). It is clearly that $-\frac{F'(r_C)}{4\pi}$ is the value of the temperature at the cosmological horizon. So the first term on lhs is in of the form $T dS$. $\delta M = \frac{1}{2}\delta r_C$ expresses the change of the Misner–Sharp mass in the universe. In dS space–times, the energy of an object is measured as a negative value, rather than its mass [31,47–49]. We identify the energy E as the negative Misner–Sharp mass, and then get $\delta E = -\delta M$. Thus, the term on rhs of the above equation denotes the change of the energy. The term $2r_C T_r^r|_{rc} \delta S$ can be written in of the form $-\sigma \delta A$ or $\sigma \delta A$ by using the relation between the entropy and the horizon area. The expression $-\sigma \delta A$ yields the same form as in Eq. (11). Considering the property of the dS space–time, we adopt the expression $\sigma \delta A$ and get $\sigma = \frac{1}{2}r_C T_r^r|_{rc} = -\frac{3}{16\pi r_C}$. Therefore, Eq. (31) is rewritten as

$$\delta E = T\delta S + \sigma \delta A, \tag{32}$$

which is the modified first law of thermodynamics at the cosmological horizon and σ is the surface tension. The inner space of a black hole horizon is an inaccessible region surrounded by the horizon. On the contrary, the outer space of the cosmological horizon is an inaccessible region where the physics isn't known. To study the property of the cosmological horizon, we only recur to the discussion of the inner space–time (normal region) of the cosmology. When a particle is absorbed by the black hole, the entropy, the horizon area and the volume of inaccessible region increase. In the dS space–time, when a particle entries the inaccessible region, the entropy, horizon area and volume of the cosmology decrease. The corresponding quantities δS, δA and δV are negative. Therefore, we should beware of the sign in the calculation. The Smarr relation,

$$E = 2(TS + \sigma A), \tag{33}$$

corresponding to Eq. (32) is also obeyed. On the other side, the modified first law gives an interpretation of the negative energy in the dS space–time.

The Gibbs free energy is $G = E - TS - \sigma A$. It is found that the Gibbs free energy is $G = -\frac{1}{4}r_C$ and there is no critical phenomenon appeared. Its differential expression is

$$\delta G = -S\delta T - A\delta\sigma, \tag{34}$$

which can be also treated as a modified horizon first law.

To make further efforts to discuss the thermodynamics of the dS space–time, we rewrite the second term on lhs of Eq. (31) as

$$T_r^r \big|_{rc} \, \delta \left(\frac{4\pi r_C^3}{3} \right) = P\delta V, \tag{35}$$

where $P = T_r^r \big|_{rc} = -\frac{\Lambda}{8\pi}$ denotes the pressure. When the cosmology radiates a particle, the volume of the inaccessible region increases, while the volume (normal region) and horizon area of the cosmology decreases. The quantity of increase equals to that of decrease. Thus, Eq. (31) is written as

$$\delta E = T\delta S + P\delta V. \tag{36}$$

Treating the cosmological constant as a variable associated to the pressure $P = -\frac{\Lambda}{8\pi}$ and ordering $E_0 = E - PV$ yield the modified law

$$\delta E_0 = T\delta S - V\delta P, \tag{37}$$

where E_0 denotes the total energy of the space–time. This result is full in consistence with that derived by other method [27]. We find $E_0 = 0$ from the expressions of E, P and V. This agree with the zero mass of the pure dS space–time. The corresponding Smarr relation

$$E_0 = 2(TS + PV), \tag{38}$$

is obeyed.

In this section, the modified first laws of thermodynamics (32) and (37) were gotten. Clearly, the expressions of the laws at the cosmological horizon of the dS space–time are different from these at the black horizon of the RN–AdS black hole.

4. Discussion and conclusion

In fact, the first law can be directly derived by the differential. Treating the cosmological constant as a variable, we carry out differential on Eq. (5) at the black hole horizon and get

$$dM = \left(\frac{M}{r_+} - \frac{Q^2}{r_+^2} - \frac{\Lambda r_+^2}{3} \right) dr_+ + \frac{Q}{r_+} dQ - \frac{r_+^3}{6} d\Lambda, \tag{39}$$

where r_+, M and Q are seen as variables and $\Phi = \frac{Q}{r_+}$ is the electromagnetic potential at the horizon. Let the cosmological constant be associated with the pressure $P = -\frac{\Lambda}{8\pi}$ and its conjugate quantity be $V = \frac{4\pi r_+^3}{3}$. It is found that Eq. (39) is just the expression of the modified first law in Eq. (1). Once again, carrying out differential on $F = 1 - \frac{r^2}{L^2}$ at the cosmological horizon yields

$$0 = dr_C + \frac{r_C^3}{6} d\Lambda. \tag{40}$$

Let the positive cosmology constant be also associated to the pressure $P = -\frac{\Lambda}{8\pi}$. Thus Eq. (40) takes on the form of

$$0 = T\delta S - V\delta P. \tag{41}$$

This result is full in accordance with that of Eq. (37), where the total energy E_0 is zero.

In this paper, using the radial Einstein's equations at the black hole horizon and the cosmological horizon, we derived the modified first laws of thermodynamics of the RN–AdS and pure dS space–times by the surface tensions. The corresponding Smarr relations were satisfied. The cosmological constants were first fixed, and then were seen as the variables associated to the pressures. For the RN–AdS black hole, the surface tension can be positive, zero and negative, and its value is determined by r_+, Q and l. Using the relation between the entropy and the horizon area, we got the effective temperature. Considering that the horizon area and the volume are associated with each other, we replaced the area with the volume and got the effective pressure. However, there is somewhat flaw when we eliminated the surface tension in terms of the pressure by introducing the volume. Namely, based on the above replacement, the differential form of the first law could be gotten, while the corresponding Smarr relation isn't obeyed. This phenomenon was also found and explained in [33]. In fact, the pressures are usually inconsistent after the surface effects of the curved surface are taken into account. To make further efforts to investigate the thermodynamics, we treated the cosmological constant as the pressure $P = -\frac{\Lambda}{8\pi}$ and rewrote the modified first law as Eq. (21). The effective surface tension was gotten. The relation between the Gibbs free energy and the temperature was discussed. For the dS space–time, we identified the energy as the negative Misner–Sharp mass and obtained the modified first laws of thermodynamics (32) and (37). In Eq. (31), the term $2r_C T_r^r \mid_{r_C} \delta S$ was expressed as $\sigma \delta A$. If we express it as $-\sigma \delta A$, the derived first law takes on the same form as Eq. (11) and obeys the corresponding Smarr relation.

Acknowledgements

We would like to thank Professors S.Q. Wu, H. Yang and P. Wang for their useful discussions. This work is supported by the National Natural Science Foundation of China (Grant No. 11205125), by Sichuan Province Science Foundation for Youths (Grant No. 2014JQ0040) and by the Innovative Research Team in China West Normal University (Grant No. 438061).

References

[1] S. Hawking, D.N. Page, Thermodynamics of black holes in anti-de Sitter space, Commun. Math. Phys. 87 (1983) 577.

[2] M. Cvetic, S.S. Gubser, Phases of R charged black holes, spinning branes and strongly coupled gauge theories, J. High Energy Phys. 9904 (1999) 024, arXiv:hep-th/9902195.

[3] M. Cvetic, S. Gubser, Thermodynamic stability and phases of general spinning branes, J. High Energy Phys. 9907 (1999) 010, arXiv:hep-th/9903132.

[4] D. Kastor, S. Ray, J. Traschen, Enthalpy and the mechanics of AdS black holes, Class. Quantum Gravity 26 (2009) 195011, arXiv:0904.2765 [hep-th].

[5] D. Kastor, S. Ray, J. Traschen, Smarr formula and an extended first law for lovelock gravity, Class. Quantum Gravity 27 (2010) 235014, arXiv:1005.5053 [hep-th].

[6] B.P. Dolan, The cosmological constant and the black hole equation of state, Class. Quantum Gravity 28 (2011) 125020, arXiv:1008.5023 [gr-qc].

[7] D. Kubiznak, R.B. Mann, P–V criticality of charged AdS black holes, J. High Energy Phys. 1207 (2012) 033, arXiv:1205.0559 [hep-th].

[8] D. Kubiznak, R.B. Mann, Black hole chemistry, Can. J. Phys. 93 (2015) 999, arXiv:1404.2126 [gr-qc].

[9] S. Wang, S.Q. Wu, F. Xie, L. Dan, The first laws of thermodynamics of the (2 + 1)-dimensional BTZ black holes and Kerr–de Sitter spacetimes, Chin. Phys. Lett. 23 (2006) 1096, arXiv:hep-th/0601147.

[10] Y. Sekiwa, Thermodynamics of de Sitter black holes: thermal cosmological constant, Phys. Rev. D 73 (2006) 084009, arXiv:hep-th/0602269.

[11] M. Urano, A. Tomimatsu, H. Saida, Mechanical first law of black hole spacetimes with cosmological constant and its application to Schwarzschild–de Sitter spacetime, Class. Quantum Gravity 26 (2009) 105010, arXiv:0903.4230v1 [gr-qc].

[12] D. Kubiznak, F. Simovic, Thermodynamics of horizons: de Sitter black holes, arXiv:1507.08630 [hep-th].

[13] S.H. Hendi, M.H. Vahidinia, Extended phase space thermodynamics and P–V criticality of black holes with non-linear source, Phys. Rev. D 88 (2013) 084045, arXiv:1212.6128 [hep-th].

[14] R.G. Cai, L.M. Cao, L. Li, R.Q. Yang, P–V criticality in the extended phase space of Gauss–Bonnet black holes in AdS space, J. High Energy Phys. 1309 (2013) 005, arXiv:1306.6233 [gr-qc].

[15] D.C. Zou, S.J. Zhang, B. Wang, Critical behavior of Born–Infeld AdS black holes in the extended phase space thermodynamics, Phys. Rev. D 89 (2014) 044002, arXiv:1311.7299 [hep-th].

[16] B.P. Dolan, Thermodynamic stability of asymptotically anti-de Sitter rotating black holes in higher dimensions, Class. Quantum Gravity 31 (2014) 165011, arXiv:1403.1507 [gr-qc].

[17] S.W. Wei, Y.X. Liu, Insight into the microscopic structure of an AdS black hole from thermodynamical phase transition, Phys. Rev. Lett. 115 (2015) 111302, arXiv:1502.00386 [gr-qc].

[18] B. Mirza, Z. Sherkatghanad, Phase transitions of hairy black holes in massive gravity and thermodynamic behavior of charged AdS black holes in an extended phase space, Phys. Rev. D 90 (2014) 084006, arXiv:1409.6839 [gr-qc].

[19] Y. Gim, W. Kim, S.H. Yi, The first law of thermodynamics in Lifshitz black holes revisited, J. High Energy Phys. 1407 (2014) 002, arXiv:1403.4704 [hep-th].

[20] J. Armas, N.A. Obers, M. Sanchioni, Gravitational tension, spacetime pressure and black hole volume, arXiv:1512.09106 [hep-th].

[21] B.P. Dolan, Black holes and Boyle's law – the thermodynamics of the cosmological constant, Mod. Phys. Lett. A 30 (2015) 1540002, arXiv:1408.4023 [gr-qc].

[22] C.V. Johnson, Holographic heat engines, Class. Quantum Gravity 31 (2014) 205002, arXiv:1404.5982 [hep-th].

[23] D. Kastor, S. Ray, J. Traschen, Chemical potential in the first law for holographic entanglement entropy, J. High Energy Phys. 1411 (2014) 120, arXiv:1409.3521 [hep-th].

[24] B.P. Dolan, Bose condensation and branes, J. High Energy Phys. 10 (2014) 179, arXiv:1406.7267 [hep-th].

[25] J.L. Zhang, R.G. Cai, H.W. Yu, Phase transition and thermodynamical geometry of Reissner–Nordstrom–AdS black holes in extended phase space, Phys. Rev. D 91 (2015) 044028, arXiv:1502.01428 [hep-th].

[26] J.L. Zhang, R.G. Cai, H.W. Yu, Phase transition and thermodynamical geometry for Schwarzschild AdS black hole in $AdS_5 \times S^5$ spacetime, J. High Energy Phys. 1502 (2015) 143, arXiv:1502.01428 [hep-th].

[27] B.P. Dolan, D. Kastor, D. Kubiznak, R.B. Mann, J. Traschen, Thermodynamic volumes and isoperimetric inequalities for de Sitter black holes, Phys. Rev. D 87 (2013) 530, arXiv:1301.5926 [hep-th].

[28] H.H. Zhao, L.C. Zhang, M.S. Ma, R. Zhao, P–V criticality of higher dimensional charged topological dilaton de Sitter black holes, Phys. Rev. D 90 (2014) 064018.

[29] H.H. Zhao, L.C. Zhang, M.S. Ma, R. Zhao, Thermodynamics of phase transition in higher dimensional Reissner–Nordstrom–de Sitter black hole, Eur. Phys. J. C 74 (2014) 3052, arXiv:1403.2151 [gr-qc].

[30] D. Kubiznak, F. Simovic, Thermodynamics of horizons: de Sitter black holes, arXiv:1507.08630 [hep-th].

[31] T. Padmanabhan, Classical and quantum thermodynamics of horizons in spherically symmetric spacetimes, Class. Quantum Gravity 19 (2012) 5387, arXiv:gr-qc/0204019.

[32] R.G. Cai, Connections between gravitational dynamics and thermodynamics, J. Phys. Conf. Ser. 484 (2014) 012003.

[33] D. Hansen, D. Kubiznak, R.B. Mann, Criticality and surface tension in rotating horizon thermodynamics, Class. Quantum Gravity 33 (2016) 165005, arXiv:1604.06312 [gr-qc].

[34] D. Kothawala, S. Sarkar, T. Padmanabhan, Einstein's equations as a thermodynamic identity: the cases of stationary axisymmetric horizons and evolving spherically symmetric horizons, Phys. Lett. B 652 (2007) 338, arXiv:gr-qc/0701002.

[35] T. Padmanabhan, Thermodynamical aspects of gravity: new insights, Rep. Prog. Phys. 73 (2010) 046901, arXiv:0911.5004 [gr-qc].

[36] T. Padmanabhan, D. Kothawala, Lanczos–Lovelock models of gravity, Phys. Rep. 531 (2013) 115, arXiv:1302.2151 [gr-qc].

[37] J.W. York Jr., Black hole thermodynamics and the Euclidean Einstein action, Phys. Rev. D 33 (1986) 2092.

[38] A. Gomberoff, C. Teitelboim, de Sitter black holes with either of the two horizons as a boundary, Phys. Rev. D 67 (2003) 104024.

[39] Y. Sekiwa, Thermodynamics of de Sitter black holes: thermal cosmological constant, Phys. Rev. D 73 (2006) 084009, arXiv:hep-th/0602269.

[40] R.G. Cai, J.Y. Ji, K.S. Soh, Action and entropy of black holes in spacetimes with a cosmological constant, Class. Quantum Gravity 15 (1998) 2783.

[41] B.B. Wang, C.G. Huang, Thermodynamics of Reissner–Nordstrom–de Sitter black hole in York's formalism, Class. Quantum Gravity 19 (2002) 2491.

[42] C.W. Misner, D.H. Sharp, Relativistic equations for adiabatic, spherically symmetric gravitational collapse, Phys. Rev. B 571 (1964) 136.

[43] M. Cahill, G. McVittie, Spherical symmetry and mass-energy in general relativity, I: general theory, J. Math. Phys. 11 (1970) 1382.

[44] S.A. Hayward, Quasi-local gravitational energy, Phys. Rev. D 49 (1994) 831, arXiv:gr-qc/9303030.

[45] Y.P. Hu, H.S. Zhang, Misner–Sharp mass and the unified first law in massive gravity, Phys. Rev. D 92 (2015) 024006, arXiv:1502.00069 [hep-th].

[46] D. Hansen, D. Kubiznak, R.B. Mann, Horizon thermodynamics from Einstein's equation of state, arXiv:1610.03079 [gr-qc].

[47] Y. Sekiwa, Decay of the cosmological constant by Hawking radiation as quantum tunneling, arXiv:0802.3266 [hep-th].

[48] M.H. Dehgham, H. KhajehAzad, Thermodynamics of a Kerr–Newman de Sitter black hole, Can. J. Phys. 81 (2003) 1363, arXiv:hep-th/0209203.

[49] V. Balasubramanian, J. de Boer, D. Minic, Mass, entropy and holography in asymptotically de Sitter spaces, Phys. Rev. D 65 (2002) 123508, arXiv:hep-th/0110108.

8

Quenching parameter in a holographic thermal QCD

Binoy Krishna Patra [a,*], Bhaskar Arya [b]

[a] *Department of Physics, Indian Institute of Technology Roorkee, Roorkee 247 667, India*
[b] *Department of Mechanical and Industrial Engineering, Indian Institute of Technology Roorkee, Roorkee 247 667, India*

Editor: Leonardo Rastelli

Abstract

We have calculated the quenching parameter, $\hat{q}$ in a model-independent way using the gauge–gravity duality. In earlier calculations, the geometry in the gravity side at finite temperature was usually taken as the pure AdS black hole metric for which the dual gauge theory becomes conformally invariant unlike QCD. Therefore we use a metric which incorporates the fundamental quarks by embedding the coincident D7 branes in the Klebanov–Tseytlin background and a finite temperature is switched on by inserting a black hole into the background, known as OKS-BH metric. Further inclusion of an additional UV cap to the metric prepares the dual gauge theory to run similar to thermal QCD. Moreover $\hat{q}$ is usually defined in the literature from the Glauber model perturbative QCD evaluation of the Wilson loop, which has no reasons to hold if the coupling is large and is thus against the main idea of gauge–gravity duality. Thus we use an appropriate definition of $\hat{q}$: $\hat{q}L^- = 1/L^2$, where L is the separation for which the Wilson loop is equal to some specific value. The above two refinements cause $\hat{q}$ to vary with the temperature as T^4 always and to depend linearly on the light-cone time L^- with an additional $(1/L^-)$ correction term in the short-distance limit whereas in the long-distance limit, $\hat{q}$ depends only linearly on L^- with no correction term. These observations agree with other holographic calculations directly or indirectly.

* Corresponding author.
 E-mail address: binoyfph@iitr.ac.in (B.K. Patra).

1. Introduction

In the initial stage of ultrarelativistic heavy-ion collisions energetic partons in the form of jets are produced from the hard collisions. After receiving a large transverse momentum, these jets plough through the fireball for a transitional period of about a few fm/c and will thus loose energy due to the interaction of the hard partons with the medium constituents, known as the jet quenching. As a result the yield of hadrons with high transverse momentum (p_T) is shown to be significantly suppressed in comparison with the cumulative yields of nucleon–nucleon collisions. There are mainly two contributions to the energy loss of the partons in the medium: one is due to the radiation emitted by the decelerated color charges, *i.e.* bremsstrahlung of gluons [1–3] and the other one is due to the collisions among the partons in the medium [4].

The experimental discoveries at RHIC revealed that the matter produced is a strongly coupled quark–gluon plasma (sQGP) unlike weakly interacting gas of partons expected from the naive asymptotic freedom, *for example*, the observed elliptic flow, the quenching of jets while traversing through the medium etc. The jet quenching is parametrized by the quenching parameter, $\hat{q}$, which is defined by the average transverse momentum square transferred from the traversing parton per unit mean free path. The extracted values of this transport coefficient in relativistic heavy-ion collisions by the JET collaboration [5] range from 1 to 25 GeV^2/fm, which are much larger than those estimated from the perturbative QCD calculations. This hints some non-perturbative mechanisms which may contribute to the jet quenching mechanism. Thus it is worthwhile to calculate the possible values of $\hat{q}$ in the strong coupling limit. The first principle lattice QCD however, cannot be applied for this purpose, which requires the real-time dynamics.

The simplest gauge–gravity duality [6–8] between the type IIB superstring theory formulated on AdS$_5 \times S^5$ space and $\mathcal{N} = 4$ supersymmetric Yang–Mills theory (SYM) in four dimensions provides a robust tool to explore the thermodynamical and transport properties of sQGP. Although the underlying dynamics, QCD is different from $\mathcal{N} = 4$ SYM but the correspondence seems feasible because some of the properties of all strongly interacting systems show some universality behavior. One of the notable observation is the universal value $(1/4\pi)$ for the η/s ratio for the quantum field theories having a holographic description [9] and thus it gives a lower bound to the ratio for sQGP. Motivated by these similarities between the $\mathcal{N} = 4$ SYM and the corresponding theory of supergravity, the jet-quenching parameter, $\hat{q}$ was related to the expectation value of the Wilson loop $W^A[\mathcal{C}]$ in adjoint representation due to the Eikonal approximation [10]:

$$\langle W^A[\mathcal{C}] \rangle \approx \exp\left(-\frac{1}{4\sqrt{2}}\hat{q}L_- L^2\right), \tag{1}$$

where $\mathcal{C}$ is a rectangular contour of size $L \times L_-$, with the sides, having the length L_- run along the light-cone. There were other calculations of $\hat{q}$ [11,30,12] using a very different setup and arriving at different conclusions. In the context of relativistic heavy ion collisions, the effects of finite 't Hooft coupling (λ) as well as chemical potential on $\hat{q}$ was studied in [31,40,39] and the jet stopping in strongly-coupled QCD-like plasmas with gravity duals have also been studied using the string α' expansion in AdS/CFT [34,35].

However, since $\hat{q}$ is related to the transverse momentum (p_T) broadening so to calculate the mean p_T, we need to Fourier transform (FT) of the Wilson loop

$$W(p_T) = \int d^2L \, e^{ip\cdot L} \, W(L). \tag{2}$$

The above FT emerges if we intend to calculate the particle production in the scattering of a quark on a target, and the target will be the medium in the jet quenching problem. It turns out that the above FT is proportional to the quark production cross section, $W(p_T) \sim d\sigma/d^2p$ [13, 14]. Let us explore the subtleties which might help us to search for the correct definition of $\hat{q}$. For example, if we define $\hat{q}$ as $\langle p_T^2 \rangle/L^-$, as some authors do. So we would then need to find $\langle p_T^2 \rangle$. But this seems easy because $\langle p_T^2 \rangle \sim \nabla_\perp^2 W(L)$ at $L = 0$. This seems consistent with getting the coefficient of the L^2 term in the exponent, as in (1). However, since our aim is to model QCD and in QCD at high p_T perturbative physics works, and $d\sigma/d^2p \sim 1/p_T^4$, so $\langle p_T \rangle$ is infinite (irrespective of what happens at lower p_T). In other words one cannot trust $W(L)$ from AdS at very small L. A way out is to define $\hat{q}$ as $\langle p_T \rangle^2/L^-$. Since $\langle p_T \rangle$ is finite even in perturbative QCD, this definition is safe. To find $\langle p_T \rangle$ we need the typical momentum scale of $W(L)$ and if one knows $W(p_T)$, then one should be able to find $\langle p_T \rangle$ exactly. Otherwise one could argue that $\langle p_T \rangle$ is given by the saturation scale Q_s, as the only scale available in the problem at high enough energy. Hence the standard prescription of finding Q_s by requiring the Wilson loop, $W(L = 1/Q_s)$ to be a constant, should probably give one a good estimate of $\langle p_T \rangle$.

In summary the above definition of $\hat{q}$ in (1) as a coefficient of the L^2 term in the Wilson line correlator may not be correct because the motivation for the definition (1) in [10] comes from the Glauber model perturbative QCD evaluation of the Wilson loop [15,16].[1] Therefore this perturbative expression has no reasons to hold when the coupling is large, which is the main idea of gauge–gravity duality. A more appropriate definition of $\hat{q}$ is then to postulate the equation

$$\hat{q}L^- = 1/L^2 , \qquad (3)$$

where L is the quark–antiquark separation for which the expectation value of the Wilson loop in adjoint representation is equal to some specific value. The above definition (3) can also be understood as follows: since $\hat{q}L^-$ behaves like the saturation scale squared in small-x physics and the saturation scale is defined by requiring that the expectation value of Wilson loop is equal to some constant at $L = 1/Q_s$ [16,15].

The calculations for $\hat{q}$ discussed so far used the geometry as the pure AdS black hole metric, for which the dual gauge theory is conformally invariant SYM theory *unlike the QCD*. This is one of the central theme of our work. Therefore, the aim of the present paper – to extend/modify the shortcomings of the above-mentioned calculation [10] – is twofold: (i) the first aim is to study the jet quenching in a gravitational background which is dual to a gauge theory with an RG flow that confines in the far IR and is asymptotically free at the far UV. Recently a gravity dual with a black hole and seven branes embedded via Ouyang embedding is constructed [23,19], which resembles the main features of strongly coupled QCD, *i.e.* is almost conformal in the UV with no Landau poles or UV divergences of the Wilson loops, but has logarithmic running of coupling in the IR. Recently one of us have explored the properties of heavy quarkonium bound states with the above geometry and the findings [20,21] can only be understood as the artifact of the correct geometry for real QCD. (ii) The second one is the appropriate definition of $\hat{q}$ as in (3) for which the Wilson loop is equal to some specific value, *say* 1/2. Our work is therefore organized as follows. Section 2 will be devoted to revisit the Ouang–Klebanov–Strassler geometry and its improvements at the UV sector. In Section 3.1, we employ the aforesaid geometry to obtain the renormalized Nambu–Goto action in both short- and long-distance limits. Thereafter we will obtain the quenching parameter in Section 3.2 and will also discuss briefly the results of other calculations. Finally we conclude in Section 4.

---

[1] In fact, it is already incorrect once someone includes perturbative QCD corrections to the Glauber formula.

2. Construction of dual geometry

A conformal gauge theory does not flow with the scale, hence it has a trivial RG flow. The AdS/CFT correspondence conjectures that a conformal theory in four dimensions can be mapped on the boundary of a pure anti-de Sitter space [6]. But if the theory has a non-trivial RG flow like QCD, which is confining in IR and conformal in UV, we cannot describe the full theory on the boundary of some higher dimensional space and hence need to envisage differently at running energy scales. One way out is to embed the D branes in the geometry and as a result the corresponding gauge theory exhibits logarithmic RG flow. Such a construction was done in the Klebanov–Strassler (KS) geometry [17] through a warped deformed conifold with three-form type IIB fluxes and the corresponding dual gauge theory is confining in the far IR limit but is not free at UV limit. The other demerits of the KS geometry are that it is devoid of quarks in the fundamental representation and cannot be generalized to finite temperature.

The inclusion of fundamental matter in string theory is possible by embedding a set of flavor branes in addition to the color branes. The strings connecting to the color and flavor branes in the adjoint representation of $U(N_c)$ group give the gauge particles and the mesons, respectively whereas those connected to both the flavor and color branes in the fundamental representation give the quarks and antiquarks, respectively. In principle one could go to large number of color (N_c) and flavor (N_f) branes in the near horizon limit and translates the branes into fluxes and then construct the gravity background which is holographically dual to gauge theory of quarks and gluons. In practice the back reaction of the probes on the background could be neglected through the probe approximation ($N_f \ll N_c$) and the flavor physics is then extracted by analyzing the effective action which describes the flavor branes in the color background [28,29]. Since the full global solution for the backreaction of D7 branes in the KS background becomes nontrivial so the insertion of the fundamental quarks in the original KS geometry [17] becomes difficult. Peter Ouyang [18] has successfully put the coincident D7 branes into the Klebanov–Tseytlin background [22], known as OKS geometry, which has all the type IIB fluxes switched on including the axio-dilaton and the local metric was then computed by incorporating the deformations of the seven branes by moving them far away from the regime of interest. Hence the axion-dilaton vanishes for the background locally, but there will be non-zero axion-dilaton globally, as a result the local back reactions on the metric modify the warp factors to the full global scenario.

For realizing the finite temperature a black hole is inserted into the OKS background, *i.e.* OKS-BH geometry, where the Hawking temperature corresponds to the gauge theory temperature. Thus the metric in OKS-BH geometry is expressed in terms of warp factor (h) [23]

$$ds^2 = \frac{1}{\sqrt{h}}\left[-g_1(u)dt^2 + dx^2 + dy^2 + dz^2\right] + \sqrt{h}\left[g_2^{-1}(u)du^2 + d\mathcal{M}_5^2\right] \tag{4}$$

where $g_i(u)$ are the black-hole factors as a function of the extra dimension, u and $d\mathcal{M}_5^2$ is due to the warped resolved–deformed conifold. The gauge theory dual to the metric (4) flows correctly at IR like QCD but the effective degrees of freedom grow indefinitely at UV limit. The situation becomes worse even in the presence of fundamental flavors because its proliferation leads to Landau poles and hence the Wilson loops diverges at UV. To circumvent the problem, one need to add the appropriate UV cap to the AdS–Schwarzschild geometry in the asymptotic UV limit. However, the additional UV caps, in general may deform the IR geometry but the far IR geometry has not been changed because the UV caps correspond to adding the non-trivial irrelevant operators in the dual gauge theory. These operators keep far IR physics completely unchanged, but the physics at not-so-small energies may be changed a bit.

Recently the IR geometry part has been suitably modified to obtain the desired dual gauge theory by the McGill group [19,23,24], where the metric (4) will receive further corrections, g_{uu}, because the unwarped metric may not remain Ricci flat due to the presence of both axio-dilaton and seven-brane sources, as:

$$ds^2 = \frac{1}{\sqrt{h}}\left[-g(u)dt^2 + dx^2 + dy^2 + dz^2\right] + \sqrt{h}\left[g(u)^{-1}g_{uu}du^2 + g_{mn}dx^m dx^n\right] \quad (5)$$

where the black hole factors $g_i(u)$ are set as $g_1(u) = g_2(u) \equiv g(u)$ and the corrections g_{uu} are of the form $1/u^n$ and may be written as a series expansion:

$$g_{uu} = 1 + \sum_{i=0}^{\infty} \frac{a_{uu,i}}{u^i}, \quad (6)$$

where the coefficients $a_{uu,i}$ are independent of the extra-dimension coordinate u and are solved exactly in [19]. Thus the warp factor, h can be extracted from the above corrections (6) as

$$h = \frac{L^4}{u^4}\left[1 + \sum_{i=1}^{\infty} \frac{a_i}{u^i}\right]$$

where the coefficients a_i are of $\mathcal{O}(g_s N_f)$ and L is the curvature of space. Thus the metric (5) reduces to OKS-BH in the IR limit and becomes $AdS_5 \times M_5$ in the UV limit, hence describes well both in IR and UV limits. Therefore, with the change of coordinates $z = 1/u$, we can rewrite the metric (5) as

$$ds^2 = g_{\mu\nu}dX^\mu dX^\nu$$
$$= A_n z^{n-2}\left[-g(z)dt^2 + d\vec{x}^2\right] + \frac{B_l z^l}{A_m z^{m+2} g(z)}dz^2 + \frac{1}{A_n z^n}ds_{M_5}^2, \quad (7)$$

where $ds_{M_5}^2$ is the metric of the internal space and the coefficients A_n can be obtained from the coefficients a_i in the warp factor (2) as follows:

$$\frac{1}{\sqrt{h}} = \frac{1}{L^2 z^2 \sqrt{a_i z^i}} \equiv A_n z^{n-2} = \frac{1}{L^2 z^2}\left[a_0 - \frac{a_1 z}{2} + \left(\frac{3a_1^2}{8a_0} - \frac{a_2}{2}\right)z^2 + \cdots\right], \quad (8)$$

which gives $A_0 = \frac{a_0}{L^2}$, $A_1 = -\frac{a_1}{2L^2}$, $A_2 = \frac{1}{L^2}\left(\frac{3a_1^2}{8a_0} - \frac{a_2}{2}\right)$ etc. Note that since a_i's for $i \geq 1$ are of $\mathcal{O}(g_s N_f)$ and $L^2 \propto \sqrt{g_s N}$, so in the limit $g_s N_f \to 0$ and $N \to \infty$ all A_i's for $i \geq 1$ are very small. The second term in the metric (7) accommodates the $1/u^n$ corrections in (5) via the series, $B_l z^l$, which is expanded further:

$$B_l z^l = 1 + a_{zz,i}z^i. \quad (9)$$

In a comprehensive study [23], the entire geometry is split into three regions. Apart from the two asymptotic regions at IR and UV denoted as regions I and III, respectively, there is an interpolating region II, where at the outermost boundary the three-forms vanish and the innermost boundary will be the outermost boundary of region I. The background in these three regions and the insertion of additional UV cap are extensively analyzed by the corresponding RG flows and the field theory realizations have been discussed in [25]. Recently another suitable model to study certain IR dynamics of QCD is the Sakai–Sugimoto model [26] in the type IIA string theory, which consists of a set of N wrapped color D4-branes on the circle and the flavor branes D8 and $\bar{D}8$ placed at the anti-nodal points of the circle to conceive the mesonic bound states. In its dual gravity, the wrapped D4-branes are replaced by an asymptotically AdS space, but the eight-branes remain and so does the circular direction. However the Sakai–Sugimoto model does not have a UV completion and had been compared recently with the aforesaid gravity dual in [27]. We shall not go into the complete details here and will use the metric (7) to obtain the Nambu–Goto action and hence the Wilson loop is computed through gauge–gravity correspondence in the next section.

3. Gauge–gravity duality

According to the gauge/gravity prescription [6], the expectation value of the Wilson loop, $W(C)$ in a strongly coupled gauge theory is related to the generating functional of the string in the bulk which has the loop C at the boundary

$$\langle W(C) \rangle \sim Z_{\text{string}} \tag{10}$$

In supergravity limit, the generating functional becomes

$$Z_{\text{string}} = e^{i S_{\text{string}}}, \tag{11}$$

where S_{string} is obtained by extremizing the string action, known as the Nambu–Goto action. So the above correspondence (10) is translated into

$$\langle W(C) \rangle \sim e^{i S_{\text{string}}} \tag{12}$$

Thus we will now evaluate the Nambu–Goto action in the next subsection.

3.1. Nambu–Goto action

By the light-cone transformation,

$$dt = \frac{dx^+ + dx^-}{\sqrt{2}}$$
$$dx_1 = \frac{dx^+ - dx^-}{\sqrt{2}} \tag{13}$$

the metric (7) is rewritten in terms of light-cone coordinates as

$$ds^2 = \left[-\frac{1}{2} A_n z^{n-2} g + \frac{1}{2} A_n z^{n-2} \right] \left[dx^{+2} + dx^{-2} \right] - (1+g) A_n z^{n-2} dx^+ dx^-$$
$$+ A_n z^{n-2} \left[dx_2^2 + dx_3^2 \right] + \frac{B_n z^n}{A_n z^{n+2} g} dz^2 + \frac{1}{A_n z^n} ds_{M_5}^2 \tag{14}$$

We parametrize the two-dimensional world sheet and their derivatives in terms of the light-cone coordinates

$$\tau = x^-, \quad \sigma = x_2 \in [-\frac{r}{2}, \frac{r}{2}],$$
$$x_2 = \text{const}, \quad x_3 = \text{const}, \quad z = z(x_2)$$
$$\partial_\alpha = \frac{\partial}{\partial \tau}, \quad \partial_\beta = \frac{\partial}{\partial \sigma}. \tag{15}$$

With the above parametrization (15), the elements of the induced metric defined by

$$g_{\alpha\beta} = G\mu\nu\frac{\partial x^\mu}{\partial \sigma^\alpha}\frac{\partial x^\nu}{\partial \sigma^\beta} \tag{16}$$

can be read off from the above metric (14)

$$g_{--} = \frac{A_n z^n (1-g)}{2z^2}$$

$$g_{-2} = g_{2-} = 0$$

$$g_{22} = \frac{A_n z^n}{z^2} + \frac{B_n z^n}{z^2 A_n z^n g} z'^2 . \tag{17}$$

Thus the determinant of the induced metric, $g_{\alpha\beta}$ can be calculated

$$\mathbf{det}\, g_{\alpha\beta} = g_{--} g_{22} = \frac{1}{2z_h^4}\left[(A_n z^n)^2 + \frac{(B_n z^n)\, z'^2}{g} \right], \tag{18}$$

hence the Nambu–Goto action can be obtained as

$$S = -\frac{1}{2\pi\alpha'}\int\int d\sigma d\tau \sqrt{-\mathbf{det}g_{\alpha\beta}}$$

$$= -\frac{1}{2\pi\alpha'}\int\int d\sigma d\tau \sqrt{-\frac{1}{2z_h^4}\left[(A_n z^n)^2 + \frac{(B_n z^n)\, z'^2}{g} \right]}, \tag{19}$$

where α' $(= \frac{R^2}{\sqrt{\lambda}}$, R is the AdS radius and λ is the 't Hooft coupling) is the string tension. Thus the equation of motion:

$$z'\frac{\partial \mathcal{L}}{\partial z'} - \mathcal{L} = C \tag{20}$$

can be written from the above Lagrangian ($\mathcal{L}$) in (19) as

$$-\left(A_n z^n\right)^2 = C\sqrt{(A_n z^n)^2 + \frac{(B_n z^n)\, z'^2}{g}}, \tag{21}$$

where C is a constant of motion and can be obtained from the condition: $z' = 0$ at $z = z_m$,

$$C^2 = \left(A_n z_m^n\right)^2 \tag{22}$$

After substituting the constant C, the equation of motion becomes finally

$$z'^2 = \frac{(A_n z^n)^2 g}{B_n z^n}\left[\frac{(A_n z^n)^2}{(A_n z_m^n)^2} - 1 \right] \tag{23}$$

Since the Lagrangian is independent of the time so after integrating over the time-like coordinate (x_-), the action becomes

$$S = -\frac{iL^-}{2\sqrt{2}\pi\alpha' z_h^2}\int_{-\frac{L}{2}}^{+\frac{L}{2}} dx_2 \sqrt{(A_n z^n)^2 + \frac{(B_n z^n)\, z'^2}{g}} \tag{24}$$

$$= -\frac{i2L^-}{2\sqrt{2}\pi\alpha' z_h^2}\int_0^{z_m} dz \sqrt{\frac{(A_n z^n)^2}{z'^2} + \frac{(B_n z^n)}{g}} \tag{25}$$

We will now substitute z'^2 from the equation of motion (23) to obtain the action. Since $(A_n z^n)^2 \ll (A_n z_m{}^n)^2$ so neglecting the higher-order terms and keeping up to the second-order term, the action (25) is simplified into (without loss of generality, $A_0 = 1$, $A_1 = 0$, and $A_2 = A$ (say), and similarly, $B_0 = 1$, $B_1 = 0$, and $B_2 = B$ (say) in units of L^2 [19][2])

$$S \simeq -\frac{\sqrt{2}L^-}{2\pi\alpha' z_h{}^2 \left(1 + A z_m{}^2\right)} \int_0^{z_m} \frac{dz}{\sqrt{g}} \left(1 + \frac{B}{2}z^2\right)\left(1 + Az^2\right) \tag{26}$$

We will now evaluate the Nambu–Goto action by solving the above integral in both short- and long-distance limits:

Case-I: In the short-distance $(z_m \ll z_h)$ limit, after performing the integration in (26) the action is written in terms of Gaussian hypergeometric functions

$$S = -\frac{L^-}{\sqrt{2}\pi\alpha' z_h{}^2 \left(1 + A z_m^2\right)} \int_0^{z_m} dz \left(\frac{1 + \frac{B+2Az^2}{2} + \frac{ABz^4}{2}}{\sqrt{1 - \frac{z^4}{z_h^4}}}\right)$$

$$= -\frac{L^- z_m}{\sqrt{2}\pi\alpha' z_h{}^2 (1 + A z_m^2)} \left[{}_2F_1\left(\frac{1}{4}, \frac{1}{2}, \frac{5}{4}; \frac{z_m^4}{z_h^4}\right) + \frac{(B+2A)z_m^2}{6} {}_2F_1\left(\frac{1}{2}, \frac{3}{4}, \frac{7}{4}; \frac{z_m^4}{z_h^4}\right)\right.$$

$$\left. + \frac{ABz_h^4}{6}\left(-\sqrt{1 - \frac{z_m^4}{z_h^4}} + {}_2F_1\left(\frac{1}{4}, \frac{1}{2}, \frac{5}{4}; \frac{z_m^4}{z_h^4}\right)\right)\right] \tag{27}$$

On expanding the hypergeometric functions in powers of $\left(\frac{z_m}{z_h}\right)$

$$ {}_2F_1\left(\frac{1}{4}, \frac{1}{2}, \frac{5}{4}; \frac{z_m^4}{z_h^4}\right) = \left(1 + \frac{z_m{}^4}{10 z_h{}^4} + \cdots\right),$$

$$ {}_2F_1\left(\frac{1}{2}, \frac{3}{4}, \frac{7}{4}; \frac{z_m^4}{z_h^4}\right) = \left(1 + \frac{3 z_m{}^4}{14 z_h{}^4} + \cdots\right),$$

$$ {}_2F_1\left(\frac{1}{4}, \frac{1}{2}, \frac{5}{4}; \frac{z_m^4}{z_h^4}\right) = \left(1 + \frac{z_m{}^4}{10 z_h{}^4} + \cdots\right), \tag{28}$$

respectively and ignoring the higher-order terms beyond the second power, the action becomes

$$S \overset{z_m \ll z_h}{\simeq} -\frac{L^- z_m}{\sqrt{2}\pi\alpha' z_h^2}\left[1 + \frac{(B-4A)z_m^2}{6} + \frac{z_m^4}{10 z_h^4}\right] \tag{29}$$

In addition to the extremal surface constructed above for the Nambu–Goto action, there is another trivial one given by the two disconnected world sheets, placed one at $x_2 = +\frac{L}{2}$ and another at $x_2 = -\frac{L}{2}$. The action for these two surfaces is

$$S_0 = -\frac{2}{2\pi\alpha'} \int dz\, dx^- \sqrt{-g_{--}g_{zz}}$$

---

[2] Such a choice is of course consistent with supergravity solution for the background we used in our work.

$$= -\frac{iL^-}{\sqrt{2\pi}\,\alpha' z_h^2} \int_0^{z_m} dz \, \frac{1 + \frac{B}{2}z^2}{\sqrt{1 - \frac{z^4}{z_h^4}}} \tag{30}$$

$$= -\frac{iL^- z_m}{\sqrt{2\pi}\,\alpha' z_h^2} \left[{}_2F_1\left(\frac{1}{4}, \frac{1}{2}, \frac{5}{4}; \frac{z_m^4}{z_h^4}\right) + \frac{B z_m^2}{6} \, {}_2F_1\left(\frac{1}{2}, \frac{3}{4}, \frac{7}{4}; \frac{z_m^4}{z_h^4}\right) \right] \tag{31}$$

Expanding the above hypergeometric functions in powers of $(\frac{z_m}{z_h})$

$${}_2F_1\left(\frac{1}{4}, \frac{1}{2}, \frac{5}{4}; \frac{z_m^4}{z_h^4}\right) = \left(1 + \frac{z_m^{\,4}}{10 z_h^{\,4}} + \dots\right), \tag{32}$$

$${}_2F_1\left(\frac{1}{2}, \frac{3}{4}, \frac{7}{4}; \frac{z_m^4}{z_h^4}\right) = \left(1 + \frac{3 z_m^{\,4}}{14 z_h^{\,4}} + \dots\right), \tag{33}$$

respectively and ignoring the higher-order terms beyond the second power, the action to be subtracted (S_0) becomes

$$S_0 \overset{z_m \ll z_h}{\simeq} -\frac{iL^- z_m}{\sqrt{2\pi}\,\alpha' z_h^2} \left[1 + \frac{B z_m^2}{6} + \frac{z_m^4}{10 z_h^4} + \cdots \right] \tag{34}$$

Therefore the renormalized action is obtained by subtracting the action (34) for the two disconnected surfaces from (29)

$$S_I \overset{z_m \ll z_h}{\simeq} S - S_0$$

$$= -\frac{L^- z_m}{\sqrt{2\pi}\,\alpha' z_h^2} \left[\left(1 + \frac{(B - 4A) z_m^2}{6} + \frac{z_m^4}{10 z_h^4}\right) - i\left(1 + \frac{B z_m^2}{6} + \frac{z_m^4}{10 z_h^4}\right) \right] \tag{35}$$

Case II: In the long-distance limit ($z_m \gg z_h$) the integral in the action (26) is split into integrations:

$$S = -\frac{L^-}{\sqrt{2\pi}\,\alpha' z_h^2 (1 + A z_m^2)} \left[\int_0^{z_h} dz \, \frac{(1 + \frac{B z^2}{2})(1 + A z^2)}{\sqrt{1 - \frac{z^4}{z_h^4}}} + \int_{z_h}^{z_m} dz \, \frac{(1 + \frac{B z^2}{2})(1 + A z^2)}{\sqrt{1 - \frac{z^4}{z_h^4}}} \right]$$

$$\equiv I + II, \tag{36}$$

where the first integral (I) becomes

$$I = -\frac{L^-}{\sqrt{2\pi}\,\alpha' z_h^2 (1 + A z_m^2)} \left[\int_0^{z_h} dz \, \frac{(1 + \frac{B z^2}{2})(1 + A z^2)}{\sqrt{1 - \frac{z^4}{z_h^4}}} \right]$$

$$\simeq -\frac{L^-}{\sqrt{2\pi}\,\alpha' z_h^2} \left[1.3 z_h + 0.3(B + 2A) z_h^3 + 0.22 A B z_h^5 \right] \tag{37}$$

and the second integral (II) becomes, after neglecting the higher-order terms in powers of $(\frac{z_h}{z_m})$ and keeping up to the second order

$$\mathrm{II} = -\frac{L^-}{\sqrt{2}\pi\alpha' z_h{}^2(1 + Az_m{}^2)} \left[\int_{z_h}^{z_m} dz \frac{(1 + \frac{Bz^2}{2})(1 + Az^2)}{\sqrt{1 - \frac{z^4}{z_h{}^4}}} \right]$$

$$\simeq \frac{-iL^-}{\sqrt{2}\pi\alpha' z_h{}^2}(1.14z_h + 0.5(B + 2A)z_m z_h{}^2 + 0.17ABz_m{}^3 z_h{}^2) \tag{38}$$

Therefore the Nambu–Goto action in this limit becomes

$$S \overset{z_m \gg z_h}{\simeq} -\frac{L^-}{\sqrt{2}\pi\alpha' z_h{}^2} \left[(1.3z_h + 0.3(B + 2A)z_h{}^3 + 0.22ABz_h{}^5) \right.$$

$$\left. + \quad i(1.14z_h + 0.5(B + 2A)z_m z_h{}^2 + 0.17ABz_m{}^3 z_h{}^2) \right] \tag{39}$$

Similarly the action to be subtracted (30) in this limit can be written as

$$S_0 = -\frac{iL^-}{\sqrt{2}\pi\alpha' z_h{}^2} \left[\int_0^{z_h} dz \frac{1 + \frac{B}{2}z^2}{\sqrt{1 - \frac{z^4}{z_h{}^4}}} + \int_{z_h}^{z_m} dz \frac{1 + \frac{B}{2}z^2}{\sqrt{1 - \frac{z^4}{z_h{}^4}}} \right] \tag{40}$$

After integrating and keeping the terms up to the second-order, the action, S_0 for two disconnected surfaces becomes

$$S_0 \overset{z_m \gg z_h}{\simeq} -\frac{L^-}{\sqrt{2}\pi\alpha' z_h{}^2} \left[-1.14z_h - 0.3Bz_h{}^3 + i(1.3z_h + 0.3Bz_h{}^3) \right] \tag{41}$$

Therefore, the renormalized action is given by

$$S_I \overset{z_m \gg z_h}{\simeq} S - S_0$$

$$= \quad -\frac{L^-}{\sqrt{2}\pi\alpha' z_h{}^2} \left[2.44z_h + 0.5Bz_m z_h{}^2 + i(-0.16z_h + 0.5(B + 2A)z_m z_h{}^2) \right] \tag{42}$$

3.2. Jet quenching parameter

We will now obtain the quenching parameter, $\hat{q}$ for which the expectation value of the Wilson loop in the adjoint representation is equal to some specific value, say, C,

$$\langle W_A \rangle = e^{i2S_I} = C \tag{43}$$

In our problem, $\langle W \rangle$ becomes complex-valued, which is a feature previously encountered in [15] as well. Since $\langle W \rangle$ is the S-matrix for a quark dipole-medium scattering, it is allowed to be complex. If we were calculating Q_s we would need the imaginary part of the forward scattering amplitude: since $S = 1 + iT$, then $\Im T = 1 - \Re S = 1 - \Re\langle W \rangle$. This was exactly done in [15]. Therefore we redefined $\hat{q}$ in (3), where L is the separation at which the real part of the Wilson loop is constant (C).

Thus decomposing the renormalized action, S_I into the real and imaginary parts, the real part of the expectation value of Wilson loop is

$$\Re\langle W_A \rangle = \Re \left[e^{i(2\Re S_I + 2i\Im S_I)} \right]$$

$$= e^{-2\Im S_I} [\cos(2\Re S_I)] = C \tag{44}$$

Now we will evaluate the quenching parameter for both long- and short-distance limits, using the actions in the respective limits.

Case I: Short-distance limit ($z_m \ll z_h$)

To write the action as a function of the separation L, we first express z_m in terms of L. For that we rewrite the equation of motion (23) in this limit ($z_m \ll z_h$)

$$z'^2 = -\frac{(A_n z^n)^2 g}{B_n z^n} \tag{45}$$

because $(A_n z^n)^2$ is much less than $(A_n z_m{}^n)^2$. Integrating both sides of the equation of motion (45)

$$\int_0^{z_m} dz \frac{\sqrt{B_n z^n}}{(A_n z^n)\sqrt{g}} = i \int_{-L/2}^0 dx_2 \tag{46}$$

the separation (L) becomes

$$
\begin{aligned}
\frac{iL}{2} &= \int_0^{z_m} dz \frac{(1 + 0.5 B z^2)(1 - A z^2)}{\sqrt{1 - \frac{z^4}{z_h{}^4}}} \\
&= z_m + \frac{z_m^5}{10 z_h^4} + 0.17(B - 2A) z_m^3 \left(1 + \frac{3 z_m^4}{14 z_h^4}\right) \\
&\quad - 0.17 A B z_h^4 z_m \left(-\sqrt{1 - \frac{z_m^4}{z_h^4}} + {}_2F_1\left(\frac{1}{4}, \frac{1}{2}, \frac{5}{4}, \frac{z_m^4}{z_h^4}\right)\right)
\end{aligned} \tag{47}
$$

Inverting the series and ignoring the higher-order terms we can express z_m as a function of L as

$$z_m = \frac{Li}{2}\left(1 + \frac{(B - 2A)}{24} L^2\right) \tag{48}$$

Thus the renormalized action can be expressed in terms of the separation (L) by replacing z_m as a function of L into (35). Ignoring the higher-order terms, the renormalized action is then given by

$$
\begin{aligned}
S_I = -\frac{\sqrt{2} L^-}{2\pi\alpha' z_h^2} \frac{Li}{2}\left(1 + \frac{(B - 2A) L^2}{24}\right)&\left[\left(1 + \frac{(B - 4A) L^2}{24} + \frac{L^4}{160 z_h^4}\right)\right. \\
&\left. - i\left(1 + \frac{B L^2}{24} + \frac{L^4}{160 z_h^4}\right)\right]
\end{aligned} \tag{49}
$$

Now the imaginary and real parts of the renormalized action can be separated, respectively as

$$\Im S_I = -\frac{\sqrt{2} L^-}{2\pi\alpha' z_h^2} \frac{L}{2}\left(1 + \frac{(B - 2A) L^2}{24}\right)\left(1 + \frac{(B - 4A) L^2}{24} + \frac{L^4}{160 z_h^4}\right) \tag{50}$$

and

$$\Re S_I = -\frac{\sqrt{2} L^-}{2\pi\alpha' z_h^2} \frac{L}{2}\left(1 + \frac{(B - 2A) L^2}{24}\right)\left(1 + \frac{B L^2}{24} + \frac{L^4}{160 z_h^4}\right). \tag{51}$$

Thus the gauge–gravity prescription (44) is reduced into

$$C = (1 - 2\Im S_I)\left(1 - 2\left(\Re S_I\right)^2\right)$$

$$= \left[1 + \frac{L^-L}{\sqrt{2}\pi\alpha'z_h{}^2}\left(1 + \frac{(B - 2A)L^2}{24}\right)\left(1 + \frac{(B - 4A)L^2}{24} + \frac{L^4}{160z_h^4}\right)\right]$$

$$\times \left[1 - \frac{L^{-2}L^2}{4\pi^2\alpha'^2z_h^4}\left(1 + \frac{(B - 2A)L^2}{12}\right)\left(1 + \frac{BL^2}{12} + \frac{L^4}{80z_h^4}\right)\right] \tag{52}$$

Let the first and the second term in the square bracket in the above equation (52) be denoted by I and II, respectively

$$\text{I} \equiv \left[1 + \frac{L^-L}{\sqrt{2}\pi\alpha'z_h{}^2}\left(1 + \frac{(B - 4A)L^2}{24} + \frac{L^4}{160z_h^4} - \frac{(B - 2A)L^2}{24}\right)\right]$$

$$= \left[1 + \frac{L^-L}{\sqrt{2}\pi\alpha'z_h{}^2}\left(1 - \frac{AL^2}{12} + \frac{L^4}{160z_h^4}\right)\right] \tag{53}$$

$$\text{II} \equiv \left[1 - \frac{L^{-2}L^2}{4\pi^2\alpha'^2z_h^4}\left(1 + \frac{(B - A)L^2}{6} + \frac{L^4}{80z_h^4}\right)\right] \tag{54}$$

Therefore the product of the terms I and II in (52) yields

$$C = \left[1 - pL - qL^2 - rL^3 - sL^4 - tL^5 - uL^6 + \text{higher order terms}\right], \tag{55}$$

where

$$p \equiv -\frac{L^-}{\pi\sqrt{2}\alpha'z_h{}^2}$$

$$q \equiv \frac{L^{-2}}{4\pi^2\alpha'^2z_h{}^4}$$

$$r \equiv \frac{AL^-}{12\pi\sqrt{2}\alpha'z_h{}^2} \tag{56}$$

By inverting the equation and ignoring the higher-order terms, the separation (L) is given by

$$L = \frac{1 - C}{p} - \frac{q(1 - C)^2}{p^2} + \frac{(1 - C)^3\left(2q^2 - pr\right)}{p^5} \tag{57}$$

Therefore the quenching parameter, $\hat{q}$ is obtained from the definition (3):

$$\hat{q} = \frac{1}{L^-L^2}$$

$$= \frac{L^-}{2\pi^2\alpha'^2z_h^4(1 - C)^2}\left[1 - \frac{L^-(1 - C)}{\sqrt{2}\pi\alpha'z_h^2} - (1 - C)^2\left(1 + \frac{A\pi^2\alpha'^2z_h^4}{3L^{-2}}\right)\right], \tag{58}$$

which finally results into for $C = \frac{1}{2}$,

$$\hat{q} = \frac{2L^-}{\pi^2\alpha'^2z_h^4}\left[\frac{3}{4} - \frac{L^-}{2\sqrt{2}\pi\alpha z_h^2} - \frac{A\pi^2\alpha'^2z_h^4}{12L^{-2}}\right] \tag{59}$$

Case II: In the long-distance limit ($z_m \gg z_h$), let us first express the separation (L) as a function of z_m. Therefore, we split up the limits of integration to the equation of motion (23) and then integrate it to yield L as a function of z_m:

$$
\begin{aligned}
\frac{iL}{2} &= \int_0^{z_m} dz \frac{(1+0.5Bz^2)(1-Az^2)}{\sqrt{1-\frac{z^4}{z_h^4}}} \\
&= \int_0^{z_h} dz \frac{(1+0.5Bz^2)(1-Az^2)}{\sqrt{1-\frac{z^4}{z_h^4}}} + \int_{z_h}^{z_m} dz \frac{(1+0.5Bz^2)(1-Az^2)}{\sqrt{1-\frac{z^4}{z_h^4}}} \\
&= 1.3z_h + 0.15(B-2A)z_h^3 - 0.22ABz_h^5 \\
&\quad + i\left[1.14z_h + 0.5(B-2A)z_m - 0.17ABz_m^3\right]
\end{aligned}
\tag{60}
$$

Inverting the series and ignoring the higher-order terms we express z_m in terms of L as

$$
z_m = \frac{L - 2.28z_h + i2.6z_h}{(B-2A)z_h^2}
\tag{61}
$$

Now the (renormalized) action (42) in this limit can be expressed as a function of L:

$$
\begin{aligned}
S_I &= -\frac{L^-}{\sqrt{2\pi}\alpha' z_h^2}\left[2.44z_h + 0.5Bz_h^2\left(\frac{L-2.28z_h+i2.6z_h}{(B-2A)z_h^2}\right) - 0.16iz_h\right. \\
&\quad \left.+ i0.5(B+2A)z_h^2\left(\frac{L-2.28z_h+i2.6z_h}{(B-2A)z_h^2}\right)\right].
\end{aligned}
\tag{62}
$$

Ignoring the higher-order terms, we get the action as a function of L,

$$
S_I \overset{z_m \gg z_h}{=} -\frac{L^-}{\sqrt{2\pi}\alpha' z_h^2}\left[\frac{0.5BL}{(B-2A)} + i\frac{0.5(B+2A)L}{(B-2A)}\right]
\tag{63}
$$

Now the real and the imaginary parts of renormalized action can be separated, respectively as

$$
\Re S_I = -\frac{BL^- L}{2\sqrt{2\pi}(B-2A)\alpha' z_h^2}
\tag{64}
$$

$$
\Im S_I = -\frac{(B+2A)L^- L}{2\sqrt{2\pi}(B-2A)\alpha' z_h^2}
\tag{65}
$$

Thus the gauge–gravity correspondence (44) in this limit is translated into:

$$
C = e^{\left[\frac{(B+2A)L^- L}{\sqrt{2\pi}\alpha'(B-2A)z_h^2}\right]} \cos\left[\frac{BL^- L}{\sqrt{2\pi}\alpha'(B-2A)z_h^2}\right]
\tag{66}
$$

Defining

$$
a \equiv \frac{(B+2A)L^-}{\sqrt{2\pi}\alpha'(B-2A)z_h^2}
$$

$$
b \equiv \frac{BL^-}{\sqrt{2\pi}\alpha'(B-2A)z_h^2},
\tag{67}
$$

the above equation (66) has been inverted to give rise the expression for the dipole separation (L) as

$$L = \frac{C-1}{a}\left[1 + \frac{(a^2-b^2)(1-C)}{2a^2} + \frac{(2a^4 - 3a^2b^2 + 3b^4)(1-C)^2}{6a^4}\right.$$
$$\left. + \frac{(6a^6 - 11a^4b^2 + 16a^2b^4 - 15b^6)(1-C)^3}{24a^6} + \cdots\right] \tag{68}$$

Using the numerical values of A and B in [19] ($A = B = 0.124$), the expressions for a and b in Eq. (67) can be rewritten as

$$a = -\frac{3\pi T^2 L^-}{\sqrt{2}\alpha'} \quad \text{and} \quad b = -\frac{\pi T^2 L^-}{\sqrt{2}\alpha'}, \tag{69}$$

and hence the separation becomes

$$L = \frac{\sqrt{2}(1-C)\alpha'}{3\pi T^2 L^-}\left[1 + \frac{4(1-C)}{9} + \frac{23(1-C)^2}{81} + \frac{301(1-C)^3}{1458} + \cdots\right] \tag{70}$$

Thus the quenching parameter $\hat{q}$ is obtained from (3) by substituting the square of the separation (70) for $C = 1/2$

$$\hat{q} = \frac{102T^4}{\alpha'^2}L^-, \tag{71}$$

which is seen to be linear in L^-.

In the study of DIS on a large nucleus in AdS/CFT set up [15], although authors did not calculate $\hat{q}$ directly but if we translate their calculation of the saturation scale, Q_s into our calculation, we could use $\hat{q} = Q_s^2/L^-$. The way Q_s depends on L is, in turn, dependent on which complex branch is chosen. In particular they took $Q_s \sim A^{1/3} \sim L^-$, since $L \sim A^{1/3}$. Hence in both cases $\hat{q}$ comes out to be $\sim L^-$, which appears to be in agreement with our calculation. Since they assumed $L^-(\sim A^{1/3})$ to be large enough, they need not keep the inverse powers of L^-. We even checked with their shock-wave metric [15], where $\hat{q}$ is $\sim L^-$ for large L^- and in agreement with our result in the respective limit.

From other perspective of jet quenching phenomena, by comparing the medium induced energy loss and the p_T-broadening in perturbative QCD with that of the trailing string picture of conformal theory in [30], they also have used $Q_s \sim L^-$, such that $\hat{q} = Q_s^2/L^- \sim L^-$ is in agreement with everything else we obtained so far in our calculations.

4. Results and discussions

We have calculated the quenching parameter, $\hat{q}$ in the holographic set-up of gauge–gravity duality, where the dual gauge theory at finite temperature is more closer to thermal QCD than the $\mathcal{N} = 4$ SYM theory usually used in the literature. Moreover we use a more appropriate definition of $\hat{q}$ compatible with the strong coupling limit of gauge–gravity duality, for which the real part of the Wilson loop expectation value is equal to some specific value (1/2). We have found that in both short- and long-distance limit, $\hat{q}$ depends linearly on L^-. However, in short-distance limit we obtain $1/L^-$ and L^{-2} correction terms.

It is however worth to mention here that it is not clear what one should do with $\hat{q}$ found in a non-perturbative AdS calculation. Since the energy loss calculations are usually done using the perturbative approximation, one cannot simply take a non-perturbative $\hat{q}$ and plug it into the perturbative energy loss expression. But then there is nothing else one can do. This is why people calculated drag force on a heavy quark without looking for $\hat{q}$ [36–38] or the instantaneous energy loss suffered by light quarks in AdS directly [32,33]. It would be interesting to see whether the drag calculation would give the same $\hat{q}$ as the one we have obtained. As far as we remember, the drag calculation in [30] obtained both $\hat{q}$ and Q_s which are in qualitative agreement with what we have gotten.

Acknowledgements

We are grateful to Yuri Kovchegov for his constant and meticulous suggestions. It would never have been possible for us to complete this work without his help. BKP is thankful to the CSIR (Grant No. 03 (1215)/12/EMR-II), Government of India for the financial assistance.

References

[1] R. Baier, Y.L. Dokshitzer, A.H. Mueller, S. Peigne, D. Schiff, Nucl. Phys. B 483 (1997) 291.
[2] B.G. Zakharov, JETP Lett. 65 (1997) 615.
[3] M. Gyulassy, P. Levai, I. Vitev, Nucl. Phys. B 594 (2001) 371.
[4] M.G. Mustafa, M.H. Thoma, Acta Phys. Hung. A 22 (2005) 93.
[5] K.M. Burke, et al., Phys. Rev. C 90 (2014) 014909.
[6] J.M. Maldacena, Phys. Rev. Lett. 80 (1998) 4859.
[7] S.S. Gubser, I.R. Klebanov, A.M. Polyakov, Phys. Lett. B 428 (1998) 105.
[8] E. Witten, Adv. Theor. Math. Phys. 2 (1998) 253;
    E. Witten, Adv. Theor. Math. Phys. 2 (1998) 505.
[9] D.T. Son, A.O. Starinets, Annu. Rev. Nucl. Part. Sci. 57 (2007) 95.
[10] H. Liu, K. Rajagopal, U.A. Wiedemann, Phys. Rev. Lett. 97 (2006) 182301.
[11] A.H. Mueller, Nucl. Phys. B 335 (1990) 115.
[12] J. Casalderrey-Solana, D.C. Gulhan, J.G. Milhano, D. Pablos, K. Rajagopal, Nucl. Phys. A 932 (2014) 421;
    J. Casalderrey-Solana, D.C. Gulhan, J.G. Milhano, D. Pablos, K. Rajagopal, J. High Energy Phys. 1410 (2014) 19;
    J. Casalderrey-Solana, D.C. Gulhan, J.G. Milhano, D. Pablos, K. Rajagopal, J. High Energy Phys. 1509 (2015) 175.
[13] F. D'Eramo, H. Liu, K. Rajagopal, Phys. Rev. D 84 (2011) 065015.
[14] R. Abir, Phys. Lett. B 748 (2015) 467.
[15] J.L. Albacete, Y.V. Kovchegov, A. Taliotis, J. High Energy Phys. 0807 (2008) 074.
[16] Y.V. Kovchegov, E. Levin, Quantum Chromodynamics at High Energy, 1st edition, Cambridge Monographs on Particle Physics, Nuclear Physics and Cosmology, 2012.
[17] I.R. Klebanov, M.J. Strassler, J. High Energy Phys. 0008 (2000) 052.
[18] P. Ouyang, Nucl. Phys. B 699 (2004) 207.
[19] M. Mia, K. Dasgupta, C. Gale, S. Jeon, Phys. Rev. D 82 (2010) 026004.
[20] Binoy Krishna Patra, H. Khanchandani, Phys. Rev. D 91 (2015) 066008.
[21] Binoy Krishna Patra, H. Khanchandani, Lata Thakur, Phys. Rev. D 92 (2015) 085034.
[22] I.R. Klebanov, A.A. Tseytlin, Nucl. Phys. B 578 (2000) 123.
[23] M. Mia, K. Dasgupta, C. Gale, S. Jeon, Nucl. Phys. B 839 (2010) 187.
[24] M. Mia, K. Dasgupta, C. Gale, S. Jeon, Phys. Lett. B 694 (2011) 460.
[25] F. Chen, L. Chen, K. Dasgupta, M. Mia, O. Trottier, Phys. Rev. D 87 (2013) 041901.
[26] T. Sakai, S. Sugimoto, Prog. Theor. Phys. 113 (2005) 843.
[27] K. Dasgupta, C. Gale, M. Mia, M. Richard, O. Trottier, J. High Energy Phys. 1507 (2015) 122.
[28] A. Karch, E. Katz, J. High Energy Phys. 0206 (2002) 043.
[29] T. Sakai, J. Sonnenschein, J. High Energy Phys. 0309 (2003) 047.
[30] F. Dominguez, C. Marquet, A.H. Mueller, B. Wu, B.W. Xiao, Nucl. Phys. A 811 (2008) 197.
[31] N. Armesto, J.D. Edelstein, J. Mas, J. High Energy Phys. 0609 (2006) 039.

[32] A. Ficnar, J. Noronha, M. Gyulassy, J. Phys. Conf. Ser. 446 (2013) 012002.
[33] S. Martins, C.B. Mariotto, J. Phys. Conf. Ser. 706 (2016) 052035.
[34] P. Arnold, P. Szepietowski, D. Vaman, G. Vong, J. High Energy Phys. 1302 (2013) 130.
[35] P. Arnold, P. Szepietowski, D. Vaman, J. High Energy Phys. 1207 (2012) 024.
[36] S.S. Gubser, Phys. Rev. D 74 (2006) 126005.
[37] S.S. Gubser, Phys. Rev. D 76 (2007) 126003.
[38] C.P. Herzog, A. Karch, P. Kovtun, C. Kozcaz, L.G. Yaffe, J. High Energy Phys. 0607 (2006) 013.
[39] S. Li, K.A. Mamo, H. Yee, arXiv:1605.00188 [hep-ph].
[40] Feng-Li Lin, Toshihiro Matsuo, Phys. Lett. B 641 (2006) 45–49.

9

D-branes from pure spinor superstring in $\text{AdS}_5 \times \text{S}^5$ background

Sota Hanazawa, Makoto Sakaguchi *

Department of Physics, Ibaraki University, Mito 310-8512, Japan

Editor: Leonardo Rastelli

Abstract

We examine the surface term for the BRST transformation of the open pure spinor superstring in an $\text{AdS}_5 \times \text{S}^5$ background. We find that the boundary condition to eliminate the surface term leads to a classification of possible configurations of 1/2 supersymmetric D-branes.

1. Introduction

Before the pure spinor formulation of the superstring was initiated by Berkovits [1], there were mainly two superstring formulations, a Ramond–Neveu–Schwarz (RNS) formulation and a Green–Schwarz (GS) formulation. The RNS superstring is described by a superconformal field theory on the two-dimensional world-sheet. So it is difficult to read off the target space geometry coupling to Ramond–Ramond fields, because spacetime supersymmetry emerges only after the GSO projection. On the other hand, for the GS superstring, spacetime supersymmetry is manifest from the outset. The action has world-sheet fermionic gauge symmetry, called κ-symmetry, instead of world-sheet supersymmetry. This makes it difficult to covariantly quantize the GS

---

* Corresponding author.

E-mail addresses: 16nd109n@vc.ibaraki.ac.jp (S. Hanazawa), makoto.sakaguchi.phys@vc.ibaraki.ac.jp (M. Sakaguchi).

superstring even in flat spacetime. In the pure spinor superstring,[1] the κ-symmetry in the GS superstring is replaced with the BRST symmetry. The pure spinor superstring can be quantized in a super-Poincaré covariant manner.

Furthermore, the pure spinor superstring in an $AdS_5 \times S^5$ background with Ramond–Ramond flux [1] is shown to be consistent even at the quantum level [5,6]. The action is composed of $\mathfrak{psu}(2,2|4)$ currents J, and the left- and right-moving ghosts, $(\lambda^\alpha, w_\alpha)$, and $(\hat{\lambda}^{\hat{\alpha}}, \hat{w}_{\hat{\alpha}})$, respectively. The ghosts satisfy the pure spinor constraints $\lambda \gamma^A \lambda = \hat{\lambda} \gamma^A \hat{\lambda} = 0$ $(A = 0, 1, \cdots, 9)$. The pure spinor superstring in the $AdS_5 \times S^5$ background, as well as the GS superstring in the $AdS_5 \times S^5$ background given in [7], is integrable in the sense that infinitely many conserved charges are constructed [8,9] (see also [10]). Nevertheless, for the detailed study of the AdS/CFT correspondence [11], covariant quantization of the superstring should be useful. Though the action of the pure spinor superstring in the $AdS_5 \times S^5$ background is bilinear in the current J, its quantization is still difficult because the J is not (anti-)holomorphic unlike the principal chiral model. We need more effort to quantize the pure spinor superstring covariantly.

The purpose of this paper is to study D-branes in the $AdS_5 \times S^5$ background. A D-brane is a solitonic object in string theory, and is characterized by the Dirichlet boundary condition of an open string. The classical BRST invariance of the open pure spinor superstring in a background implies that the background fields satisfy full non-linear equations of motion for a supersymmetric Born–Infeld action [12]. This is the open string version of [13] in which the classical BRST invariance of the closed pure spinor superstring in a curved background was shown to imply that the background fields satisfy full non-linear equations of motion for the type-II supergravity. For D-branes in the $AdS_5 \times S^5$ background, supersymmetric D-brane configurations are derived in [14] by examining equations of motion for a Dirac–Born–Infeld action for each D-brane embedding ansatz.

In the present paper, we will examine D-branes in the $AdS_5 \times S^5$ background by using the open pure spinor superstring. Especially, we concentrate ourselves on the BRST invariance in the presence of the boundary. Namely, we examine the surface term for the BRST transformation of the open pure spinor superstring in the $AdS_5 \times S^5$ background. We will find that the boundary condition to eliminate the surface term leads to a classification of possible configurations of 1/2 supersymmetric D-branes. This approach is the pure spinor superstring version of [15,18].[2] In [15,18], the boundary condition for the κ-symmetry surface term of the GS superstring in the $AdS_5 \times S^5$ background was shown to lead to a classification of possible configurations of 1/2 supersymmetric D-branes. We find that our result is consistent with those obtained in [14,15,18]. One of the main advantages in our approach is that the derivation is much simpler than the one by using the Dirac–Born–Infeld action and the GS superstring action. This is because the pure spinor superstring action is bilinear in the currents, and because we don't need to deal with the κ-symmetry variation which is highly non-linear.

This paper is organized as follows. In section 2, after introducing the pure spinor superstring in the $AdS_5 \times S^5$ background, we examine the BRST invariance of the open superstring action and extract the surface term. For the BRST invariance to be preserved even in the presence of the boundary, the surface term must be eliminated by a certain boundary condition. In section 3, we fix the boundary conditions by examining a few terms contained in the surface term. The

---

[1] The extended versions of the pure spinor superstring were proposed in [2,3] which introduced new ghosts to relax the pure spinor constraint, and in [4] which introduced doubled spinor degrees of freedom with a compensating local supersymmetry.

[2] A covariant approach to study D-branes in flat spacetime was proposed by Lambert and West in [19].

boundary conditions lead us to a classification of possible 1/2 supersymmetric D-branes in the $AdS_5 \times S^5$ background. In section 4, the boundary conditions fixed above are shown to eliminate all terms contained in the surface term. The last section is devoted to a summary and discussions. Our notation and convention are summarized in Appendix.

2. Pure spinor superstring in $AdS_5 \times S^5$ background

The manifestly covariant action of the pure spinor superstring in the $AdS_5 \times S^5$ background [1,5,16] (see e.g. [17] for reviews) is composed of three parts

$$S = S_\sigma + S_{\text{WZ}} + S_{\text{gh}} , \tag{2.1}$$

with

$$S_\sigma = \frac{1}{2} \langle J_2 \bar{J}_2 + J_1 \bar{J}_3 + J_3 \bar{J}_1 \rangle , \tag{2.2}$$

$$S_{\text{WZ}} = \frac{1}{4} \langle J_3 \bar{J}_1 - J_1 \bar{J}_3 \rangle , \tag{2.3}$$

$$S_{\text{gh}} = \langle w \bar{\partial} \lambda + \hat{w} \partial \hat{\lambda} + N \bar{J}_0 + \hat{N} J_0 - N \hat{N} \rangle , \tag{2.4}$$

where $\langle \cdots \rangle$ stands for $\frac{1}{\pi \alpha'} \int d^2\sigma \, \text{Str}(\cdots)$. Here $J = \frac{1}{2}(J_\tau + J_\sigma)$ and $\bar{J} = \frac{1}{2}(J_\tau - J_\sigma)$, are the left- and right-moving currents, respectively. The J_ξ ($\xi = \tau, \sigma$) stands for the pullback of the Cartan one-form on the worldsheet, $J_\xi = g^{-1}\partial_\xi g$, with $g \in \text{PSU}(2,2|4)/(\text{SO}(1,4) \times \text{SO}(5))$. Furthermore J_ξ is decomposed into four parts, under the $\mathbb{Z}_4$-graded decomposition of $\mathfrak{psu}(2, 2|4)$, namely

$$J_\xi = J_{0\xi} + J_{1\xi} + J_{2\xi} + J_{3\xi},$$

$$J_{0\xi} = \frac{1}{2} J_{0\xi}^{ab} M_{ab} + \frac{1}{2} J_{0\xi}^{a'b'} M_{a'b'} , \quad J_{1\xi} = q_\alpha J_{1\xi}^\alpha , \quad J_{2\xi} = J_{2\xi}^a P_a + J_{2\xi}^{a'} P_{a'} , \quad J_{3\xi} = \hat{q}_{\hat{\alpha}} J_{3\xi}^{\hat{\alpha}} . \tag{2.5}$$

The set of generators $\{M_{ab}, M_{a'b'}, P_a, P_{a'}, q_\alpha, \hat{q}_{\hat{\alpha}}\}$ of $\mathfrak{psu}(2, 2|4)$ satisfies (anti-)commutation relations given in (A.1) and (A.14). The pure spinor variables are defined as

$$\lambda = \lambda^\alpha q_\alpha , \quad \hat{\lambda} = \hat{\lambda}^{\hat{\alpha}} \hat{q}_{\hat{\alpha}} , \quad w = w_\alpha (\tilde{\gamma}^{0\cdots4})^{\alpha\hat{\alpha}} \hat{q}_{\hat{\alpha}} , \quad \hat{w} = \hat{w}_{\hat{\alpha}} (\tilde{\gamma}^{0\cdots4})^{\hat{\alpha}\alpha} q_\alpha , \tag{2.6}$$

where $(\lambda^\alpha, w_\alpha)$ and $(\hat{\lambda}^{\hat{\alpha}}, \hat{w}_{\hat{\alpha}})$ are left- and right-moving ghosts, respectively. In terms of these ghosts the Lorentz currents are given as $N = -\{w, \lambda\}$ and $\hat{N} = -\{\hat{w}, \hat{\lambda}\}$.

2.1. BRST invariance

The BRST transformation of the action is examined below. We will not drop any surface term here. In the next section we will consider the boundary condition for the surface term to be eliminated, and show that the condition leads us to a classification of possible configurations of 1/2 supersymmetric D-branes in the $AdS_5 \times S^5$ background.

First we examine S_σ. The BRST transformation law of currents with a Grassmann odd parameter ε is given as [5]

$$\varepsilon Q(J_1) = \nabla(\varepsilon\lambda) + [J_2, \varepsilon\hat{\lambda}] , \quad \varepsilon Q(\bar{J}_1) = \bar{\nabla}(\varepsilon\lambda) + [\bar{J}_2, \varepsilon\hat{\lambda}] ,$$

$$\varepsilon Q(J_2) = [J_1, \varepsilon\lambda] + [J_3, \varepsilon\hat{\lambda}] , \quad \varepsilon Q(\bar{J}_2) = [\bar{J}_1, \varepsilon\lambda] + [\bar{J}_3, \varepsilon\hat{\lambda}] ,$$

$$\varepsilon Q(J_3) = \nabla(\varepsilon\hat{\lambda}) + [J_2, \varepsilon\lambda] \,, \quad \varepsilon Q(\bar{J}_3) = \bar{\nabla}(\varepsilon\hat{\lambda}) + [\bar{J}_2, \varepsilon\lambda] \,,$$

$$\varepsilon Q(J_0) = [J_3, \varepsilon\lambda] + [J_1, \varepsilon\hat{\lambda}] \,, \quad \varepsilon Q(\bar{J}_0) = [\bar{J}_3, \varepsilon\lambda] + [\bar{J}_1, \varepsilon\hat{\lambda}] \,, \tag{2.7}$$

where $\nabla A \equiv \partial A + [J_0, A]$ and $\bar{\nabla} A \equiv \bar{\partial} A + [\bar{J}_0, A]$. By using (2.7), we obtain

$$\varepsilon Q(S_\sigma) = \frac{1}{2} \left\langle J_1 \bar{\nabla}(\varepsilon\hat{\lambda}) + \nabla(\varepsilon\lambda)\bar{J}_3 + J_3 \bar{\nabla}(\varepsilon\lambda) + \nabla(\varepsilon\hat{\lambda})\bar{J}_1 \right\rangle \,. \tag{2.8}$$

To derive this expression we have used the cyclicity of the Str, for example $\mathrm{Str}(J_2[\bar{J}_1, \varepsilon\lambda]) = \mathrm{Str}(\varepsilon\lambda[J_2, \bar{J}_1])$.

Next we consider S_{WZ}. Using (2.7), one derives

$$\varepsilon Q(S_{\mathrm{WZ}}) = \frac{1}{4} \Big\langle \nabla(\varepsilon\hat{\lambda})\bar{J}_1 + J_3 \bar{\nabla}(\varepsilon\lambda) - \nabla(\varepsilon\lambda)\bar{J}_3 - J_1 \bar{\nabla}(\varepsilon\hat{\lambda})$$

$$+ \varepsilon\lambda([\bar{J}_1, J_2] - [J_1, \bar{J}_2]) + \varepsilon\hat{\lambda}([J_3, \bar{J}_2] - [\bar{J}_3, J_2]) \Big\rangle \,. \tag{2.9}$$

By using Maurer–Cartan equations

$$\nabla \bar{J}_3 - \bar{\nabla} J_3 = [\bar{J}_1, J_2] - [J_1, \bar{J}_2] \,, \quad \nabla \bar{J}_1 - \bar{\nabla} J_1 = [\bar{J}_3, J_2] - [J_3, \bar{J}_2] \,, \tag{2.10}$$

the second line of the right-hand side of (2.9) may be rewritten as

$$\varepsilon Q(S_{\mathrm{WZ}}) = \frac{1}{4} \Big\langle \nabla(\varepsilon\hat{\lambda})\bar{J}_1 + J_3 \bar{\nabla}(\varepsilon\lambda) - \nabla(\varepsilon\lambda)\bar{J}_3 - J_1 \bar{\nabla}(\varepsilon\hat{\lambda})$$

$$+ \varepsilon\lambda(\nabla\bar{J}_3 - \bar{\nabla}J_3) - \varepsilon\hat{\lambda}(\nabla\bar{J}_1 - \bar{\nabla}J_1) \Big\rangle \,. \tag{2.11}$$

Finally we examine S_{gh}. The BRST transformation law of ghosts

$$\varepsilon Q(w) = -J_3\varepsilon \,, \quad \varepsilon Q(\hat{w}) = -\bar{J}_1\varepsilon \,, \quad \varepsilon Q(\lambda) = \varepsilon Q(\hat{\lambda}) = 0 \tag{2.12}$$

implies that

$$\varepsilon Q(N) = [J_3, \varepsilon\lambda] \,, \quad \varepsilon Q(\hat{N}) = [\bar{J}_1, \varepsilon\hat{\lambda}] \,. \tag{2.13}$$

Further noting that

$$\mathrm{Str}(N[\bar{J}_3, \varepsilon\lambda]) = \mathrm{Str}(-\bar{J}_3\varepsilon[\lambda\lambda, w]) = 0 \,, \quad \mathrm{Str}(\hat{N}[J_1, \varepsilon\hat{\lambda}]) = \mathrm{Str}(-J_1\varepsilon[\hat{\lambda}\hat{\lambda}, \hat{w}]) = 0 \,, \tag{2.14}$$

which follow from the pure spinor conditions $\{\lambda, \lambda\} = \{\hat{\lambda}, \hat{\lambda}\} = 0$, we can derive

$$\varepsilon Q(S_{\mathrm{gh}}) = \left\langle -J_3 \bar{\nabla}(\varepsilon\lambda) - \bar{J}_1 \nabla(\varepsilon\hat{\lambda}) \right\rangle \,. \tag{2.15}$$

Gathering all results obtained above together, we find that the BRST transformation of S is

$$\varepsilon Q(S) = \frac{1}{4} \Big\langle \nabla(\varepsilon\lambda\bar{J}_3 - \varepsilon\hat{\lambda}\bar{J}_1) - \bar{\nabla}(\varepsilon\lambda J_3 - \varepsilon\hat{\lambda}J_1) \Big\rangle$$

$$= \frac{1}{4} \Big\langle \partial(\varepsilon\lambda\bar{J}_3 - \varepsilon\hat{\lambda}\bar{J}_1) - \bar{\partial}(\varepsilon\lambda J_3 - \varepsilon\hat{\lambda}J_1) \Big\rangle \,. \tag{2.16}$$

In the second equality, we have used the fact that $\mathrm{Str}([J_0, \varepsilon\lambda\bar{J}_3]) = 0$ and the similar relations.

We can conclude that S is BRST invariant as long as this surface term vanishes. For a closed string, the surface term always vanishes. For an open string, however, appropriate boundary conditions are required. In the next section we will examine these boundary conditions.

3. Boundary BRST invariance to D-brane configurations

In this section we will examine boundary conditions for the surface term to be eliminated, and show that they lead us to a classification of possible 1/2 supersymmetric D-brane configurations in the $AdS_5 \times S^5$ background.

The surface term (2.16) turns to[3]

$$
\varepsilon Q(S) = \frac{1}{4\pi\alpha'} \int d^2\sigma \, \text{Str}\left[\partial_\sigma (\varepsilon\lambda J_3 - \varepsilon\hat{\lambda}J_1)_\tau + \partial_\tau (\varepsilon\lambda J_3 - \varepsilon\hat{\lambda}J_1)_\sigma \right]
$$
$$
= \frac{1}{4\pi\alpha'} \int d\tau \, \text{Str}\left[(\varepsilon\lambda J_3 - \varepsilon\hat{\lambda}J_1)_\tau \right]\bigg|_{\sigma=\sigma_*}
\tag{3.1}
$$

where we have assumed that the surface term at $\tau = \pm\infty$ vanishes as usual. The open string boundaries are at $\sigma = \sigma_*$ with $\sigma_* = 0, \pi$. As seen in (A.21), $J_{1\tau}$ and $J_{3\tau}$ correspond to $q_\alpha L_\tau^{1\alpha}$ and $\hat{q}_{\hat{\alpha}} L_\tau^{2\hat{\alpha}}$, respectively, where the corresponding Cartan one-form L^I is given in (A.18). It follows that $J_{1\tau}$ and $J_{3\tau}$ are polynomials in θ^I. We should note that the surface terms do not cancel out each other. So we may examine each surface term separately without loss of generality. Our strategy is as follows. First we examine a few terms contained in J_1 and J_3, and fix the boundary condition. Next we will show that the boundary condition we have fixed would eliminate all terms in (3.1).

First, we shall fix the boundary condition by examining the following three terms contained in $J_{1\tau}$ and those in $J_{3\tau}$

$$
J_{1\tau} = q_\alpha L_\tau^{1\alpha} = q_\alpha \left(\partial_\tau\theta + \frac{1}{2}e_\tau^A \tilde{\gamma}^{0\cdots4}\gamma_A\hat{\theta} - \frac{i}{6}\tilde{\gamma}^{0\cdots4}\gamma_A\hat{\theta} (\theta\gamma^A\partial_\tau\theta + \hat{\theta}\gamma^A\partial_\tau\hat{\theta}) \right) + \cdots ,
\tag{3.2}
$$

$$
J_{3\tau} = \hat{q}_{\hat{\alpha}} L_\tau^{2\hat{\alpha}} = \hat{q}_{\hat{\alpha}} \left(\partial_\tau\hat{\theta} - \frac{1}{2}e_\tau^A \tilde{\gamma}^{0\cdots4}\gamma_A\theta + \frac{i}{6}\tilde{\gamma}^{0\cdots4}\gamma_A\theta (\theta\gamma^A\partial_\tau\theta + \hat{\theta}\gamma^A\partial_\tau\hat{\theta}) \right) + \cdots .
\tag{3.3}
$$

The first two terms in the most right-hand sides in the above equations come from $D\theta$ defined in (A.19), while the last one is contained in $\frac{m^2}{3!}D\theta$ where m^2 is defined in (A.20). After fixing the boundary condition we shall show that the boundary condition eliminates all surface terms in section 4.

Substituting (3.2) and (3.3) into (3.1), we obtain the corresponding surface terms, which we shall denote as $\delta_0 S$,

$$
\delta_0 S = \frac{1}{4\pi\alpha'} \int d\tau \, \varepsilon \left[\lambda\gamma^{0\cdots4} \left(\partial_\tau\hat{\theta} - \frac{1}{2}e_\tau^A \tilde{\gamma}^{0\cdots4}\gamma_A\theta + \frac{i}{6}\tilde{\gamma}^{0\cdots4}\gamma_A\theta (\theta\gamma^A\partial_\tau\theta + \hat{\theta}\gamma^A\partial_\tau\hat{\theta}) \right) \right.
$$
$$
\left. + \hat{\lambda}\gamma^{0\cdots4} \left(\partial_\tau\theta + \frac{1}{2}e_\tau^A \tilde{\gamma}^{0\cdots4}\gamma_A\hat{\theta} - \frac{i}{6}\tilde{\gamma}^{0\cdots4}\gamma_A\hat{\theta} (\theta\gamma^A\partial_\tau\theta + \hat{\theta}\gamma^A\partial_\tau\hat{\theta}) \right) \right]\bigg|_{\sigma=\sigma_*}
\tag{3.4}
$$

where we have used $\text{Str}(q_\alpha\hat{q}_{\hat{\beta}}) = (\gamma^{0\cdots4})_{\alpha\hat{\beta}}$ and $\text{Str}(\hat{q}_{\hat{\alpha}}q_\beta) = -(\gamma^{0\cdots4})_{\hat{\alpha}\beta}$. For our purpose, it is convenient to rewrite (3.4) in a 32-component notation. Defining

---

[3] $\partial\bar{J} - \bar{\partial}J = \partial_\tau J_\sigma + \partial_\sigma J_\tau$, as $\partial = \partial_\tau + \partial_\sigma$ and $\bar{\partial} = \partial_\tau - \partial_\sigma$.

$$\lambda^I \equiv \begin{pmatrix} \lambda^I \\ 0 \end{pmatrix} , \quad \theta^I \equiv \begin{pmatrix} \theta^I \\ 0 \end{pmatrix} , \tag{3.5}$$

where $(\lambda^1, \lambda^2) = (\lambda, \hat{\lambda})$ and $(\theta^1, \theta^2) = (\theta, \hat{\theta})$, and using the 32 component spinor notation given in Appendix A, we find that (3.4) may be simplified to

$$\delta_0 S = \frac{1}{4\pi\alpha'} \int d\tau\, \varepsilon \left[\bar{\lambda} I \sigma_1 \partial_\tau \theta + \frac{1}{2} e_\tau^A \bar{\lambda} \Gamma_A \sigma_3 \theta - \frac{i}{6} \bar{\lambda} \Gamma_A \sigma_3 \theta\, \bar{\theta} \Gamma^A \partial_\tau \theta \right] \Bigg|_{\sigma=\sigma_*} . \tag{3.6}$$

To show this, the following relations are useful

$$\bar{\lambda}^I I(\sigma_1)_{IJ} \partial \theta^J = \lambda^1 \gamma^{0\cdots4} \partial \theta^2 + \lambda^2 \gamma^{0\cdots4} \partial \theta^1 , \tag{3.7}$$

$$\bar{\lambda}^I \Gamma_A (\sigma_3)_{IJ} \theta^J = \lambda^1 \gamma_A \theta^1 - \lambda^2 \gamma_A \theta^2 , \tag{3.8}$$

$$\bar{\theta} \Gamma^A \partial_\tau \theta = \theta^1 \gamma^A \partial_\tau \theta^1 + \theta^2 \gamma^A \partial_\tau \theta^2 . \tag{3.9}$$

For bosonic coordinates, we impose the boundary conditions as follows: Neumann boundary condition $e_\sigma^{\bar{A}} = \partial_\sigma x^\mu e_\mu^{\bar{A}} = 0$ for $\bar{A} = \bar{A}_0, \cdots, \bar{A}_p$, or Dirichlet boundary condition $e_\tau^{\underline{A}} = \partial_\tau x^\mu e_\mu^{\underline{A}} = 0$ for $\underline{A} = \underline{A}_{p+1}, \cdots, \underline{A}_9$. This boundary condition eliminates the surface term $\delta x^\mu e_\mu^A e_\sigma^B \eta_{AB}$ at $\sigma = \sigma_*$. In order to delete $\delta_0 S$, we must impose boundary conditions on θ and λ. The boundary condition we shall impose on θ is

$$\theta = M\theta , \quad M = s\Gamma^{\bar{A}_0 \cdots \bar{A}_p} \otimes \rho \tag{3.10}$$

where $\rho_{IJ} \in \{1, \sigma_1, i\sigma_2, \sigma_3\}$ is a two-by-two matrix acting on θ^I ($I = 1, 2$). For the reality of θ, we choose s as $s = \pm 1$. The boundary condition leads to 1/2 supersymmetric D-branes. As θ^I are a pair of Majorana–Weyl spinors satisfying $\theta = \Gamma_{11}\theta$, we find $p = $ odd for consistency, $[M, \Gamma_{11}] = 0$. For the boundary condition on λ, we must impose $\lambda = M\lambda$. This is necessary for the BRST transformation to be non-trivial even at the boundary. In fact, the BRST transformation of J_1 in (2.7), namely $\varepsilon Q(\partial \theta^\alpha) = \varepsilon \partial \lambda^\alpha + \cdots$, is consistent if we impose the same boundary condition on θ and λ.

We shall examine each term contained in (3.6) below so that we will fix p and ρ. Let us begin with examining the second term in the right hand side of (3.6). Because $e_\tau^{\bar{A}} = 0$,

$$\bar{\lambda} \Gamma_{\bar{A}} \sigma_3 \theta = 0 \tag{3.11}$$

must be satisfied. It follows from (3.11) that in order to delete the third term in the right hand side of (3.6),

$$\bar{\theta} \Gamma^{\underline{A}} \partial_\tau \theta = 0 \tag{3.12}$$

must be satisfied. We examine (3.12) first. Noting that $CM = \mp \alpha M^T C$ with $\rho = \alpha \rho^T$ for $p = \{ {}^1_3$ mod 4, respectively, we derive

$$\bar{\theta} \Gamma^{\underline{A}} \partial_\tau \theta = \bar{\theta} \Gamma^{\underline{A}} M \partial_\tau \theta = \theta^T C M \Gamma^{\underline{A}} \partial_\tau \theta = \mp \alpha \overline{M\theta} \Gamma^{\underline{A}} \partial_\tau \theta = \mp \alpha \bar{\theta} \Gamma^{\underline{A}} \partial_\tau \theta , \tag{3.13}$$

so that α is fixed as $\alpha = \pm 1$ for (3.12). It means that $\rho = \pm \rho^T$ for $p = \{ {}^1_3$ mod 4.

Now, we return to (3.11). We derive, defining β by $\sigma_3 \rho = \beta \rho \sigma_3$,

$$\bar{\lambda} \Gamma_{\bar{A}} \sigma_3 \theta = \bar{\lambda} \Gamma_{\bar{A}} \sigma_3 M\theta = \beta \bar{\lambda} \Gamma_{\bar{A}} M \sigma_3 \theta = -\beta \bar{\lambda} M \Gamma_{\bar{A}} \sigma_3 \theta = \beta \overline{M\lambda} \Gamma_{\bar{A}} \sigma_3 \theta = \beta \bar{\lambda} \Gamma_{\bar{A}} \sigma_3 \theta . \tag{3.14}$$

Table 1
1/2 supersymmetric D-brane configurations in AdS$_5 \times$ S^5.

p	-1	1	3	5	7	9
D-brane	–	$(2, 0)$	$(1, 3), (3, 1)$	$(2, 4), (4, 2)$	$(3, 5), (5, 3)$	–

It implies that β is fixed as $\beta = -1$ for (3.11). This means that $\rho = \sigma_1$ or $i\sigma_2$. Combining this with the result obtained from (3.11), we can conclude that $\rho = \sigma_1$ for $p = 1$ mod 4, and that $\rho = i\sigma_2$ for $p = 3$ mod 4. For consistency we require that $M^2 = 1$. This implies that the time direction 0 is a Neumann direction since $s^2 = 1$. The results so far coincide with the boundary condition for 1/2 supersymmetric D-branes in flat spacetime.

Finally, we examine the first term in the right hand side of (3.6) which leads to the additional condition specific to the AdS$_5 \times$ S^5 background. One may show that

$$\bar{\lambda} I \sigma_1 \partial_\tau \theta = \bar{\lambda} I \sigma_1 M \partial_\tau \theta = \pm \bar{\lambda} I M \sigma_1 \partial_\tau \theta = \pm (-1)^n \bar{\lambda} M I \sigma_1 \partial_\tau \theta = \mp (-1)^n \bar{\lambda} I \sigma_1 \partial_\tau \theta .$$

(3.15)

In the second equality we have used $\sigma_1 \rho = \pm \rho \sigma_1$ for $p = \{ {1 \atop 3}$ mod 4. The third equality follows from $IM = (-1)^n MI$ where n is the number of Neumann directions contained in AdS$_5$ spanned by $\{0, 1, 2, 3, 4\}$. As a result, for $\bar{\lambda} I \sigma_1 \partial_\tau \theta = 0$ we must impose $n = $ even for $p = 1$ mod 4 and $n = $ odd for $p = 3$ mod 4. We summarize the result in the Table 1 where (n, n') means a D-brane of which world-volume is extended along AdS$_n \times$ S$^{n'}$. This gives a classification of 1/2 supersymmetric D-brane configurations in the AdS$_5 \times$ S^5 background. This result is consistent with the ones obtained by using the κ-symmetry variation of the Green–Schwarz superstring [15, 18] and by examining Dp-brane field equations [14].

Summarizing the results, we find that the surface term (3.6) vanishes if we impose the boundary conditions

$$\theta = M\theta , \quad \lambda = M\lambda ,$$

(3.16)

with

$$M = \left\{ \begin{array}{l} s\Gamma^{\bar{A}_0 \cdots \bar{A}_p} \otimes \sigma_1 \\ s\Gamma^{\bar{A}_0 \cdots \bar{A}_p} \otimes i\sigma_2 \end{array} \right. , \quad n = \left\{ \begin{array}{l} \text{even} \\ \text{odd} \end{array} \right. \quad \text{for} \quad p = \left\{ \begin{array}{l} 1 \\ 3 \end{array} \right. \text{mod 4} , \quad s = \pm 1$$

(3.17)

respectively.

4. Proof of validity in eliminating all surface terms

In the previous section, we have derived the boundary conditions (3.16) with (3.17) which eliminate a certain terms (3.6) contained in the surface term (3.1). In this section, we shall show that the boundary conditions fixed in the previous section may eliminate all terms contained in the surface term (3.1).

For this purpose, it is convenient to rewrite the surface term (3.1) in the 32-component notation as

$$\varepsilon Q(S) = -\frac{\varepsilon}{4\pi \alpha'} \int d\tau \left[\bar{\lambda} I \sigma_1 L_\tau \right] \big|_{\sigma = \sigma_*}$$

(4.1)

where the corresponding Cartan one-form L is given in (A.6). We will show that the validity of the boundary conditions for $p = 1$ mod 4 and for $p = 3$ mod 4, in turn.

4.1. $p = 1 \bmod 4$

We shall show that the surface term (4.1) is eliminated by the boundary conditions for $p = 1 \bmod 4$: $\theta = P_+\theta$ and $\lambda = P_+\lambda$ where $P_+ = \frac{1}{2}(1 + M)$ with $M = s\Gamma^{\bar{A}_0\cdots\bar{A}_p} \otimes \sigma_1$ and $n = $ even. First we examine $D_\tau\theta$ defined in (A.7). It follows from $\theta = P_+\theta$ that

$$D_\tau\theta = P_+ D_\tau\theta \ . \tag{4.2}$$

In order to derive this relation, we have used

$$\partial_\tau\theta = P_+\partial_\tau\theta \ , \tag{4.3}$$

$$\frac{1}{2}e_\tau^{\bar{A}}I\Gamma_{\bar{A}}\epsilon\theta = \frac{1}{2}e_\tau^{\bar{A}}I\Gamma_{\bar{A}}\epsilon P_+\theta = \frac{1}{2}e_\tau^{\bar{A}}I\Gamma_{\bar{A}}P_-\epsilon\theta = \frac{1}{2}e_\tau^{\bar{A}}IP_+\Gamma_{\bar{A}}\epsilon\theta = P_+\frac{1}{2}e_\tau^{\bar{A}}I\Gamma_{\bar{A}}\epsilon\theta \ , \tag{4.4}$$

$$\frac{1}{4}w_\tau^{AB}\Gamma_{AB}\theta = \frac{1}{4}w_\tau^{\bar{A}\bar{B}}\Gamma_{\bar{A}\bar{B}}P_+\theta + \frac{1}{4}w_\tau^{\underline{AB}}\Gamma_{\underline{AB}}P_+\theta = P_+\left(\frac{1}{4}w_\tau^{\bar{A}\bar{B}}\Gamma_{\bar{A}\bar{B}}\theta + \frac{1}{4}w_\tau^{\underline{AB}}\Gamma_{\underline{AB}}\theta\right). \tag{4.5}$$

Here we have assumed that $w^{\bar{A}\underline{B}} = 0$. This is because a D-brane breaks rotational invariance in the plane spanned by one of Neumann directions and one of Dirichlet directions.

Next we will examine $\mathcal{M}^2 P_+$ where $\mathcal{M}^2$ is defined in (A.8). Noting that $CP_\pm = P_\mp^T C$ one derives

$$-iI\Gamma_A\epsilon\theta\,\bar{\theta}\Gamma^A P_+ = -iI\Gamma_{\bar{A}}\epsilon\theta\,\bar{\theta}P_-\Gamma^{\bar{A}}P_+ = -iP_+IP_+\Gamma_{\bar{A}}P_-\epsilon P_+\theta\,\bar{\theta}P_-\Gamma^{\bar{A}}P_+ \ , \tag{4.6}$$

$$\frac{i}{2}\Gamma_{AB}\theta\,\bar{\theta}\hat{\Gamma}^{AB}\epsilon P_+ = \frac{i}{2}\Gamma_{AB}\theta\,\bar{\theta}\hat{\Gamma}^{AB}P_-\epsilon P_+$$

$$= \frac{i}{2}\Gamma_{\bar{A}\bar{B}}\theta\,\bar{\theta}P_-\hat{\Gamma}^{\bar{A}\bar{B}}P_-\epsilon P_+ + \frac{i}{2}\Gamma_{\underline{AB}}\theta\,\bar{\theta}P_-\hat{\Gamma}^{\underline{AB}}P_-\epsilon P_+$$

$$= P_+\frac{i}{2}\Gamma_{\bar{A}\bar{B}}\theta\,\bar{\theta}P_-\hat{\Gamma}^{\bar{A}\bar{B}}P_-\epsilon P_+ + P_+\frac{i}{2}\Gamma_{\underline{AB}}\theta\,\bar{\theta}P_-\hat{\Gamma}^{\underline{AB}}P_-\epsilon P_+ \ . \tag{4.7}$$

It follows that

$$\mathcal{M}^2 P_+ = P_+\mathcal{M}^2 P_+ \ . \tag{4.8}$$

Gathering the results (4.2) and (4.8) together we obtain $\boldsymbol{L}_\tau = P_+\boldsymbol{L}_\tau$. Using this we may derive

$$\bar{\lambda}I\sigma_1\boldsymbol{L}_\tau = \bar{\lambda}I\sigma_1 P_+\boldsymbol{L}_\tau = \bar{\lambda}P_+IP_+\sigma_1 P_+\boldsymbol{L}_\tau = \overline{P_-\lambda}P_+IP_+\sigma_1 P_+\boldsymbol{L}_\tau = 0 \ . \tag{4.9}$$

This shows that the boundary conditions for $p = 1 \bmod 4$ eliminate the surface term (4.1).

4.2. $p = 3 \bmod 4$

We show that the surface term (4.1) is eliminated by the boundary conditions for $p = 3 \bmod 4$: $\theta = P_+\theta$ and $\lambda = P_+\lambda$ where $P_+ = \frac{1}{2}(1 + M)$ with $M = \Gamma^{\bar{A}_0\cdots\bar{A}_p} \otimes i\sigma_2$ and $n = $ odd. First we examine $D_\tau\theta$. The calculation similar to the one in the case with $p = 1 \bmod 4$, except for

$$\frac{1}{2}e_\tau^{\bar{A}}I\Gamma_{\bar{A}}\epsilon\theta = \frac{1}{2}e_\tau^{\bar{A}}I\Gamma_{\bar{A}}\epsilon P_+\theta = \frac{1}{2}e_\tau^{\bar{A}}I\Gamma_{\bar{A}}P_+\epsilon\theta = \frac{1}{2}e_\tau^{\bar{A}}IP_-\Gamma_{\bar{A}}\epsilon\theta = P_+\frac{1}{2}e_\tau^{\bar{A}}I\Gamma_{\bar{A}}\epsilon\theta \ , \tag{4.10}$$

leads us to

$$D_\tau \theta = P_+ D_\tau \theta \ . \tag{4.11}$$

Next we will examine $\mathcal{M}^2 P_+$. Noting that $CP_\pm = P_\mp^T C$ one derives

$$-i\Gamma_A \epsilon \theta \, \bar{\theta} \Gamma^A P_+ = -i P_+ I P_- \Gamma_{\bar{A}} P_+ \epsilon P_+ \theta \, \bar{\theta} P_- \Gamma^{\bar{A}} P_+ \ , \tag{4.12}$$

$$\frac{i}{2}\Gamma_{AB} \theta \, \bar{\theta} \hat{\Gamma}^{AB} \epsilon P_+ = P_+ \frac{i}{2} \Gamma_{\bar{A}\bar{B}} \theta \, \bar{\theta} P_- \hat{\Gamma}^{\bar{A}\bar{B}} P_+ \epsilon P_+ + P_+ \frac{i}{2} \Gamma_{\underline{AB}} \theta \, \bar{\theta} P_- \hat{\Gamma}^{\underline{AB}} P_+ \epsilon P_+ \ , \tag{4.13}$$

so that

$$\mathcal{M}^2 P_+ = P_+ \mathcal{M}^2 P_+ \ . \tag{4.14}$$

Gathering the results (4.11) and (4.14) together we obtain $L_\tau = P_+ L_\tau$. Using this we may derive

$$\bar{\lambda} I \sigma_1 L_\tau = \bar{\lambda} I \sigma_1 P_+ L_\tau = \bar{\lambda} P_+ I P_- \sigma_1 P_+ L_\tau = \overline{P_- \lambda} P_+ I P_- \sigma_1 P_+ L_\tau = 0 \ . \tag{4.15}$$

It implies that the boundary condition for $p = 3 \bmod 4$ eliminates the surface term (4.1).

Summarizing we have shown that the boundary condition (3.16) with (3.17) eliminates the surface term (3.1) of the BRST transformation $\varepsilon Q(S)$.

5. Summary and discussions

We examined the BRST invariance of the open pure spinor superstring action in the $AdS_5 \times S^5$ background. In order for the BRST symmetry to be preserved even in the presence of the boundary, the surface term of the BRST transformation must be eliminated by appropriate boundary conditions. We determined such boundary conditions and found that the boundary conditions lead to a classification of possible configurations of 1/2 supersymmetric D-branes in the $AdS_5 \times S^5$ background. Our result is summarized in the Table 1. This is consistent with the results obtained by the other approaches [14,15,18].

We have used an exponential parametrization of the coset representative g throughout this paper. In [21] a GS superstring action in the $AdS_5 \times S^5$ background was derived based on an alternate version of the coset superspace construction in terms of GL(4|4). The pure spinor superstring action in this coset superspace construction was given in [22]. This action is expected to make it more transparent to examine the surface term for the BRST transformation of the action and to derive possible D-brane configurations in the $AdS_5 \times S^5$ background.

The method used in this paper can be applied easily to a superstring in the other background, for example the superstring in the type IIB pp-wave background [23,24]. The result will be consistent with the one obtained by the boundary κ-invariance of the open GS superstring [15, 20] and by examining equations of motion for a D-brane [14].

It is also known that in the presence of a constant flux, the boundary condition to ensure the κ-invariance of the GS superstring action leads to possible (non-commutative) D-branes [25]. Furthermore the boundary condition to ensure the κ-invariance of the supermembrane action leads to the self-duality condition for the three-form flux on the M5-brane world-volume [26]. The same result is expected to be obtained by using the pure spinor supermembrane action [27]. We hope to report this issue in another place [28].

Finally let us comment on a characterization of the Wess–Zumino (WZ) action. The WZ action is necessary for the κ-symmetry of the action, and then halves fermionic degrees of freedom on the world-volume so as to match bosonic and fermionic degrees of freedom. It is shown that the WZ term of a (D)p-brane in flat spacetime is characterized as a non-trivial element of the Chevalley–Eilenberg (CE) cohomology in [29] for p-branes, and in [30,31] for Dp-branes. In [32], Dp-brane actions in the extended pure spinor formalism [2] are characterized as a non-trivial element of the BRST cohomology of the extended BRST symmetry. It is interesting for us to extend this analysis to the Dp-brane action in the AdS$_5 \times$ S^5 background.[4]

Acknowledgements

The authors would like to thank Takanori Fujiwara, Yoshifumi Hyakutake and Kentaroh Yoshida for useful comments.

Appendix A. Notation and convention

The super-isometry algebra of the AdS$_5 \times$ S^5 background is $\mathfrak{psu}(2,2|4)$ of which (anti-)commutation relations are[5]

$$[P_a, P_b] = M_{ab}, \quad [P_{a'}, P_{b'}] = -M_{a'b'},$$

$$[M_{AB}, P_C] = \eta_{BC} P_A - \eta_{AC} P_B, \quad [M_{AB}, M_{CD}] = \eta_{BC} M_{AD} + 3\text{-terms}, \tag{A.1}$$

and

$$[Q_I, M_{AB}] = -\frac{1}{2} Q_I \Gamma_{AB}, \quad [Q_I, P_A] = \frac{1}{2}\epsilon_{IJ} Q_J I \Gamma_A,$$

$$\{Q_I, Q_J\} = -2iC\Gamma^A P_A \delta_{IJ} h_+ + \epsilon_{IJ}\left(iC\Gamma^{ab} I M_{ab} - iC\Gamma^{a'b'} I M_{a'b'}\right) h_+, \tag{A.2}$$

where Γ^A ($A = 0, 1, \cdots, 9$) are 32×32 gamma matrices, and $Q_I = Q_I h_+$ ($I = 1, 2$) are a pair of Majorana–Weyl spinors. We introduced $\epsilon_{IJ} = \begin{pmatrix} 0 & 1 \\ -1 & 0 \end{pmatrix}$. We have defined $I = \Gamma^{01234}$ and $h_\pm = \frac{1}{2}(1 \pm \Gamma_{11})$ with $\Gamma_{11} \equiv \Gamma^{012\cdots 9}$. The charge conjugation matrix C satisfies $C\Gamma_A = -\Gamma_A^T C$. The AdS$_5$ isometry is generated by P_a an M_{ab} ($a, b = 0, 1, \cdots, 4$), while the S^5 isometry is by $P_{a'}$ and $M_{a'b'}$ ($a', b' = 5, 6, \cdots, 9$).

The left-invariant Cartan one-form L is defined by

$$L = g^{-1}dg = L^A P_A + \frac{1}{2}L^{AB} M_{AB} + Q_I L^I \tag{A.3}$$

where $g \in \text{PSU}(2,2|4)/(\text{SO}(1,4) \times \text{SO}(5))$. Parametrizing g by $g = g_x(x)e^{Q_I \theta^I}$ and introducing the vielbein $e^A(x)$ and spin-connection $w^{AB}(x)$ by $g_x^{-1}dg_x = e^A P_A + \frac{1}{2}w^{AB} M_{AB}$, we obtain

$$L^A = e^A - 2i\bar{\theta}\Gamma^A\left(\frac{1}{2} + \frac{\mathcal{M}^2}{4!} + \frac{\mathcal{M}^4}{6!} + \cdots\right)D\theta, \tag{A.4}$$

$$L^{AB} = w^{AB} + 2i\bar{\theta}\hat{\Gamma}^{AB}\epsilon\left(\frac{1}{2} + \frac{\mathcal{M}^2}{4!} + \frac{\mathcal{M}^4}{6!} + \cdots\right)D\theta, \tag{A.5}$$

$$L^I = \left(1 + \frac{\mathcal{M}^2}{3!} + \frac{\mathcal{M}^4}{5!} + \cdots\right)D\theta, \tag{A.6}$$

---

[4] On a CE cohomology classification of Dp-brane actions in the AdS$_5 \times$ S^5 background see e.g. [33,34].
[5] We follow the notation given in [35] except for Γ_{11}.

where we have defined $\bar{\theta} \equiv \theta^T C$, $\hat{\Gamma}^{AB} \equiv (\Gamma^{ab}I, -\Gamma^{a'b'}I)$ and

$$D\theta = d\theta + \frac{1}{2}e^A I\Gamma_A \epsilon \theta + \frac{1}{4}w^{AB}\Gamma_{AB}\theta \, , \tag{A.7}$$

$$\mathcal{M}^2 = -i I\Gamma_A \epsilon \theta \cdot \bar{\theta}\Gamma^A + \frac{i}{2}\Gamma_{AB}\theta \cdot \bar{\theta}\hat{\Gamma}^{AB}\epsilon \, . \tag{A.8}$$

A.1. 16-component spinor notation

Let us rewrite (A.2) in terms of 16×16 gamma-matrices γ^A. We decompose Γ^A as

$$\Gamma^0 = \mathbf{1}_{16} \otimes i\sigma_2 \, , \quad \Gamma^i = \gamma^i \otimes \sigma_1 \, , \quad \Gamma^9 = \gamma \otimes \sigma_1 \tag{A.9}$$

where $\gamma \equiv \gamma^{1\cdots 8}$ and $i = 1, 2, \cdots, 8$. It implies that[6]

$$\Gamma^A = \begin{pmatrix} 0 & \tilde{\gamma}^A \\ \gamma^A & 0 \end{pmatrix} \, , \quad \tilde{\gamma}^A \equiv (1, \gamma^i, \gamma) \, , \quad \gamma^A \equiv (-1, \gamma^i, \gamma) \, . \tag{A.10}$$

The anti-commutation relation $\{\Gamma^A, \Gamma^B\} = 2\eta^{AB}\mathbf{1}_{32}$ turns out to

$$\tilde{\gamma}^A \gamma^B + \tilde{\gamma}^B \gamma^A = 2\eta^{AB}\mathbf{1}_{16} \, , \quad \gamma^A \tilde{\gamma}^B + \gamma^B \tilde{\gamma}^A = 2\eta^{AB}\mathbf{1}_{16} \, . \tag{A.11}$$

It follows that

$$\Gamma_{AB} = \begin{pmatrix} \tilde{\gamma}_{AB} & 0 \\ 0 & \gamma_{AB} \end{pmatrix} \, , \quad I = \begin{pmatrix} 0 & \tilde{\gamma}^{0\cdots 4} \\ \gamma^{0\cdots 4} & 0 \end{pmatrix} \, , \quad I\Gamma_A = \begin{pmatrix} \tilde{\gamma}^{0\cdots 4}\gamma_A & 0 \\ 0 & \gamma^{0\cdots 4}\tilde{\gamma}_A \end{pmatrix} \, ,$$

$$\Gamma_{11} = \begin{pmatrix} \tilde{\gamma}^0 \gamma^1 \cdots \gamma^9 & 0 \\ 0 & \gamma^0 \tilde{\gamma}^1 \cdots \gamma^9 \end{pmatrix} = \begin{pmatrix} 1 & 0 \\ 0 & -1 \end{pmatrix} \, , \quad C = \Gamma_0^\dagger = \begin{pmatrix} 0 & 1 \\ -1 & 0 \end{pmatrix} \, ,$$

$$C\Gamma^A = \begin{pmatrix} \gamma^A & 0 \\ 0 & -\tilde{\gamma}^A \end{pmatrix} \, , \quad C\Gamma^{AB}I = \begin{pmatrix} \gamma^{AB}\gamma^{0\cdots 4} & 0 \\ 0 & -\tilde{\gamma}^{AB}\tilde{\gamma}^{0\cdots 4} \end{pmatrix} \, , \tag{A.12}$$

where we have defined the following objects

$$\tilde{\gamma}_{AB} \equiv \frac{1}{2}(\tilde{\gamma}_A \gamma_B - \tilde{\gamma}_B \gamma_A) \, , \quad \gamma_{AB} \equiv \frac{1}{2}(\gamma_A \tilde{\gamma}_B - \gamma_B \tilde{\gamma}_A) \, ,$$

$$\tilde{\gamma}^{0\cdots 4} = \tilde{\gamma}^0 \gamma^1 \tilde{\gamma}^2 \gamma^3 \tilde{\gamma}^4 \, , \quad \gamma^{0\cdots 4} = \gamma^0 \tilde{\gamma}^1 \gamma^2 \tilde{\gamma}^3 \gamma^4 \, . \tag{A.13}$$

As $I^2 = -1$, we have $\gamma^{0\cdots 4}\tilde{\gamma}^{0\cdots 4} = \tilde{\gamma}^{0\cdots 4}\gamma^{0\cdots 4} = -1$. $Q_I h_+ = Q_I$ implies that $Q_I = (q_I, 0)$ with q_I ($I = 1, 2$) being a pair of 16-component spinors. Similarly, $h_+\theta = \theta$ implies that $\theta^I = \begin{pmatrix} \theta^I \\ 0 \end{pmatrix}$ with θ^I ($I = 1, 2$) being a pair of 16-component spinors. By using q_I, (A.2) takes of the form

$$[q_I, M_{AB}] = -\frac{1}{2}q_I \tilde{\gamma}_{AB} \, , \quad [q_I, P_A] = \frac{1}{2}\epsilon_{IJ}q_J \tilde{\gamma}^{0\cdots 4}\gamma_A \, ,$$

$$\{q_I, q_J\} = -2i\gamma^A P_A \delta_{IJ} + \epsilon_{IJ}(i\gamma^{ab}\gamma^{0\cdots 4}M_{ab} - i\gamma^{a'b'}\gamma^{0\cdots 4}M_{a'b'}) \, . \tag{A.14}$$

The fact that $C\Gamma^{A_1\cdots A_n}$ is symmetric iff $n = 1, 2$ mod 4 implies that γ_A, $\tilde{\gamma}_A$, $\gamma_{A_1\cdots A_5}$ and $\tilde{\gamma}_{A_1\cdots A_5}$ are symmetric and that γ_{ABC} and $\tilde{\gamma}_{ABC}$ are antisymmetric.

---

[6] In the literature, γ^A and $\tilde{\gamma}^A$ are distinguished each other by putting spinor indices. In this paper, however, we use a matrix notation without spinor indices to make our presentation simpler.

In this notation, the left-invariant Cartan one-forms are given as

$$L = L^A P_A + \frac{1}{2} L^{AB} M_{AB} + q_I L^I , \qquad (A.15)$$

$$L^A = e^A - 2i\theta\gamma^A \left(\frac{1}{2} + \frac{m^2}{4!} + \frac{m^4}{6!} + \cdots \right) D\theta , \qquad (A.16)$$

$$L^{AB} = w^{AB} + 2i\theta\hat{\gamma}^{AB} \epsilon \left(\frac{1}{2} + \frac{m^2}{4!} + \frac{m^4}{6!} + \cdots \right) D\theta , \qquad (A.17)$$

$$L^I = \left(1 + \frac{m^2}{3!} + \frac{m^4}{5!} + \cdots \right) D\theta , \qquad (A.18)$$

where $\hat{\gamma}^{AB} = (\gamma^{ab}\gamma^{0\cdots4}, -\gamma^{a'b'}\gamma^{0\cdots4})$ and

$$D\theta = d\theta + \frac{1}{2} e^A \tilde{\gamma}^{0\cdots4}\gamma_A\epsilon\theta + \frac{1}{4} w^{AB}\tilde{\gamma}_{AB}\theta , \qquad (A.19)$$

$$m^2 = -i\tilde{\gamma}^{0\cdots4}\gamma_A\epsilon\theta \cdot \theta\gamma^A + \frac{i}{2}\tilde{\gamma}_{AB}\theta \cdot \theta\epsilon\hat{\gamma}^{AB} . \qquad (A.20)$$

The currents used in (2.5) are related to the above objects by

$$J_{0\xi} = \frac{1}{2} L_\xi^{AB} M_{AB} , \quad J_{1\xi} = q_\alpha L_\xi^{1\alpha} , \quad J_{2\xi} = L_\xi^A P_A , \quad J_{3\xi} = \hat{q}_{\hat{\alpha}} L_\xi^{2\hat{\alpha}} , \qquad (A.21)$$

where we have replaced (q^1, q^2) with $(q_\alpha, \hat{q}_{\hat{\alpha}})$ and correspondingly (θ^1, θ^2) with $(\theta^\alpha, \hat{\theta}^{\hat{\alpha}})$.

References

[1] N. Berkovits, Super-Poincaré covariant quantization of the superstring, J. High Energy Phys. 0004 (2000) 018, arXiv:hep-th/0001035.

[2] P.A. Grassi, G. Policastro, M. Porrati, P. Van Nieuwenhuizen, Towards covariant quantization of superstrings without pure spinor constraints, J. High Energy Phys. 0210 (2002) 054, arXiv:hep-th/0112162;
P.A. Grassi, G. Policastro, P. van Nieuwenhuizen, On the BRST cohomology of superstrings with/without pure spinors, Adv. Theor. Math. Phys. 7 (3) (2003) 499, arXiv:hep-th/0206216.

[3] Y. Aisaka, Y. Kazama, A new first class algebra, homological perturbation and extension of pure spinor formalism for superstring, J. High Energy Phys. 0302 (2003) 017, arXiv:hep-th/0212316;
Y. Aisaka, Y. Kazama, Operator mapping between RNS and extended pure spinor formalisms for superstring, J. High Energy Phys. 0308 (2003) 047, arXiv:hep-th/0305221;
Y. Aisaka, Y. Kazama, Relating Green–Schwarz and extended pure spinor formalisms by similarity transformation, J. High Energy Phys. 0404 (2004) 070, arXiv:hep-th/0404141.

[4] Y. Aisaka, Y. Kazama, Origin of pure spinor superstring, J. High Energy Phys. 0505 (2005) 046, arXiv:hep-th/0502208.

[5] N. Berkovits, Quantum consistency of the superstring in AdS$_5$ × S^5 background, J. High Energy Phys. 0503 (2005) 041, arXiv:hep-th/0411170.

[6] B.C. Vallilo, One loop conformal invariance of the superstring in an AdS$_5$ × S^5 background, J. High Energy Phys. 0212 (2002) 042, arXiv:hep-th/0210064.

[7] R.R. Metsaev, A.A. Tseytlin, Type IIB superstring action in AdS$_5$ × S^5 background, Nucl. Phys. B 533 (1998) 109, arXiv:hep-th/9805028.

[8] I. Bena, J. Polchinski, R. Roiban, Hidden symmetries of the AdS$_5$ × S^5 superstring, Phys. Rev. D 69 (2004) 046002, arXiv:hep-th/0305116.

[9] B.C. Vallilo, Flat currents in the classical AdS$_5$ × S^5 pure spinor superstring, J. High Energy Phys. 0403 (2004) 037, arXiv:hep-th/0307018;
I. Adam, A. Dekel, L. Mazzucato, Y. Oz, Integrability of type II superstrings on Ramond–Ramond backgrounds in various dimensions, J. High Energy Phys. 0706 (2007) 085, arXiv:hep-th/0702083.

[10] M. Hatsuda, K. Yoshida, Classical integrability and super Yangian of superstring on $AdS_5 \times S^5$, Adv. Theor. Math. Phys. 9 (5) (2005) 703, arXiv:hep-th/0407044.

[11] J.M. Maldacena, The large N limit of superconformal field theories and supergravity, Int. J. Theor. Phys. 38 (1999) 1113, Adv. Theor. Math. Phys. 2 (1998) 231, arXiv:hep-th/9711200;
S.S. Gubser, I.R. Klebanov, A.M. Polyakov, Gauge theory correlators from noncritical string theory, Phys. Lett. B 428 (1998) 105, arXiv:hep-th/9802109;
E. Witten, Anti-de Sitter space and holography, Adv. Theor. Math. Phys. 2 (1998) 253, arXiv:hep-th/9802150.

[12] N. Berkovits, V. Pershin, Supersymmetric Born–Infeld from the pure spinor formalism of the open superstring, J. High Energy Phys. 0301 (2003) 023, arXiv:hep-th/0205154.

[13] N. Berkovits, P.S. Howe, Ten-dimensional supergravity constraints from the pure spinor formalism for the superstring, Nucl. Phys. B 635 (2002) 75, arXiv:hep-th/0112160.

[14] K. Skenderis, M. Taylor, Branes in AdS and pp wave space–times, J. High Energy Phys. 0206 (2002) 025, arXiv:hep-th/0204054.

[15] M. Sakaguchi, K. Yoshida, D-branes of covariant AdS superstrings, Nucl. Phys. B 684 (2004) 100, arXiv:hep-th/0310228.

[16] N. Berkovits, O. Chandia, Superstring vertex operators in an $AdS_5 \times S^5$ background, Nucl. Phys. B 596 (2001) 185, arXiv:hep-th/0009168.

[17] N. Berkovits, ICTP lectures on covariant quantization of the superstring, arXiv:hep-th/0209059;
Y. Oz, The pure spinor formulation of superstrings, Class. Quantum Gravity 25 (2008) 214001, arXiv:0910.1195 [hep-th];
O.A. Bedoya, N. Berkovits, GGI lectures on the pure spinor formalism of the superstring, arXiv:0910.2254 [hep-th];
L. Mazzucato, Superstrings in AdS, Phys. Rep. 521 (2012) 1, arXiv:1104.2604 [hep-th].

[18] M. Sakaguchi, K. Yoshida, Notes on D-branes of type IIB string on $AdS_5 \times S^5$, Phys. Lett. B 591 (2004) 318, arXiv:hep-th/0403243.

[19] N.D. Lambert, P.C. West, D-branes in the Green–Schwarz formalism, Phys. Lett. B 459 (1999) 515, arXiv:hep-th/9905031.

[20] P. Bain, K. Peeters, M. Zamaklar, D-branes in a plane wave from covariant open strings, Phys. Rev. D 67 (2003) 066001, arXiv:hep-th/0208038.

[21] R. Roiban, W. Siegel, Superstrings on $AdS_5 \times S^5$ supertwistor space, J. High Energy Phys. 0011 (2000) 024, arXiv:hep-th/0010104.

[22] I. Ramirez, B.C. Vallilo, Supertwistor description of the AdS pure spinor string, Phys. Rev. D 93 (8) (2016) 086008, arXiv:1510.08823 [hep-th].

[23] R.R. Metsaev, Type IIB Green–Schwarz superstring in plane wave Ramond–Ramond background, Nucl. Phys. B 625 (2002) 70, arXiv:hep-th/0112044;
R.R. Metsaev, A.A. Tseytlin, Exactly solvable model of superstring in Ramond–Ramond plane wave background, Phys. Rev. D 65 (2002) 126004, arXiv:hep-th/0202109.

[24] N. Berkovits, Conformal field theory for the superstring in a Ramond–Ramond plane wave background, J. High Energy Phys. 0204 (2002) 037, arXiv:hep-th/0203248.

[25] M. Sakaguchi, K. Yoshida, Noncommutative D-brane from covariant AdS superstring, Nucl. Phys. B 797 (2008) 179, arXiv:hep-th/0604039.

[26] M. Sakaguchi, K. Yoshida, Noncommutative M-branes from covariant open supermembranes, Phys. Lett. B 642 (2006) 400, arXiv:hep-th/0608099;
M. Sakaguchi, K. Yoshida, Intersecting noncommutative M5-branes from covariant open supermembrane, Nucl. Phys. B 781 (2007) 85, arXiv:hep-th/0702062;
M. Sakaguchi, K. Yoshida, A covariant approach to noncommutative M5-branes, Prog. Theor. Phys. Suppl. 171 (2007) 275, arXiv:hep-th/0702132.

[27] N. Berkovits, Towards covariant quantization of the supermembrane, J. High Energy Phys. 0209 (2002) 051, arXiv:hep-th/0201151.

[28] S. Hanazawa, M. Sakaguchi, in preparation.

[29] J.A. De Azcarraga, P.K. Townsend, Superspace geometry and classification of supersymmetric extended objects, Phys. Rev. Lett. 62 (1989) 2579.

[30] C. Chryssomalakos, J.A. de Azcarraga, J.M. Izquierdo, J.C. Perez Bueno, The geometry of branes and extended superspaces, Nucl. Phys. B 567 (2000) 293, arXiv:hep-th/9904137.

[31] M. Sakaguchi, IIB branes and new space–time superalgebras, J. High Energy Phys. 0004 (2000) 019, arXiv:hep-th/9909143.

[32] L. Anguelova, P.A. Grassi, Super D-branes from BRST symmetry, J. High Energy Phys. 0311 (2003) 010, arXiv:hep-th/0307260.

[33] M. Sakaguchi, K. Yoshida, Non-relativistic AdS branes and Newton–Hooke superalgebra, J. High Energy Phys. 0610 (2006) 078, arXiv:hep-th/0605124.

[34] M. Hatsuda, K. Kamimura, Wess–Zumino terms for AdS D-branes, Nucl. Phys. B 703 (2004) 277, arXiv:hep-th/0405202.

[35] M. Hatsuda, K. Kamimura, M. Sakaguchi, From super AdS$_5 \times$ S^5 algebra to super pp wave algebra, Nucl. Phys. B 632 (2002) 114, arXiv:hep-th/0202190.

10

A general holographic insulator/ superconductor model with dark matter sector away from the probe limit

Yan Peng [a,b,*], Qiyuan Pan [c], Yunqi Liu [d]

[a] *School of Mathematical Sciences, Qufu Normal University, Qufu, Shandong 273165, China*
[b] *School of Mathematics and Computer Science, Shaanxi Sci-Tech University, Hanzhong, Shaanxi 723000, China*
[c] *Department of Physics, Key Laboratory of Low Dimensional Quantum Structures and Quantum Control of Ministry of Education, Hunan Normal University, Changsha, Hunan 410081, China*
[d] *School of Physics, Huazhong University of Science and Technology, Wuhan, Hubei 430074, China*

Editor: Stephan Stieberger

Abstract

We investigate holographic phase transitions with dark matter sector in the AdS soliton background away from the probe limit. In cases of weak backreaction, we find that the larger coupling parameter α makes the gap of condensation shallower and the critical chemical potential keeps as a constant. In contrast, for very heavy backreaction, the dark matter sector could affect the critical chemical potential and the order of phase transitions. We also find the jump of the holographic topological entanglement entropy corresponds to a first order transition between superconducting states in this model with dark matter sector. More importantly, for certain sets of parameters, we observe novel phenomenon of retrograde condensation. In a word, the dark matter sector provides richer physics in the phase structure and the holographic superconductor properties are helpful in understanding dark matter.

* Corresponding author.
E-mail addresses: yanpengphy@163.com (Y. Peng), panqiyuan@126.com (Q. Pan), liuyunqi@hust.edu.cn (Y. Liu).

1. Introduction

The anti-de Sitter/conformal field theories (AdS/CFT) correspondence relates the strongly correlated conformal field theory on the boundary with a weakly interacting gravity system in the bulk [1–3]. According to this novel idea, the gauge/gravity duality has been successfully employed to gain a better understanding of the low energy physics in condensed matter systems from a higher dimensional gravitational dual. The simplest holographic metal/superconductor model dual to gravity theories was constructed by applying a scalar field and a Maxwell field coupled in the AdS black hole background [4–6]. And then the holographic insulator/superconductor transition was also established in the AdS soliton spacetime [7,8]. At present, a lot of more complete holographic superconductor models have been widely studied to model conductivity and other condensed matter physics properties in various gravity theories, such as the Einstein–Gauss–Bonnet gravity, Hořava–Lifshitz gravity, nonlinear electrodynamics gravity and so on, see Refs. [9–34].

According to contemporary astronomical observations, almost 24 percent of the total energy density in our universe is in the form of dark matter, whose configuration is still unclear. In order to model the dark matter, a new gravity was established by introducing an additional $U(1)$ gauge field coupled to the normal Maxwell field [35,36]. This dark matter gravity is strongly supported by astrophysical observation of 511 keV gamma rays [37] and the electron positron excess in galaxy [38,39]. Moreover, this dark matter gravity reveals new physics allowing the 3.6σ discrepancy between measured value of the muon anomalous magnetic moment and its prediction in the Standard Model [40]. In order to understand dark matter from properties of holographic systems, new holographic metal/superconductor transition models were constructed in the background of this dark matter gravity, for references see [41,42]. The interesting topics of effects of dark matter on holographic vortices and holographic fluid viscosity were also carried out in [43,44]. Since the holographic superconductor provides a lot of qualitative characteristic properties shared by real superconductor, it is believed that holographic superconductor theory may be applied to superconductor in the laboratory. As a further step along this line, some possible methods to directly detect the dark matter in the lab were proposed according to holographic theory [45]. Another interest to study this model is related to the question of possible matter configurations in AdS spacetime [46,47]. On the sides of CFT on the boundary, it was shown that the dark matter sector really provides richer physics in the critical temperature and also the stability of metal/superconductor transitions on the boundary. Most surprisingly, for a certain set of parameters, there is a novel phenomenon of retrograde condensation in holographic transition models with dark matter sector in Refs. [41,42], which was already observed in other metal/superconductor systems [48,49]. We further obtained the general conditions for retrograde condensations and proved retrograde condensations to be unstable in the s-wave metal/superconductor transitions [50]. In this work, we mainly focus on the effects of the dark matter in holographic insulator/superconductor model, which is interesting because the method provides the insight into the strong coupling system and possibly it enables findings of some 'experimental' facts connected with dark sector.

Most of holographic models were constructed in the background of AdS black hole or described metal/superconductor transitions. So the holographic insulator/superconductor transition in the AdS soliton spacetime needs much more research. Along this line, the dark matter sector was also considered in the AdS soliton spacetime while neglecting the matter fields' backreaction on the metric [51–53]. It was shown that the critical chemical potential is independent of the dark matter sector parameter without backreaction. In this work, we will show that the dark matter sector could affect the critical chemical potential and also the order of transitions when including backreaction. In contrast to retrograde condensation in holographic metal/superconductor models [41,42,48,49], we will also manage to obtain the novel retrograde condensation in our s-wave holographic insulator/superconductor model.

On the other hand, the entanglement entropy is usually applied to keep track of the degrees of freedom for strongly coupled system while other traditional methods might not be available. According to the AdS/CFT correspondence, it was proposed that the holographic entanglement entropy of a strongly interacting system on the boundary can be calculated from a weakly coupled gravity dual in the bulk [54,55]. With this elegant approach, the holographic entanglement entropy has recently been used to study properties of phase transitions in various holographic models and it provides us richer insights into the phase transitions [56–74]. For example, it was argued that the entanglement entropy is a good probe to the critical phase transition point and also the order of phase transitions. As a further step, we would like to examine whether the holographic entanglement entropy is still useful in disclosing properties of transitions in this general holographic insulator/superconductor model with dark matter sector.

This paper is organized as follows. In the next section, we introduce the holographic model with dark matter sector in the background of AdS soliton away from the probe limit. In section 3.1, we study the stability of holographic phase transitions. And in section 3.2 and section 3.3, we disclose properties of insulator/superconductor transitions by examining in detail behaviors of the scalar operator and the holographic entanglement entropy of the system. We will summarize our main results in the last section.

2. Equations of motion and boundary conditions

In this work, we consider holographic phase transitions with dark matter sector in the 5-dimensional AdS soliton spacetime. The general Lagrange density constructed by a scalar field and two $U(1)$ gauge fields coupled in the gravitational background reads [41]

$$\mathcal{L} = R + \frac{12}{L^2} - \left(\frac{1}{4} F^{MN} F_{MN} + |\nabla_M \psi - iq A_M|^2 + m^2 \psi^2 + \frac{1}{4} B^{MN} B_{MN} \right.$$
$$\left. + \frac{\alpha}{4} B^{MN} F_{MN} \right), \tag{1}$$

where $\psi(r)$ is the complex scalar field, A_M stands for the ordinary Maxwell field and another additional $U(1)$ gauge field B_M corresponds to the dark matter field, which is not completely decoupled with the visible matter. $-6/L^2$ is the negative cosmological constant, where L is the radius of AdS spacetime which will be scaled unity in the following numerical calculation. m is the mass of the scalar field, q is the scalar charge and α is the coupling parameter between the two $U(1)$ gauge fields.

The Einstein equations for the system can be written in the form

$$R_{MN} - \frac{1}{2} g_{MN} R - 6 g_{MN} = \frac{1}{2} \tilde{T}_{MN}, \tag{2}$$

where $\tilde{T}_{MN}$ is the energy-momentum tensor expressed as

$$\tilde{T}_{MN} = F_{M\beta}F_N^\beta + B_{M\beta}B_N^\beta + \alpha B_{M\beta}F_N^\beta + 2\nabla_M\psi\nabla_N\psi + 2q^2 A_M A_N \psi^2$$
$$+ g_{MN}\left(-\frac{1}{4}F_{MN}F^{MN} - \frac{1}{4}B_{MN}B^{MN} - \frac{\alpha}{4}F_{MN}B^{MN} - \nabla_M\psi\nabla_M\psi\right.$$
$$\left. - q^2 A_M A^M \psi^2 - m^2\psi^2\right). \tag{3}$$

With the variation of the considered matter fields, we obtain the corresponding independent equations of motion in the form

$$\nabla_M F^{MN} - 2q^2 A^N \psi^2 + \frac{\alpha}{2}\nabla_M B^{MN} = 0, \tag{4}$$

$$\nabla_M\nabla^M\psi - q^2 A_M A^M \psi - m^2\psi = 0, \tag{5}$$

$$\nabla_M B^{MN} + \frac{\alpha}{2}\nabla_M F^{MN} = 0. \tag{6}$$

Putting (6) into (4), we can eliminate the dark matter field and obtain the equation

$$\nabla_M F^{MN} - \frac{2q^2\psi^2 A^N}{\tilde{\alpha}} = 0, \tag{7}$$

where $\tilde{\alpha} = 1 - \frac{\alpha^2}{4}$.

Substituting (7) into (6), we arrive at

$$\nabla_M B^{MN} + \frac{\alpha q^2\psi^2 A^N}{\tilde{\alpha}} = 0. \tag{8}$$

From Eqs. (7) and (8), we find that the ordinary Maxwell field A_M and the $U(1)$ gauge field B_M which corresponds to the dark matter field share similar features for the holographic superconductor system in AdS soliton background, so we can still use the metric ansatz given in Ref. [8] where the authors took the backreaction into account (without dark matter sector), i.e.,

$$ds^2 = -r^2 e^{C(r)}dt^2 + \frac{dr^2}{r^2 B(r)} + r^2 dx^2 + r^2 dy^2 + r^2 B(r)e^{D(r)}d\chi^2, \tag{9}$$

$$A = \phi(r)dt, \qquad B = \eta(r)dt, \qquad \psi = \psi(r). \tag{10}$$

In order to get smooth solutions at the tip r_s satisfying $B(r_s) = 0$, we have to impose on the coordinate χ a period Γ as

$$\Gamma = \frac{4\pi e^{-D(r_s)/2}}{r_s^2 B'(r_s)}. \tag{11}$$

We also need $C(r \to \infty) = 0$ and $D(r \to \infty) = 0$ to recover the AdS boundary.

From above assumptions, we can obtain the equations of motion as

$$\psi'' + \left(\frac{5}{r} + \frac{B'}{B} + \frac{C'}{2} + \frac{D'}{2}\right)\psi' + \frac{q^2\phi^2 e^{-C}}{r^4 B}\psi - \frac{m^2}{r^2 B}\psi = 0, \tag{12}$$

$$\phi'' + \left(\frac{3}{r} + \frac{B'}{B} - \frac{C'}{2} + \frac{D'}{2}\right)\phi' - \frac{2q^2\psi^2}{\tilde{\alpha}r^2 B}\phi = 0, \tag{13}$$

$$\eta'' + \left(\frac{3}{r} + \frac{B'}{B} - \frac{C'}{2} + \frac{D'}{2}\right)\eta' + \frac{\alpha q^2\psi^2\phi}{\tilde{\alpha}r^2 B} = 0, \tag{14}$$

$$C'' + \frac{1}{2}C'^2 + \left(\frac{5}{r} + \frac{A'}{2} + \frac{B'}{B}\right)C' - \frac{e^{-C}}{r^2}\left(\phi'^2 + \eta'^2 + \alpha\phi'\eta'\right)$$

$$- \frac{2q^2\phi^2\psi^2 e^{-C}}{r^4 B} = 0, \tag{15}$$

$$B'\left(\frac{3}{r} - \frac{C'}{2}\right) + B\left[\psi'^2 - \frac{1}{2}A'C' + \frac{e^{-C}}{2r^2}\left(\phi'^2 + \eta'^2 + \alpha\phi'\eta'\right) + \frac{12}{r^2}\right]$$

$$+ \frac{q^2\phi^2\psi^2 e^{-C}}{r^4} + \frac{m^2\psi^2}{r^2} - \frac{12}{r^2} = 0, \tag{16}$$

$$D' = \frac{2r^2 C'' + r^2 C'^2 + 4rC' - 2e^{-C}(\phi'^2 + \eta'^2 + \alpha\phi'\eta') + 4r^2\psi'^2}{r(6 + rC')}, \tag{17}$$

where the prime denotes the derivative with respect to r. Since the equations are nonlinear and coupled to each other, we have to solve these equations by using the numerical shooting method which will integrate the equations of motion from the tip of the soliton out to the infinity. Thus, we have to specify the boundary conditions for this system. At the tip, we can impose proper boundary conditions as

$$\psi(r) = \psi_0 + \psi_1(r - r_s) + \cdots, \quad \phi(r) = \phi_0 + \phi_1(r - r_s) + \cdots,$$

$$\eta(r) = \eta_0 + \eta_1(r - r_s) + \cdots, \quad B(r) = B_0(r - r_s) + \cdots,$$

$$C(r) = C_0 + C_1(r - r_s) + \cdots, \quad D(r) = D_0 + D_1(r - r_s) + \cdots, \tag{18}$$

where the dots denote higher order terms.

After putting expansions (18) into (12)–(17) and considering leading terms of these equations, we are left with six independent parameters r_s, ψ_0, ϕ_0, η_0, C_0 and D_0 at the tip. Near the AdS boundary ($r \to \infty$), the asymptotic behaviors of the solutions are

$$\psi \to \frac{\psi_-}{r^{\lambda_-}} + \frac{\psi_+}{r^{\lambda_+}} + \cdots, \quad \phi \to \mu - \frac{\rho}{r^2} + \cdots, \quad \eta \to \xi - \frac{\varpi}{r^2} + \cdots,$$

$$B \to 1 + \frac{B_4}{r^4} + \cdots, \quad C \to \frac{C_4}{r^4} + \cdots, \quad D \to \frac{D_4}{r^4} + \cdots, \tag{19}$$

with $\lambda_\pm = (2 \pm \sqrt{4 + m^2})$. μ and ρ can be interpreted as the chemical potential and charge density in the dual theory respectively. The other two operators ξ and ϖ are dual to the $U(1)$ gauge field $\eta(r)$.

From the equations of motion for the system, we obtain the scaling symmetry

$$r \to ar, \quad (\chi, x, y, t) \to (\chi, x, y, t)/a, \quad \phi \to a\phi, \quad \eta \to a\eta, \tag{20}$$

which can be used to set $r_s = 1$ in the following calculation.

We will impose four constraint conditions at infinity. The asymptotic expressions (19) imply two constraint conditions at infinity: $C(\infty) = 0$ and $D(\infty) = 0$. We also have to impose one falloff of the scalar fields vanishes in order to get a stable CFT on the boundary. Choosing $m^2 = -\frac{15}{4} > -4$ above the BF bound [75], the second mode ψ_+ is always normalizable. In this paper, we will fix $\psi_- = 0$ and use the operator $\psi_+ = <O_+>$ to describe the phase transition in the dual CFT. For different values of ψ_0, we can rely on the independent parameters ϕ_0, η_0, C_0 and D_0 as the shooting parameter to search for the solutions with the boundary conditions $\psi_- = 0$, $C(r \to \infty) = 0$, $D(r \to \infty) = 0$ and $\frac{\xi}{\mu}$ fixed.

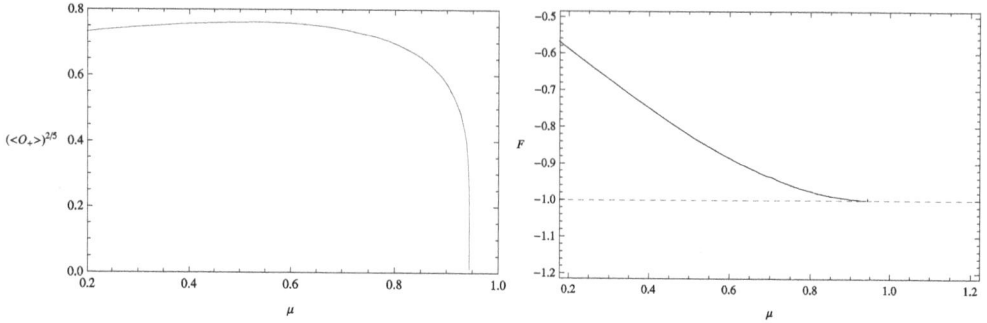

Fig. 1. The phase transition in cases of $\Gamma = \pi$, $m^2 = -\frac{15}{4}$, $q = 2$, $\alpha = 2.5$ and $\frac{\xi}{\mu} = -1$. The left panel shows the behavior of the scalar condensation. The right panel represents the free energy of the system, where the solid line corresponds to the superconducting phase and the dashed line is with the normal phase. (Color online.)

3. Holographic phase transitions in AdS soliton background

3.1. Stability of the scalar condensation

In this part, we investigate the stability of holographic insulator/superconductor phase transitions through the behaviors of the scalar operator. We choose the example $\Gamma = \pi$, $m^2 = -\frac{15}{4}$, $q = 2$, $\alpha = 2.5$ and $\frac{\xi}{\mu} = -1$ in Fig. 1. It is surprising in the left panel that the condensed phases appear at small chemical potential $\mu < \mu_c = 0.944$, which is different from the normal results in [7,21]. We refer this phenomenon as the retrograde condensation, which was also observed in the holographic metal/superconductor transitions in the background of AdS black hole [41,50]. By choosing different sets of parameters, we find that there are retrograde condensations for all superconducting solutions satisfying $\frac{\xi}{\mu} = -1$ and $\alpha > 2$.

In order to determine whether the retrograde condensation is thermodynamically favored, we should calculate the free energy of the system for both normal phase and condensed phase. We show the corresponding free energy of the system in the right panel of Fig. 1. It shows that the free energy of this hairy AdS soliton is larger than the free energy of the soliton in normal phase. Since the physical procedure corresponds to the phases with the lowest free energy, we arrive at a conclusion that the retrograde condensation superconducting solutions are thermodynamically unstable and the unusual behaviors of the scalar operator can be used to detect the thermodynamical instability of phase transitions. In the retrograde condensation, the soliton superconductor phase is not thermodynamically favored. In other words, there is no soliton/soliton superconductor transition as usual. In this case, we expect that there maybe interesting soliton/black hole and soliton/black hole superconductor transitions with the increase of chemical potential. And we plan to draw the complete diagram of the soliton/black hole/black hole superconductor system in the next work. And on the aspects of bulk theory, the unstable condensation also means the scalar field can't condense in the background of AdS soliton, which corresponds to an approach from the CFT on the boundary to the AdS gravity in the bulk.

Now we study the case of $\Gamma = \pi$, $m^2 = -\frac{15}{4}$, $q = 2$, $\alpha = 0.5$ and $\frac{\xi}{\mu} = 1$ in Fig. 2. In the left panel, we find a critical chemical potential $\mu_c = 0.944$ above which there is scalar condensation. In order to further study the phase transition, we plot the free energy in the right panel of Fig. 2. The free energy of superconducting state lies below the free energy of the normal state, which suggests that the superconducting solutions are thermodynamically stable. More calculations

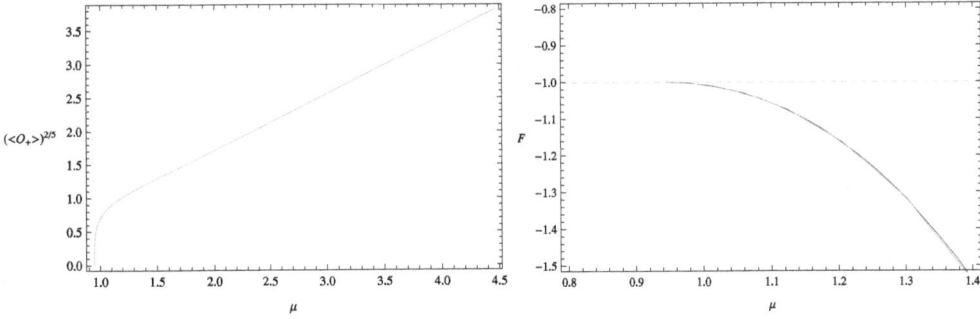

Fig. 2. The phase transitions in cases of $\Gamma = \pi$, $m^2 = -\frac{15}{4}$, $q = 2$, $\alpha = 0.5$ and $\frac{\xi}{\mu} = 1$. The left panel represents the scalar operator with respect to the chemical potential. The solid green line in the right panel shows the free energy of the superconducting phase and the dashed line corresponds to normal phases. As a comparison, we also plot the results of equation (21) with solid blue line in the right panel. (Color online.)

show that phase transitions are thermodynamically stable for all $\alpha > 0$ and $\frac{\xi}{\mu} = 1$. When $\frac{\xi}{\mu} = -1$, there are thermodynamically stable solutions for $0 \leqslant \alpha < 2$.

With fitting methods, we obtain approximate formulas for the free energy of normal state (F_{SL}) and superconducting state (F_{SC}) that:

$$F_{SL} \approx -1.00, \quad F_{SC} \approx -1.00 - 2.21(\mu - \mu_c)^2 - 0.86(\mu - \mu_c)^3. \tag{21}$$

In the right panel, the solid blue line corresponding to formula (21) almost coincides with the solid green line obtained from the original data. That means our fitting formula is valid. More importantly, the front formulas suggest that: $F_{SL}|_{\mu=\mu_c} = F_{SC}|_{\mu=\mu_c}$, $\frac{\partial F_{SL}}{\partial \mu}|_{\mu=\mu_c} = \frac{\partial F_{SC}}{\partial \mu}|_{\mu=\mu_c}$, $\frac{\partial^2 F_{SL}}{\partial \mu^2}|_{\mu=\mu_c} \neq \frac{\partial^2 F_{SC}}{\partial \mu^2}|_{\mu=\mu_c}$. In other words, the curve representing the physical phases with the lowest free energy is smooth at the critical point μ_c or the insulator/superconductor transition is of the second order.

3.2. Properties of the scalar condensation with weak backreaction

According to the transformation in [26], larger scalar charge q corresponds to cases of smaller backreaction parameter. In the following discussion, we take $q = 2$ as phases with weak backreaction in this part and $q = 1$ as the strong backreaction cases in the next part. We show the scalar operator $\langle O_+ \rangle^{1/\lambda_+}$ as a function of μ with $\Gamma = \pi$, $m^2 = -\frac{15}{4}$ and $q = 2$ in Fig. 3. It is shown that there are phase transitions at critical chemical potentials, above which the charged scalar condensation turns on. We exhibit the condensation of the scalar operator by choosing various α from top to bottom as: $\alpha = 0$, $\alpha = 0.5$, $\alpha = 1.0$ and $\frac{\xi}{\mu} = 1$. The increase of the parameter α develops shallower condensation gap. And in both curves, the critical chemical potential keeps as a constant $\mu_c = 0.944$ for different values of α. In all, the effects of the dark matter sector on the critical phase transition points and condensation gap are very different from those reported in the metal/superconductor system [41,42,50]. We conclude that the effects of the dark matter sector on transitions depend on backgrounds.

In the following, we pay attention to the holographic entanglement entropy (HEE) of the transition system. The authors in Refs. [54,55] have provided a method to compute the entanglement entropy of conformal field theories (CFTs) from the gravity side. For simplicity, we consider the entanglement entropy for a half space corresponding to a subsystem $\bar{A}$ defined by

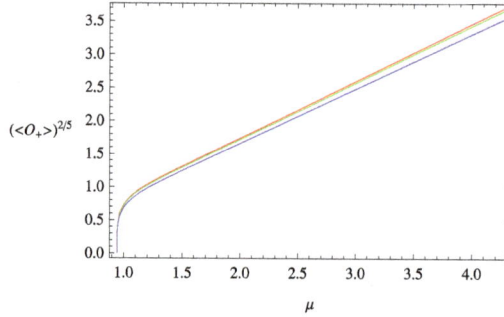

Fig. 3. The phase transition in the cases of $\Gamma = \pi$, $m^2 = -\frac{15}{4}$ and $q = 2$. It shows the behavior of scalar operator with $\frac{\xi}{\mu} = 1$ and various α from top to bottom as: $\alpha = 0$ (red), $\alpha = 0.5$ (green)and $\alpha = 1$ (blue). (Color online.)

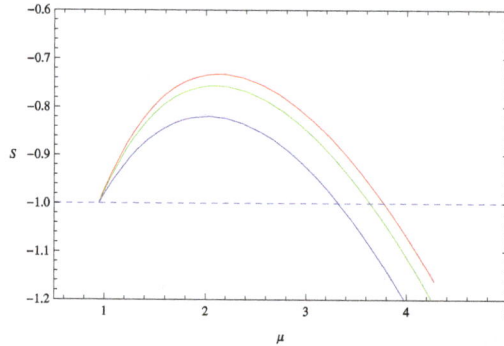

Fig. 4. The holographic entanglement entropy as a function of the chemical potential μ for $\Gamma = \pi$, $m^2 = -\frac{15}{4}$ and $q = 2$. The dashed lines correspond to the entanglement entropy of a pure AdS soliton and the solid lines correspond to cases of superconducting solutions. We have choose $\frac{\xi}{\mu} = 1$ and various α from top to bottom as: $\alpha = 0$ (red), $\alpha = 0.5$ (green) and $\alpha = 1.0$ (blue). (Color online.)

$x > 0$, $-\frac{R}{2} < y < \frac{R}{2}$ $(R \to \infty)$, $0 \leqslant \chi \leqslant \Gamma$. Then the entanglement entropy can be expressed as [67–69]:

$$S_A^{half} = \frac{R\Gamma}{4G_N} \int_{r_0}^{\frac{1}{\varepsilon}} r e^{\frac{D(r)}{2}} dr = \frac{R\pi}{8G_N} (\frac{1}{\varepsilon^2} + S),$$
(22)

where $r = \frac{1}{\varepsilon}$ is the UV cutoff. The first term is divergent as $\varepsilon \to 0$. In contrast, the second term does not depend on the cutoff and thus is physical important. As a matter of fact, this finite term is the difference between the entanglement entropy in the pure AdS soliton and the pure AdS space. $S = -1$ corresponds to the pure AdS soliton.

We present the holographic entanglement entropy as a function of the chemical potential μ in Fig. 4 with $\Gamma = \pi$, $m^2 = -\frac{15}{4}$, $q = 2$, $\frac{\xi}{\mu} = 1$ and with various α. For all set of parameters, we find a threshold chemical potential $\mu = 0.944$, above which the hairy soliton appears. The jump of the slop of the entanglement entropy at $\mu = 0.944$ signals that some kind of new degrees of freedom like the Cooper pair would emerge in the new phase. It also can be easily seen from the pictures that when the parameters are fixed, the entanglement entropy first increases and then decreases monotonously as we choose a larger chemical potential. It means that there is firstly an increase and then a reduction in the number of degrees of freedom due to the condensate generated in the phase transitions [73]. We also see that larger α corresponds to smaller maximum entanglement entropy. Most importantly, we mention that compared with the scalar operator, the entanglement entropy is more sensitive to the change of the coupling parameter α.

The critical phase transition point $\mu = 0.944$ obtained from the behaviors of holographic entanglement entropy is equal to the threshold chemical potential $\mu_c = 0.944$ obtained from the behaviors of the free energy and the scalar operator. That means the holographic entanglement entropy can be used to search for the critical chemical potential. With detailed analysis of the holographic entanglement entropy, we conclude that the critical chemical potential is independent of the dark matter sector parameter. This numerical result is nontrivial since when considering the matter fields' backreaction on the metric, the equations of motion depend on the coupling parameter α even at the phase transition points where the scalar field is zero. And the jump of the slope of holographic entanglement entropy corresponds to second order phase transitions in the general holographic superconductor model with dark matter sector. In summary, we conclude that the entanglement entropy can be used to determine the critical phase transition point, the order of the phase transition and the values of the coupling parameter α. So the entanglement entropy is indeed a good probe of insulator/superconductor phase transitions with dark matter sector.

3.3. Various phase transitions with strong backreaction

Now we pay attention to the holographic superconductor with small charge $q = 1$ or strong backreaction. We exhibit the free energy as a function of the chemical potential in Fig. 5 with $\Gamma = \pi$, $m^2 = -\frac{15}{4}$ and various α. It can be seen from the top panel that, for the small model parameter $\alpha = 0.1$, F develops a discontinuity in the first derivative of the free energy with respect to the chemical potential at a critical value $\mu_c = 1.82$, which implies the first order phase transition. In the middle panel with $\alpha = 1.0$, F decreases smoothly near the critical point $\mu_c = 1.89$ indicating the second order phase transitions from normal state into superconducting state. What's more, besides the second order phase transition at $\mu_c = 1.89$, the free energy develops a "swallow tail" at $\mu = 2.08$ within the superconducting phase, a typical signal for a first order phase transition. The bottom panel shows that when the coupling parameter is larger as $\alpha = 1.5$, there is only second order insulator/superconductor phase transitions at the critical phase transition point $\mu_c = 1.89$. We conclude that the dark matter sector can affect the critical phase transition points and also the order of phase transitions.

We also detect properties of phase transitions by studying the corresponding behaviors of the scalar operator in Fig. 6. In the top panel with $\alpha = 0.1$, at the critical chemical potential $\mu_c = 1.82$, the scalar operator $\langle O_+ \rangle^{1/\lambda_+}$ has a jump from insulator state to superconductor state indicating a first order phase transition. When choosing $\alpha = 1.0$ in the middle panel, the curves firstly increase continuously around the insulator/superconductor points $\mu_c = 1.89$ and then have a jump at $\mu = 2.08$ in the superconducting state. It means there are second order phase transitions at $\mu_c = 1.89$ and then first order phase transitions at $\mu = 2.08$. In cases of lager parameter $\alpha = 1.5$ in the bottom panel, $\langle O_+ \rangle^{1/\lambda_+}$ increases continuously with chemical potential, which is a classical performance of the second order phase transition.

At last, we turn to study the holographic phase transition through the entanglement entropy approach. We plot the corresponding entanglement entropy with respect to the chemical potential μ in Fig. 7 with $\alpha = 1.0$ as an example. The blue dashed line $S = -1$ describes the entanglement entropy of the normal phase, while the blue solid line corresponds to the entanglement entropy of

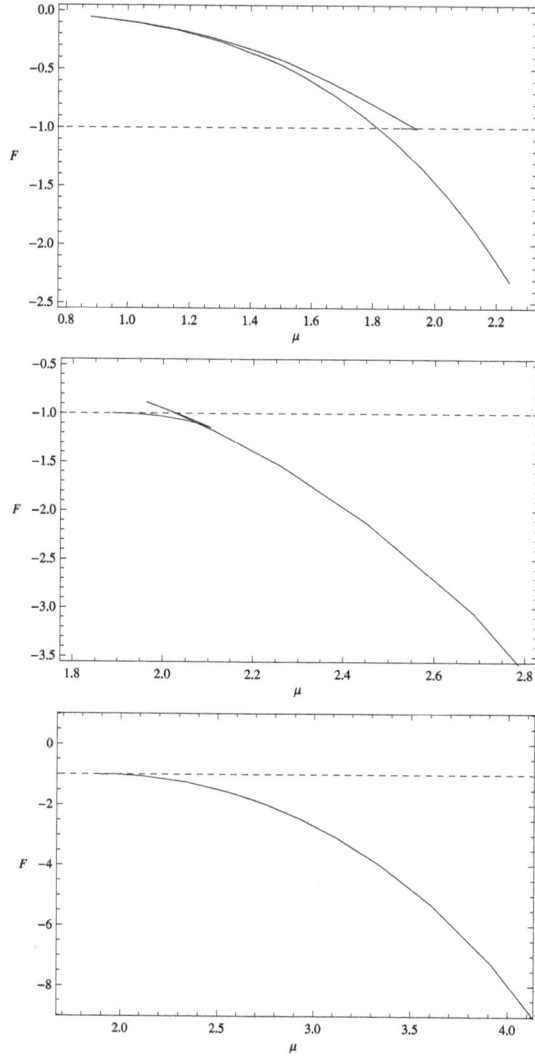

Fig. 5. The free energy with respect to the chemical potential in cases of $\Gamma = \pi$, $m^2 = -\frac{15}{4}$, $q = 1$ and $\frac{\xi}{\mu} = 1$. The panels from top to bottom represent the cases of $\alpha = 0.1$, $\alpha = 1.0$ and $\alpha = 1.5$. The solid line corresponds to the superconducting phase and the dashed line of $S = -1$ is with the normal phase. (Color online.)

the superconducting phase. The holographic entanglement entropy is continuously at the critical chemical potential $\mu = 1.89$ and there are discontinuous slops at this transition point, which implies the phase transitions at $\mu = 1.89$ are of the second order. When we go on to increase the chemical potential, the entanglement entropy has a jump around $\mu = 2.08$ which corresponds to the "swallow tail" of the free energy F in Fig. 5 and the dump of the scalar operator in Fig. 6. The front phenomenon implies a first order phase transition around $\mu = 2.08$ in the superconducting phase. We state that the entanglement entropy can be used to study the critical phase transition points and the order of phase transitions in our general holographic superconductor model. It is clear that there is numerical noise in the picture. Since the numerical noise is small compared to

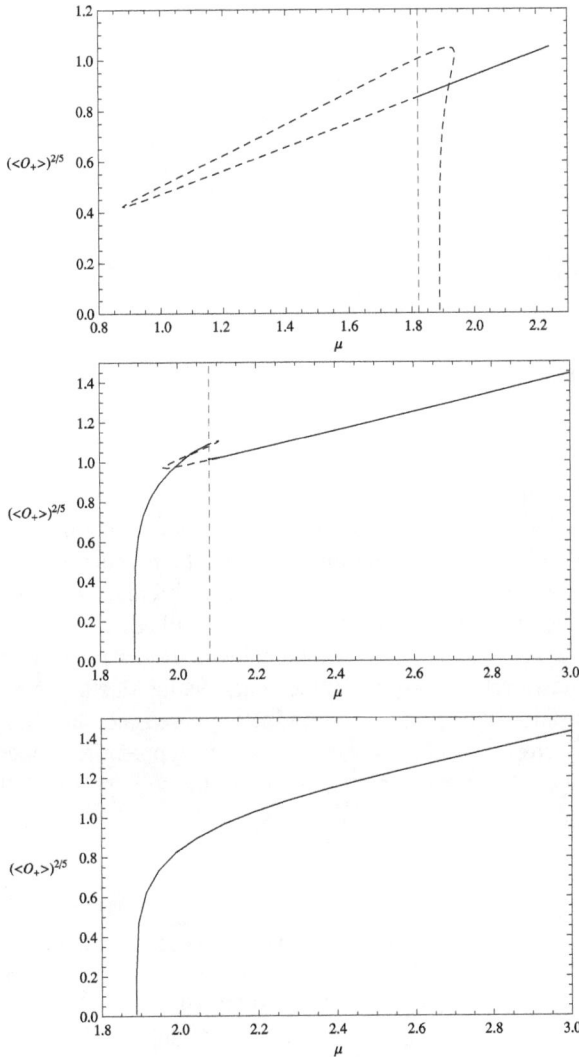

Fig. 6. The behaviors of scalar operator in cases of $\Gamma = \pi$, $m^2 = -\frac{15}{4}$, $q = 1$ and $\frac{\xi}{\mu} = 1$. The panels from top to bottom show the case of $\alpha = 0.1$, $\alpha = 1.0$ and $\alpha = 1.5$. The blue solid lines correspond to the physical superconducting states. (Color online.)

the jump of the holographic entanglement entropy at the red dashed line, it will not change our results.

4. Conclusions

We studied holographic insulator/superconductor transition model with backreaction in the presence of dark matter sector. We disclosed properties of transitions through analyzing behaviors of the scalar operator and the holographic entanglement entropy of the system. For the case of weak backreaction, it was shown that the phase transition is of the second order and the larger

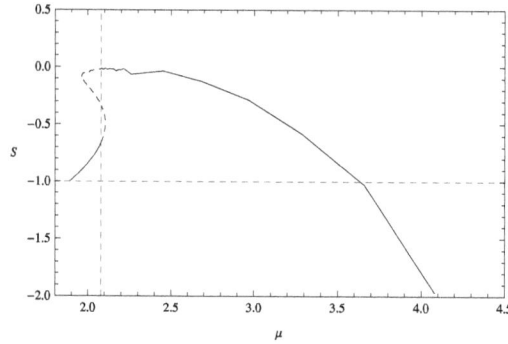

Fig. 7. The entanglement entropy in cases of $\Gamma = \pi$, $m^2 = -\frac{15}{4}$, $q = 1$, $\alpha = 1.0$ and $\frac{\xi}{\mu} = 1$. The blue solid line shows the entanglement entropy of the superconducting phase and the blue dashed line $S = -1$ corresponds to normal phases. (Color online.)

coupling parameter α makes the gap of condensation lower. And the critical phase transition chemical potential is always a constant for different values of α similar to cases without back-reaction. When including strong backreaction, the parameter α could affect the critical chemical potential and the order of phase transitions, which is very different from cases in the probe limit. In this case, we also found the jump of the holographic topological entanglement entropy corresponds to a first order transition in this general model with dark matter sector. More importantly, we observed novel phases referred as retrograde condensation due to the dark matter sector. In the retrograde condensation, the soliton superconductor phase is not thermodynamically favored. The front holographic properties of dark matter sector have potential application in dark matter detection and we plan to draw the complete diagram of the soliton/black hole/black hole super-conductor/soliton superconductor system in the next work.

Acknowledgements

This work was supported by the National Natural Science Foundation of China under Grant Nos. 11305097, 11275066 and 11505066; the Shaanxi Province Science and Technology Department Foundation of China under Grant Nos. 2016JQ1039 and 2016JM1028; and the Hunan Provincial Natural Science Foundation of China under Grant No. 2016JJ1012. This work was also partly finished during the International Conference on holographic duality for condensed matter physics at Kavli Institute for Theoretical Physics China (KITPC), Chinese Academy of Sciences on July 6–31, 2015.

References

[1] J.M. Maldacena, The large-N limit of superconformal field theories and supergravity, Adv. Theor. Math. Phys. 2 (1998) 231.
[2] S.S. Gubser, I.R. Klebanov, A.M. Polyakov, Gauge theory correlators from non-critical string theory, Phys. Lett. B 428 (1998) 105.
[3] E. Witten, Anti-de Sitter space and holography, Adv. Theor. Math. Phys. 2 (1998) 253.
[4] S.A. Hartnoll, Lectures on holographic methods for condensed matter physics, Class. Quantum Gravity 26 (2009) 224002.
[5] C.P. Herzog, Lectures on holographic superfluidity and superconductivity, J. Phys. A 42 (2009) 343001.
[6] G.T. Horowitz, Introduction to Holographic Superconductors, Lect. Notes Phys. 828 (2011) 313, arXiv:1002.1722 [hep-th].

[7] T. Nishioka, S. Ryu, T. Takayanagi, Holographic superconductor/insulator transition at zero temperature, J. High Energy Phys. 1003 (2010) 131, arXiv:0911.0962 [hep-th].

[8] G.T. Horowitz, B. Way, Complete phase diagrams for a holographic superconductor/insulator system, J. High Energy Phys. 1011 (2010) 011, arXiv:1007.3714 [hep-th].

[9] R. Gregory, S. Kanno, J. Soda, Holographic superconductors with higher curvature corrections, J. High Energy Phys. 0910 (2009) 010.

[10] L. Barclay, R. Gregory, S. Kanno, P. Sutcliffe, Gauss–Bonnet holographic superconductors, J. High Energy Phys. 1012 (2010) 029.

[11] Q. Pan, B. Wang, E. Papantonopoulos, J. Oliviera, A. Pavan, Holographic superconductors with various condensates in einstein–Gauss–Bonnet gravity, Phys. Rev. D 81 (2010) 106007.

[12] F. Aprile, J.G. Russo, Models of holographic superconductivity, Phys. Rev. D 81 (2010) 026009.

[13] A. Salvio, Holographic superfluids and superconductors in dilaton gravity, J. High Energy Phys. 1209 (2012) 134.

[14] R. Cai, H. Zhang, Holographic superconductors with Hořava–Lifshitz black holes, Phys. Rev. D 81 (2010) 066003.

[15] J. Jing, Q. Pan, S. Chen, Holographic superconductors with Power-Maxwell field, J. High Energy Phys. 1111 (2011) 045.

[16] G.T. Horowitz, M.M. Roberts, Holographic superconductors with various condensates, Phys. Rev. D 78 (2008) 126008.

[17] J. Sonner, A rotating holographic superconductor, Phys. Rev. D 80 (2009) 084031.

[18] Q. Pan, B. Wang, General holographic superconductor models with backreactions, arXiv:1101.0222 [hep-th].

[19] S.A. Hartnoll, C.P. Herzog, G.T. Horowitz, Holographic superconductors, J. High Energy Phys. 0812 (2008) 015.

[20] Y. Liu, Q. Pan, B. Wang, Holographic superconductor developed in BTZ black hole background with backreactions, Phys. Lett. B 702 (2011) 94.

[21] Y. Peng, Q. Pan, B. Wang, Various types of phase transitions in the AdS soliton background, Phys. Lett. B 699 (2011) 383, arXiv:1104.2478 [hep-th].

[22] J.P. Gauntlett, J. Sonner, T. Wiseman, Holographic superconductivity in M-Theory, Phys. Rev. Lett. 103 (2009) 151601.

[23] J. Jing, S. Chen, Holographic superconductors in the Born–Infeld electrodynamics, Phys. Lett. B 686 (2010) 68.

[24] K. Maeda, M. Natsuume, T. Okamura, Universality class of holographic superconductors, Phys. Rev. D 79 (2009) 126004.

[25] X.H. Ge, B. Wang, S.F. Wu, G.H. Yang, Analytical study on holographic superconductors in external magnetic field, J. High Energy Phys. 1008 (2010) 108.

[26] Y. Brihaye, B. Hartmann, Holographic superconductors in 3 + 1 dimensions away from the probe limit, Phys. Rev. D 81 (2010) 126008.

[27] C.P. Herzog, P.K. Kovtun, D.T. Son, Holographic model of superfluidity, Phys. Rev. D 79 (2009) 066002.

[28] D. Roychowdhury, AdS/CFT superconductors with Power Maxwell electrodynamics: reminiscent of the Meissner effect, Phys. Lett. B 718 (2013) 1089.

[29] S. Franco, A.M. Garcia-Garcia, D. Rodriguez-Gomez, A general class of holographic superconductors, J. High Energy Phys. 1004 (2010) 092.

[30] S. Franco, A.M. Garcia-Garcia, D. Rodriguez-Gomez, A holographic approach to phase transitions, Phys. Rev. D 81 (2010) 041901(R).

[31] Q. Pan, B. Wang, General holographic superconductor models with Gauss–Bonnet corrections, Phys. Lett. B 693 (2010) 159.

[32] Y. Peng, Q. Pan, Stückelberg holographic superconductor models with backreactions, Commun. Theor. Phys. 59 (2013) 110.

[33] D. Arean, L.A. Pando Zayas, I.S. Landea, A. Scardicchio, The holographic disorder-driven superconductor–metal transition, arXiv:1507.02280 [hep-th].

[34] M. Baggioli, M. Goykhman, Phases of holographic superconductors with broken translational symmetry, J. High Energy Phys. 1507 (2015) 035.

[35] H. Davoudiasl, H.-S. Lee, I. Lewis, W.J. Marciano, "Dark" Z implications for parity violation, rare meson decays, and Higgs physics, Phys. Rev. D 85 (2012) 115019;
C.-F. Chang, E. Ma, T.-C. Yuan, Multilepton Higgs decays through the dark portal, J. High Energy Phys. 1403 (2014) 054.

[36] A. Achucarro, T. Vachaspati, Semilocal and electroweak strings, Phys. Rep. 327 (2000) 347;
T. Vachaspati, Dark strings, Phys. Rev. D 80 (2009) 063502;
B. Hartmann, F. Arbabzadah, Cosmic strings interacting with dark strings, J. High Energy Phys. 07 (2009) 068;
Y. Brihaye, B. Hartmann, Effect of dark strings on semilocal strings, Phys. Rev. D 80 (2009) 123502.

[37] P. Jean, et al., Early SPI/INTEGRAL measurements of 511 keV line emission from the 4th quadrant of the Galaxy, Astron. Astrophys. 407 (2003) L55.

[38] J. Chang, et al., An excess of cosmic ray electrons at energies of 300–800 GeV, Nature (London) 456 (2008) 362.

[39] O. Adriani, et al., PAMELA Collaboration, An anomalous positron abundance in cosmic rays with energies 1.5–100 GeV, Nature 458 (2009) 607.

[40] G.W. Bennett, et al., Final report of the E821 muon anomalous magnetic moment measurement at BNL, Phys. Rev. D 73 (2006) 072003.

[41] L. Nakonieczny, M. Rogatko, Analytic study on backreacting holographic superconductors with dark matter sector, Phys. Rev. D 90 (2014) 106004.

[42] L. Nakonieczny, M. Rogatko, K.I. Wysokinski, Magnetic field in holographic superconductor with dark matter sector, Phys. Rev. D 91 (2015) 046007.

[43] M. Rogatko, K.I. Wysokinski, Holographic vortices in the presence of dark matter sector, J. High Energy Phys. 1512 (2015) 041.

[44] M. Rogatko, K.I. Wysokinski, Viscosity of holographic fluid in the presence of dark matter sector, J. High Energy Phys. 1608 (2016) 124.

[45] M. Rogatko, K.I. Wysokinski, Holographic superconductivity in the presence of dark matter: basic issues, Acta Phys. Pol. A 130 (2016) 558, arXiv:1603.02068 [physics].

[46] T. Shiromizu, S. Ohashi, R. Suzuki, No-go on strictly stationary spacetimes in four/higher dimensions, Phys. Rev. D 86 (2012) 064041.

[47] B. Bakon, M. Rogatko, Complex scalar field in strictly stationary Einstein–Maxwell-axion-dilaton spacetime with negative cosmological constant, Phys. Rev. D 87 (2013) 084065.

[48] F. Aprile, D. Roest, J.G. Russo, Holographic superconductors from gauged supergravity, J. High Energy Phys. 1106 (2011) 040.

[49] R.-G. Cai, L. Li, L.-F. Li, A holographic p-wave superconductor model, J. High Energy Phys. 1401 (2014) 032.

[50] Y. Peng, Holographic entanglement entropy in superconductor phase transition with dark matter sector, Phys. Lett. B 750 (2015) 420.

[51] L. Nakonieczny, M. Rogatko, K.I. Wysokinski, Analytic investigation of holographic phase transitions influenced by dark matter sector, Phys. Rev. D 92 (2015) 066008.

[52] M. Rogatko, K.I. Wysokinski, P-wave holographic superconductor/insulator phase transitions affected by dark matter sector, J. High Energy Phys. 1603 (2016) 215.

[53] M. Rogatko, K.I. Wysokinski, Condensate flow in holographic models in the presence of dark matter, J. High Energy Phys. 1610 (2016) 152.

[54] S. Ryu, T. Takayanagi, Holographic derivation of entanglement entropy from AdS/CFT, Phys. Rev. Lett. 96 (2006) 181602.

[55] S. Ryu, T. Takayanagi, Aspects of holographic entanglement entropy, J. High Energy Phys. 0608 (2006) 045.

[56] T. Nishioka, T. Takayanagi, Entropy and closed string tachyons, J. High Energy Phys. 0701 (2007) 090.

[57] I.R. Klebanov, D. Kutasov, A. Murugan, Entanglement as a probe of confinement, Nucl. Phys. B 796 (2008) 274.

[58] A. Pakman, A. Parnachev, Topological entanglement entropy and holography, J. High Energy Phys. 0807 (2008) 097.

[59] T. Nishioka, S. Ryu, T. Takayanagi, Holographic entanglement entropy: an overview, J. Phys. A 42 (2009) 504008.

[60] L.-Y. Hung, R.C. Myers, M. Smolkin, On holographic entanglement entropy and higher curvature gravity, J. High Energy Phys. 1104 (2011) 025.

[61] J. de Boer, M. Kulaxizi, A. Parnachev, Holographic entanglement entropy in Lovelock gravities, J. High Energy Phys. 1107 (2011) 109.

[62] N. Ogawa, T. Takayanagi, Higher derivative corrections to holographic entanglement entropy for AdS solitons, J. High Energy Phys. 1110 (2011) 147.

[63] T. Albash, C.V. Johnson, Holographic entanglement entropy and renormalization group flow, J. High Energy Phys. 1202 (2012) 095.

[64] R.C. Myers, A. Singh, Comments on holographic entanglement entropy and RG flows, J. High Energy Phys. 1204 (2012) 122.

[65] X.M. Kuang, E. Papantonopoulos, B. Wang, Entanglement entropy as a probe to the proximity effect in holographic superconductors, J. High Energy Phys. 1405 (2014) 130.

[66] Y. Peng, Q. Pan, Holographic entanglement entropy in general holographic superconductor models, J. High Energy Phys. 1406 (2014) 011.

[67] R.-G. Cai, S. He, L. Li, Y.-L. Zhang, Holographic entanglement entropy in insulator/superconductor transition, J. High Energy Phys. 1207 (2012) 088, arXiv:1203.6620 [hep-th].

[68] R.-G. Cai, S. He, L. Li, L.-F. Li, Entanglement entropy and Wilson loop in Stückelberg holographic insulator/superconductor model, J. High Energy Phys. 1210 (2012) 107, arXiv:1209.1019 [hep-th].

[69] W.P. Yao, J.L. Jing, Holographic entanglement entropy in metal/superconductor phase transition with Born–Infeld electrodynamics, Nucl. Phys. B 889 (2014) 109.

[70] Y. Peng, Y. Liu, A general holographic metal/superconductor phase transition model, J. High Energy Phys. 1502 (2015) 082.

[71] T. Albash, C.V. Johnson, Holographic studies of entanglement entropy in superconductors, J. High Energy Phys. 1205 (2012) 079, arXiv:1202.2605 [hep-th].

[72] B. Swingle, T. Senthil, Universal crossovers between entanglement entropy and thermal entropy, Phys. Rev. B 87 (2013) 045123.

[73] R.-G. Cai, S. He, L. Li, Y.-L. Zhang, Holographic entanglement entropy on p-wave superconductor phase transition, J. High Energy Phys. 1207 (2012) 027.

[74] L.-F. Li, R.-G. Cai, L. Li, C. Shen, Entanglement entropy in a holographic p-wave superconductor model, Nucl. Phys. B 894 (2015) 15.

[75] P. Breitenlohner, D.Z. Freedman, Positive energy in Anti-de Sitter backgrounds and gauged extended supergravity, Phys. Lett. B 115 (1982) 197.

11

Double quarkonium production at high Feynman-x

Sergey Koshkarev, Stefan Groote *

Institute of Physics, University of Tartu, Tartu 51010, Estonia

Editor: Stephan Stieberger

Abstract

In this paper we give estimates for the proton–proton cross sections into pairs of quarkonium states J/ψ, $\psi(2S)$, $\Upsilon(1S)$ and $\Upsilon(2S)$ at the scheduled AFTER@LHC energy of 115 GeV. The estimates are based on the intrinsic heavy quark mechanism which is observable for high values of x_F, a range outside the dominance of single parton and double parton scattering.

1. Introduction

In the era of high luminosity and high energy accelerators the associated heavy quarkonium production plays a special role as a testing ground to study multiple parton scattering in a single hadron collision. Significant progress on the Double Parton Scattering (DPS) has been provided by the Tevatron and the LHC in measuring the productions of $J/\psi + W$ [1], $J/\psi + Z$ [2], $J/\psi + $ charm [3] and $J/\psi + J/\psi$ [4–6]. Therefore and for many other reasons, heavy quarkonium production is always a hot topic in high energy physics, as this kind of physics is an ideal probe for testing quantum chromodynamics.

Current colliders provide access only to the physics at low values of the Feynman parameter x_F. However, significant interest is given also for physics at high x_F [7–11]. This region will be accessible at a future fixed-target experiment at the LHC (AFTER@LHC). In a recent paper, Jean-Philippe Lansberg and Hua-Sheng Shao discussed contributions of the DPS to the double-

---

* Corresponding author.
 E-mail address: groote@ut.ee (S. Groote).

quarkonium production in the kinematic region of the AFTER@LHC [12]. However, as we learned from the low statistics NA3 experiment measurements of the double J/ψ production [13, 14] and the observation of the double charmed baryons by the SELEX collaboration [15–17], the double intrinsic heavy quark mechanism can be the leading production mechanism [18,19].

The existence of a non-perturbative intrinsic heavy quark component in the nucleon is a rigorous prediction of QCD. Intrinsic charm and bottom quarks are contained in the wavefunction of a light hadron – from diagrams where the heavy quarks are multiply attached via gluons to the valence quarks. In detail, the intrinsic heavy quark components are contributed by the twist-six contribution of the operator product expansion proportional to $1/m_Q^2$ [20,21]. In this case, the frame-independent light-front wavefunction of the light hadron has maximum probability if the Fock state is minimally off-shell. This means that all the constituents are at rest in the hadron rest frame and thus have the same rapidity y if the hadron is boosted. Equal rapidity occurs if the light-front momentum fractions $x = k^+/P^+$ of the Fock state constituents are proportional to their transverse masses, $x_i \propto m_{T,i} = (m_i^2 + k_{T,i}^2)^{1/2}$, i.e. if the heavy constituents have the largest momentum fractions. This features the BHPS model given by Brodsky, Hoyer, Peterson and Sakai for the distribution of intrinsic heavy quarks [22,23].

In the BHPS model the wavefunction of a hadron in QCD can be represented as a superposition of Fock state fluctuations, e.g. $|h\rangle \sim |h_l\rangle + |h_l g\rangle + |h_l Q\bar{Q}\rangle \ldots$, where h_l is the light quark content, and $Q = c, b$. If the projectile interacts with the target, the coherence of the Fock components is broken and the fluctuation can hadronize. The intrinsic heavy quark Fock components are generated by virtual interactions such as $gg \to Q\bar{Q}$ where the gluons couple to two or more valence quarks of the projectile. The probability to produce such $Q\bar{Q}$ fluctuations scales as $\alpha_s^2(m_Q^2)/m_Q^2$ relative to the leading-twist production.

Following Refs. [18,22,23], the general formula for the probability distribution of an n-particle intrinsic heavy quark Fock state as a function of the momentum fractions x_i and the transfer momenta $\vec{k}_{T,i}$ can be written as

$$\frac{dP_{iQ}}{\prod_{i=1}^{n} dx_i d^2 k_{T,i}} \propto \alpha_s^4(M_{Q\bar{Q}}) \frac{\delta\left(\sum_{i=1}^{n} \vec{k}_{T,i}\right)\delta\left(1 - \sum_{i=1}^{n} x_i\right)}{\left(m_h^2 - \sum_{i=1}^{n} m_{T,i}^2/x_i\right)^2}, \tag{1}$$

where m_h is the mass of the initial hadron. The probability distribution for the production of two heavy quark pairs is given by

$$\frac{dP_{iQ_1 Q_2}}{\prod_{i=1}^{n} dx_i d^2 k_{T,i}} \propto \alpha_s^4(M_{Q_1 \bar{Q}_1})\alpha_s^4(M_{Q_2 \bar{Q}_2}) \frac{\delta\left(\sum_{i=1}^{n} \vec{k}_{T,i}\right)\delta\left(1 - \sum_{i=1}^{n} x_i\right)}{\left(m_h^2 - \sum_{i=1}^{n} m_{T,i}^2/x_i\right)^2}. \tag{2}$$

If one is interested in the calculation of the x distribution, one can simplify the formula by replacing $m_{T,i}$ by the effective mass $\hat{m}_i = (m_i^2 + \langle k_{T,i}^2 \rangle)^{1/2}$ and neglecting the masses of the light quarks,

$$\frac{dP_{iQ_1 Q_2}}{\prod_{i=1}^{n} dx_i} \propto \alpha_s^4(M_{Q_1 \bar{Q}_1})\alpha_s^4(M_{Q_2 \bar{Q}_2}) \frac{\delta\left(1 - \sum_{i=1}^{n} x_i\right)}{\left(\sum_{i=1}^{n} \hat{m}_{T,i}^2/x_i\right)^2}. \tag{3}$$

The x_F distribution for the double quarkonium production $X_1 + X_2$ (with $X_i = J/\psi, \psi(2S)$, $\Upsilon(1S), \Upsilon(2S), \ldots$) is then given by [18]

$$\frac{dP_{iQ_1 Q_2}}{dx_{X_1 X_2}} = \int \prod_{i=1}^{n} dx_i dx_{X_1} dx_{X_2} \frac{dP_{iQ_1 Q_2}}{\prod_{i=1}^{n} dx_i} \delta(x_{X_1 X_2} - x_{X_1} - x_{X_2})$$
$$\times \delta(x_{X_1} - x_{Q_1} - x_{\bar{Q}_1})\delta(x_{X_2} - x_{Q_2} - x_{\bar{Q}_2}). \tag{4}$$

The BHPS model assumes that the vertex function in the intrinsic heavy quark wavefunction is varying relatively slowly. The particle distributions are then controlled by the light-cone energy denominator and the phase space. The Fock states can be materialized by a soft collision in the target which brings the state on shell. The distribution of produced open and hidden charm states will reflect the underlying shape of the Fock state wavefunction.

In this paper we investigate the double intrinsic heavy quark mechanism for the double-quarkonium production in the high Feynman-x region at the AFTER@LHC experiment. In this particular case the production of the double quarkonium plays a special role as it provides the direct access to extract the double heavy quark probabilities P_{icc}, P_{icb} and P_{ibb}. To the best of our knowledge the x_F distribution for double-quarkonium production in proton beam events has not yet been measured (cf. also a comment at the end of the third paragraph in the Introduction of Ref. [18]). Therefore, our estimates cannot be compared to existing data but wait for future confirmation by experiments like AFTER@LHC, for which we give numerical values. As an innovative element, for our analysis we use the color evaporation model, applied also to excited $2S$ states. Finally, in the conclusions we discuss why existing LHC measurements cannot be interpreted as non-evidence of the intrinsic heavy quark mechanism.

2. Double-quarkonium production cross section

The production cross section of the quarkonium can be obtained as an application of the quark–hadron duality principle known as color evaporation model (CEM) [24]. In this model the cross section of quarkonium are obtained by calculating the production of a $Q\bar{Q}$ in the small invariant mass interval between $2m_Q$ and the threshold to produce open heavy-quark hadrons, $2m_H$. The $Q\bar{Q}$ pair has $3 \times \bar{3} = (1 + 8)$ color components, consisting of a color-singlet and a color-octet. Therefore, the probability that a color-singlet is formed and produces a quarkonium state is $1/(1 + 8)$, and the model predicts

$$\sigma(Q\bar{Q}) = \frac{1}{9} \int_{2m_Q}^{2m_H} dM_{Q\bar{Q}} \frac{d\sigma_{Q\bar{Q}}}{dM_{Q\bar{Q}}} = \frac{1}{9} \int_{4m_Q^2}^{4m_H^2} dM_{Q\bar{Q}}^2 \frac{d\sigma_{Q\bar{Q}}}{dM_{Q\bar{Q}}^2}, \tag{5}$$

where $\sigma_{Q\bar{Q}}$ is the production cross section of the heavy quark pairs and $\sigma(Q\bar{Q})$ is a sum of production cross sections of all quarkonium states in the duality interval. For example, in case of charmonium states one has $\sigma(Q\bar{Q}) = \sigma(J/\psi) + \sigma(\psi(2S)) + \ldots$. According to a simple statistical counting, the fraction of the total color-singlet cross section into a quarkonium state is given by

$$\sigma(X) = \rho_X \cdot \sigma(Q\bar{Q}) \tag{6}$$

$(X = J/\psi, \psi(2S), \ldots)$ with

$$\rho_X = \frac{2J_X + 1}{\sum_i (2J_i + 1)}, \tag{7}$$

where J_X is the spin of the quarkonium state X and the sum runs over all quarkonium states. In case of the J/ψ meson the calculation gives

$$\rho_{J/\psi} \simeq 0.2. \tag{8}$$

This statistical counting rule works well for J/ψ but not so well for other charmonium states, even not for $\psi(2S)$. Instead, in this paper we use the fact that a quarkonium production matrix element is proportional to the absolute square of the radial wave function at the origin [25], so that

$$\sigma(J/\psi) : \sigma(\psi(2S)) \approx |R_{J/\psi}(0)|^2 : |R_{\psi(2S)}(0)|^2. \tag{9}$$

The absolute square of the radial wave function $R_X(0)$ of the quarkonium state $X = J/\psi$, $\psi(2S), \ldots$ at the origin is determined by the leptonic decay rate [26]

$$\Gamma(X \to e^+e^-) = \frac{4N_c \alpha_{\text{em}}^2 e_Q^2}{3} \frac{|R_X(0)|^2}{M_X^2} \left(1 - \frac{16\alpha_s}{3\pi}\right), \tag{10}$$

where $N_c = 3$ is the number of quark colors, e_Q is the electric charge of the heavy quark, and M_X is the mass of the quarkonium state X. Splitting $\sigma(Q\bar{Q})$ up into the different quarkonium states one can obtain the corresponding production cross sections.

According to the intrinsic heavy quark mechanism the production cross section $\sigma(Q\bar{Q})$ of a $Q\bar{Q}$ pair in the duality interval is given by [18]

$$\sigma^{iQ}(Q\bar{Q}) = f_{Q\bar{Q}/p}^{iQ} \cdot P_{iQ} \cdot \sigma_{pp}^{inel} \cdot \frac{1}{9} \frac{\mu^2}{4\hat{m}_Q}, \tag{11}$$

where $\mu \approx 0.2 \text{ GeV}$ denotes the soft interaction scale parameter, $f_{Q\bar{Q}/p}^{iQ}$ is the fragmentation ratio of the $Q\bar{Q}$ pair written as

$$f_{Q\bar{Q}/p}^{iQ} = \int\limits_{4m_Q^2}^{4m_H^2} dM_{Q\bar{Q}}^2 \frac{dP_{iQ}}{dM_{Q\bar{Q}}^2} \bigg/ \int\limits_{4m_Q^2}^{s} dM_{Q\bar{Q}}^2 \frac{dP_{iQ}}{dM_{Q\bar{Q}}^2}, \tag{12}$$

and the inelastic proton–proton cross section σ_{pp}^{inel} in the region of $\sqrt{s} \geq 100 \text{ GeV}$ is obtained by the approximation [27]

$$\sigma_{pp}^{inel} = 62.59 \hat{s}^{-0.5} + 24.09 + 0.1604 \ln(\hat{s}) + 0.1433 \ln^2(\hat{s}) \text{ mb}, \tag{13}$$

where $\hat{s} = s/2m_p^2$. At the AFTER@LHC energy $\sqrt{s} = 115 \text{ GeV}$, one obtains $\sigma_{pp}^{inel} = 28.4 \text{ mb}$.

2.1. Double-charmonium production from $|uudc\bar{c}c\bar{c}\rangle$

The double-charmonium production cross section $\sigma(c\bar{c} + c\bar{c})$ from the Fock state $|uudc\bar{c}c\bar{c}\rangle$ can be written obviously as

$$\sigma^{icc}(c\bar{c} + c\bar{c}) = (f_{c\bar{c}/p}^{icc})^2 P_{icc} \sigma_{pp}^{inel} \frac{1}{9} \frac{1}{9} \frac{\mu^2}{4\hat{m}_c}, \tag{14}$$

where the fragmentation ratio $f_{Q\bar{Q}/p}^{iQ_1Q_2}$ is obtained as

$$f_{Q\bar{Q}/p}^{iQ_1Q_2} = \int\limits_{4m_Q^2}^{4m_H^2} dM_{Q\bar{Q}}^2 \frac{dP_{iQ_1Q_2}}{dM_{Q\bar{Q}}^2} \bigg/ \int\limits_{4m_Q^2}^{s} dM_{Q\bar{Q}}^2 \frac{dP_{iQ_1Q_2}}{dM_{Q\bar{Q}}^2}. \tag{15}$$

In this case ($Q = c$, $H = D$) we use $m_c \approx 1.3$ GeV for the mass of c quark, $\hat{m}_c = 1.5$ GeV for the effective transverse c-quark mass, and $m_D = 1.87$ GeV for the mass of the D meson. For the integrated probability distribution we take the value $P_{icc} \simeq 0.002$ [18].

Combining Eqs. (14) and (15), we may expect the double-charmonium production cross section to be

$$\sigma^{icc}(c\bar{c} + c\bar{c}) \approx 1.5 \times 10^2 \text{ pb.}$$

Analyzing the values of the radial wave functions at the origin [26], one finds

$$\sigma(J/\psi + J/\psi) : \sigma(J/\psi + \psi(2S)) : \sigma(\psi(2S) + \psi(2S)) \approx 1 : 0.65 : 0.43.$$

Taking into account Eq. (8) and the generalization of Eq. (6),

$$\sigma(X_1 + X_2) = \rho_{X_1}\rho_{X_2} \cdot \sigma(Q\bar{Q} + Q\bar{Q}), \tag{16}$$

one obtains

$$\sigma^{icc}(J/\psi + J/\psi) \approx 6.0 \text{ pb}$$
$$\sigma^{icc}(J/\psi + \psi(2S)) \approx 3.9 \text{ pb}$$
$$\sigma^{icc}(\psi(2S) + \psi(2S)) \approx 2.6 \text{ pb.} \tag{17}$$

2.2. Associated charmonium–bottomonium production from $|uudc\bar{c}b\bar{b}\rangle$

Following Refs. [28,29], the associated charmonium–bottomonium production cross section is given by

$$\sigma^{icb}(c\bar{c} + b\bar{b}) = f^{icb}_{c\bar{c}/p} \, f^{icb}_{b\bar{b}/p} \, P_{icb} \, \sigma^{incl}_{pp} \frac{1}{9}\frac{1}{9}\frac{\mu^2}{4\hat{m}_b^2}\left(\frac{\hat{m}_c}{\hat{m}_b}\frac{\alpha_s(M_{b\bar{b}})}{\alpha_s(M_{c\bar{c}})}\right)^4. \tag{18}$$

Applying Eq. (15) to this case ($Q = b$, $H = B$) we use $m_b \approx 4.2$ GeV for the mass of the b quark, $\hat{m}_b = 4.6$ GeV for the effective transverse b-quark mass, and $m_B = 5.3$ GeV for the mass of the B meson. The value of P_{icb} is unknown at this moment but we assume it to be approximately equal to P_{icc}. Finally, we calculate the associated charmonium–bottomonium production cross section to be

$$\sigma^{icb}(c\bar{c} + b\bar{b}) = 0.35 \text{ pb.} \tag{19}$$

In this section we calculate only the production cross section for the ground states,

$$\sigma^{icb}(J/\psi + \Upsilon(1S)) \approx 14 \text{ fb.} \tag{20}$$

2.3. Double-bottomonium production from $|uudb\bar{b}b\bar{b}\rangle$

We already have all ingredients for the calculation of the production cross section of the double-bottomonium states except for $P_{ibb} = (\hat{m}_c/\hat{m}_b)^2 \cdot P_{icb}$, so the numerical value will be

$$\sigma^{ibb}(b\bar{b} + b\bar{b}) = 0.03 \text{ pb,} \tag{21}$$

and the cross sections for the particular double-bottomonium states are given by

$$\sigma^{ibb}(\Upsilon(1S) + \Upsilon(1S)) \approx 1.2 \text{ fb}$$
$$\sigma^{ibb}(\Upsilon(1S) + \Upsilon(2S)) \approx 0.6 \text{ fb}$$
$$\sigma^{ibb}(\Upsilon(2S) + \Upsilon(2S)) \approx 0.3 \text{ fb.} \tag{22}$$

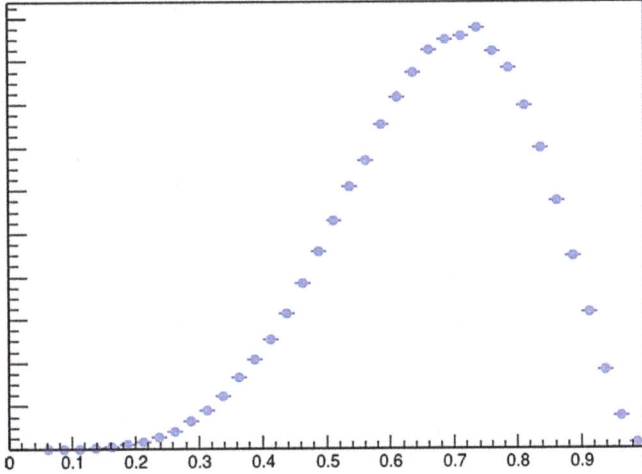

Fig. 1. The histogram shows the x_F distribution of the J/ψ pair due to the double intrinsic heavy quark mechanism in arbitrary units.

3. Conclusions

In this paper we investigated the contribution of the double intrinsic heavy quark mechanism to the production of a quarkonium pair. It is clear that Single Parton Scattering (SPS) and Double Parton Scattering (DPS) provide the main contributions to the double quarkonium production cross section. However, both these contributions are vanishing fast with increasing Feynman parameter x_F. On the other hand, the contribution from the double intrinsic heavy quark mechanism mainly grows with x_F (see Fig. 1). If one considers proton–proton collisions in the center-of-mass frame, one can distinguish between charm production at positive x_F coming from the intrinsic heavy in the beam proton and negative x_F coming from the intrinsic heavy in the nucleons of the target. As it is shown in Ref. [12], the DPS contribution starts at $x_F = -0.5$. This is the region where the double intrinsic heavy quark mechanism is from the target and contributes on the average. On the other hand, the double intrinsic charm becomes the leading production mechanism at high x_F, $\langle x_{\psi\psi} \rangle \simeq 0.64$ [18].

Another interesting aspect to be discussed is $\sigma(J/\psi + J/\psi)/\sigma(J/\psi)$. The only result for this ratio with access to high values of x_F was provided by the NA3 experiment and was found to be $(3 \pm 1) \times 10^{-4}$ with 150 and 280 GeV/c pion [13] and 400 GeV/c proton [14] beams. The same ratio measured by the LHCb Collaboration is found to be $(5.1 \pm 1.0 \pm 0.6^{+1.2}_{-1.0}) \times 10^{-4}$ [4]. This result can be interpreted wrongly as non-evidence for the intrinsic heavy quark mechanism. However, the traditional $q\bar{q}$ annihilation mechanism and the leading gluon–gluon fusion mechanism for LHCb are not in good agreement with the NA3 data (cf. the discussion in Ref. [29]) which shows that perturbative QCD can explain neither the NA3 cross section nor the x_F distribution. Compared to this, the double intrinsic heavy quark mechanism reproduces x_F dependencies very well [18], at least for the case measured by NA3, namely the case of pion–nucleon scattering [13] (cf. Fig. 2).

Current experimental knowledge does not give us much information about the main contribution of the double intrinsic heavy quark mechanism. In our calculations we use P_{icc} from data with low statistics and provide other formal assumptions. However, the key feature of

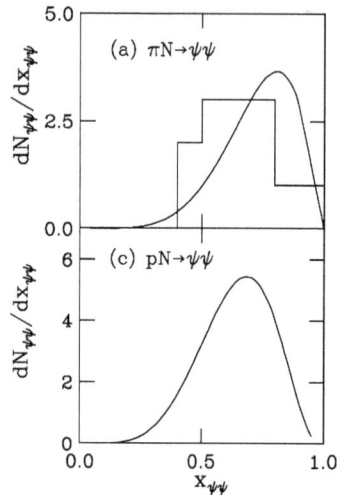

Fig. 2. x_F distribution for (a) $\pi N \to \psi\psi$ and (c) $pN \to \psi\psi$ ($x_F \in [0, 1]$). The plots are taken from Ref. [18]. Shown are NA3 $\pi^- N$ data at 150 and 280 GeV/c [13] (histograms) and estimates of the intrinsic heavy quark mechanism (solid curve).

AFTER@LHC is the access to high Feynman-x. Therefore, the measurement of the double quarkonium production can provide more accurate data and shed more light on the role of the intrinsic heavy quark mechanism.

Acknowledgements

This work was supported by the Estonian Research Council under Grant No. IUT2-27.

References

[1] G. Aad, et al., ATLAS Collaboration, J. High Energy Phys. 1404 (2014) 172.
[2] G. Aad, et al., ATLAS Collaboration, Eur. Phys. J. C 75 (5) (2015) 229.
[3] R. Aaij, et al., LHCb Collaboration, J. High Energy Phys. 1206 (2012) 141; J. High Energy Phys. 1403 (2014) 108 (Addendum).
[4] R. Aaij, et al., LHCb Collaboration, Phys. Lett. B 707 (2012) 52.
[5] V.M. Abazov, et al., D0 Collaboration, Phys. Rev. D 90 (11) (2014) 111101.
[6] V. Khachatryan, et al., CMS Collaboration, J. High Energy Phys. 1409 (2014) 094.
[7] S.J. Brodsky, F. Fleuret, C. Hadjidakis, J.P. Lansberg, Phys. Rep. 522 (2013) 239.
[8] J.P. Lansberg, S.J. Brodsky, F. Fleuret, C. Hadjidakis, Few-Body Syst. 53 (2012) 11.
[9] J.P. Lansberg, et al., EPJ Web Conf. 66 (2014) 11023.
[10] A. Rakotozafindrabe, et al., PoS DIS 2013 (2013) 250.
[11] S.J. Brodsky, A. Kusina, F. Lyonnet, I. Schienbein, H. Spiesberger, R. Vogt, Adv. High Energy Phys. 2015 (2015) 231547.
[12] J.P. Lansberg, H.S. Shao, Nucl. Phys. B 900 (2015) 273.
[13] J. Badier, et al., NA3 Collaboration, Phys. Lett. B 114 (1982) 457.
[14] J. Badier, et al., NA3 Collaboration, Phys. Lett. B 158 (1985) 85.
[15] M. Mattson, et al., SELEX Collaboration, Phys. Rev. Lett. 89 (2002) 112001.
[16] M. Mattson, Ph.D. thesis, Carnegie Mellon University, 2002.
[17] A. Ocherashvili, et al., SELEX Collaboration, Phys. Lett. B 628 (2005) 18.
[18] R. Vogt, S.J. Brodsky, Phys. Lett. B 349 (1995) 569.
[19] S. Koshkarev, V. Anikeev, arXiv:1605.03070 [hep-ph].

[20] S.J. Brodsky, J.C. Collins, S.D. Ellis, J.F. Gunion, A.H. Mueller, DOE/ER/40048-21 P4, SLAC-PUB-15471.
[21] M. Franz, M.V. Polyakov, K. Goeke, Phys. Rev. D 62 (2000) 074024.
[22] S.J. Brodsky, P. Hoyer, C. Peterson, N. Sakai, Phys. Lett. B 93 (1980) 451.
[23] S.J. Brodsky, C. Peterson, N. Sakai, Phys. Rev. D 23 (1981) 2745.
[24] J.F. Amundson, O.J.P. Eboli, E.M. Gregores, F. Halzen, Phys. Lett. B 372 (1996) 127.
[25] B. Humpert, P. Méry, Z. Phys. C 20 (1983) 83.
[26] E.J. Eichten, C. Quigg, Phys. Rev. D 52 (1995) 1726.
[27] M.M. Block, F. Halzen, Phys. Rev. D 86 (2012) 014006.
[28] R. Vogt, S.J. Brodsky, Nucl. Phys. B 438 (1995) 261.
[29] R. Vogt, Nucl. Phys. B 446 (1995) 159.

12

ODE/IM correspondence for modified $B_2^{(1)}$ affine Toda field equation

Katsushi Ito, Hongfei Shu *

Department of Physics, Tokyo Institute of Technology, Tokyo, 152-8551, Japan

Editor: Stephan Stieberger

Abstract

We study the massive ODE/IM correspondence for modified $B_2^{(1)}$ affine Toda field equation. Based on the ψ-system for the solutions of the associated linear problem, we obtain the Bethe ansatz equations. We also discuss the T–Q relations, the T-system and the Y-system, which are shown to be related to those of the $A_3/\mathbf{Z}_2$ integrable system. We consider the case that the solution of the linear problem has a monodromy around the origin, which imposes nontrivial boundary conditions for the T-/Y-system. The high-temperature limit of the T- and Y-system and their monodromy dependence are studied numerically.

1. Introduction

It has been recognized that the relation between classical and quantum integrable systems is useful for studying non-perturbative properties of supersymmetric gauge theories and the AdS/CFT correspondence [1–3]. The ODE/IM correspondence [4–6] provides an interesting example of this classical/quantum correspondence, which relates the spectral determinants of certain ordinary differential equations (ODE) to the Bethe ansatz equations in the massless limit of certain integrable models (IM). It is an interesting problem to make a complete list of this

---

* Corresponding author.
E-mail address: h.shu@th.phys.titech.ac.jp (H. Shu).

ODE/IM correspondence. The ordinary differential equations for the integrable models related to classical Lie algebras have been proposed in [7]. The Wronskian of the solutions obeys the functional relations called the ψ-system, which leads to the Bethe ansatz equations of the related quantum integrable system. The ψ-system for classical Lie algebra has been reformulated in the form of the matrix valued linear differential equations [8], where the Bethe ansatz equations of the integrable models associated with the untwisted affine Lie algebra $X^{(1)}$ of a classical Lie algebra X are related to the linear differential equations associated with the Langlands dual $(X^{(1)})^\vee$.

The ODE/IM correspondence has been generalized to massive integrable models. It was found that for the classical sinh-Gordon equation modified by a conformal transformation, the spectral problem for the associated linear problem leads to the functional relations of the quantum sine-Gordon model [9]. By taking the conformal limit, it reduces to the ODE/IM correspondence for the Schrödinger type differential equation [4,5].

Recently the massive ODE/IM correspondence has been generalized to a class of modified affine Toda field equations [10–15]. In particular, Locke and one of the present authors studied the modified affine Toda equations for affine Lie algebra $\hat{\mathfrak{g}}^\vee$, where $\hat{\mathfrak{g}}$ is an untwisted affine Lie algebra including exceptional type [13]. It has been shown that from their associated linear problems one obtains the ψ-system which leads to the Bethe ansatz equations for the affine Lie algebra $\hat{\mathfrak{g}}$.

It would be an interesting problem to explore the modified affine Toda field equation with an affine Lie algebra $\hat{\mathfrak{g}}$ which is not of the form of the Langlands dual of an untwisted one, where the corresponding integrable models are not identified yet. In this paper we will work with the modified affine Toda field equation associated with the affine Lie algebra $B_2^{(1)}$ (or $C_2^{(1)}$), which provides the simplest and nontrivial example. This equation also appears in the study of the area of minimal surface with a null-polygonal boundary in AdS_4 spacetime [16–19], which is dual to the gluon scattering amplitudes with specific momentum configurations. The equation of motion of strings is described by the $B_2^{(1)}$ affine Toda field equation modified by the conformal transformation. The Stokes problem of the associated linear system determines the functional equations for the cross-rations of external momenta. These functional relations are known to be the same as the Y-system of the homogeneous sine-Gordon model [20,21] and the free energy of the Y-system determines the area of the minimal surface.

The purpose of this paper is to apply the massive ODE/IM correspondence to the modified $B_2^{(1)}$ affine Toda field equation and to investigate the functional relations for the Stokes coefficients of the linear problem, which include the Bethe ansatz equations, the T–Q relations, the T-system and the Y-system. We study the boundary condition of the T-system arising from the nontrivial monodromy of the linear problem solution around the origin. This monodromy condition also appears in the study of the form factors via the AdS/CFT correspondence [22,23].

This paper is organized as follows: In sect. 2, we introduce modified $B_2^{(1)}$ affine Toda equation and the associated linear problem. In sect. 3, we discuss the ψ-system and derive the Bethe ansatz equations. In sect. 4, we discuss the spinor representation of B_2 in detail and study the quantum Wronskian and the T–Q relations. In sect. 5, we argue the T-system and Y-system and their boundary conditions which come from the monodromy of the solution of the linear system around the origin. In sect. 6, we investigate the high-temperature limit of the Y-system in the presence of monodromy. Sect. 7 is devoted for conclusions and discussion. In the appendix, we summarize the auxiliary T-functions and their functional relations used in this paper.

2. Modified $B_2^{(1)}$ affine Toda field equation

The Lie algebra $B_2 = so(5)$ has simple roots $\alpha_1 = e_1 - e_2$ and $\alpha_2 = e_2$, where e_i $(i = 1, 2)$ is an orthonormal basis of $\mathbf{R}^2$. We denote the highest root by θ, which is given by $\theta = \alpha_1 + 2\alpha_2$, and define the extended root $\alpha_0 = -\theta$. $\omega_1 = e_1$ and $\omega_2 = \frac{1}{2}(e_1 + e_2)$ are the fundamental weights satisfying $2\omega_i \cdot \alpha_j / \alpha_j^2 = \delta_{ij}$. Let $\{H^i, E_\alpha\}$ $(i = 1, 2, \alpha \in \Delta)$ be the Chevalley basis of B_2, where Δ is the set of roots.

Let $\phi = (\phi^1, \phi^2)$ be the two-component scalar field on the complex plane with coordinates $(z, \bar{z})$. We define the modified affine Toda field equation for $B_2^{(1)}$ by

$$\partial\bar{\partial}\phi - \frac{m^2}{\beta}\left(\alpha_1 e^{\beta\alpha_1\cdot\phi} + 2\alpha_2 e^{\beta\alpha_2\cdot\phi} + p(z)\bar{p}(\bar{z})\alpha_0 e^{\beta\alpha_0\cdot\phi}\right) = 0, \tag{2.1}$$

where $\partial = \frac{\partial}{\partial z}, \bar{\partial} = \frac{\partial}{\partial\bar{z}}$, m is a mass parameter and β is a coupling parameter. $p(z)$ is a holomorphic function of z and is chosen as

$$p(z) = z^{4M} - s^{4M}, \tag{2.2}$$

with $M > \frac{1}{3}$ and s is a complex parameter. This equation is obtained by the conformal transformation $z \to w$ with $\frac{\partial w}{\partial z} = p^{\frac{1}{4}}$ and the field redefinition $\phi \to \phi - \frac{1}{4\beta}\rho^\vee \log(p\bar{p})$, where $\rho^\vee = \omega_1 + 2\omega_2$ is the co-Weyl vector.[1] Note that the Coxeter number of B_2 is 4. Eq. (2.1) can be written in the form of the compatibility condition $[\partial + A_z, \bar{\partial} + A_{\bar{z}}] = 0$ of the linear differential equations defined in a B_2-module:

$$(\partial + A_z)\Psi = 0, \quad (\bar{\partial} + A_{\bar{z}})\Psi = 0, \tag{2.3}$$

where the connections are defined by

$$A_z = \frac{\beta}{2}\partial\phi \cdot H + me^\lambda\left(e^{\beta\alpha_1\cdot\phi/2}E_{\alpha_1} + e^{\beta\alpha_2\cdot\phi/2}E_{\alpha_2} + p(z)e^{\beta\alpha_0\cdot\phi/2}E_{\alpha_0}\right),$$

$$\bar{A}_{\bar{z}} = -\frac{\beta}{2}\bar{\partial}\phi \cdot H + me^{-\lambda}\left(e^{\beta\alpha_1\cdot\phi/2}E_{-\alpha_1} + e^{\beta\alpha_2\cdot\phi/2}E_{-\alpha_2} + \bar{p}(\bar{z})e^{\beta\alpha_0\cdot\phi/2}E_{-\alpha_0}\right). \tag{2.4}$$

Here λ is the spectral parameter. We are interested in the special class of solutions of (2.1), which satisfy the periodicity condition $\phi(\rho, \theta + \frac{\pi}{4M}) = \phi(\rho, \theta)$ and the boundary conditions at infinity and the origin of the complex plane:

$$\phi(\rho, \theta) = \frac{2M\rho^\vee}{\beta}\log\rho + \cdots, \quad (\rho \to \infty) \tag{2.5}$$

$$\phi(\rho, \theta) = 2g\log\rho + \cdots, \quad (\rho \to 0) \tag{2.6}$$

where we have introduced the polar coordinate (ρ, θ) by $z = \rho e^{i\theta}$ and g is a 2-vector satisfying $\beta\alpha_a \cdot g + 1 > 0$ $(a = 0, 1, 2)$. Due to the special form (2.2) of $p(z)$, (2.1) and the linear problem are invariant under the Symanzik rotation

$$\hat{\Omega}_k : (z, s, \lambda) \to (ze^{\frac{2\pi i k}{4M}}, se^{\frac{2\pi i k}{4M}}, \lambda - \frac{2\pi i k}{4M}), \tag{2.7}$$

---

[1] Note that the modified equation in [17] is $\partial\bar{\partial}\phi - \frac{m^2}{\beta}(\alpha_1\sqrt{p\bar{p}}e^{\beta\alpha_1\cdot\phi} + 2\alpha_2 e^{\beta\alpha_2\cdot\phi} + \sqrt{p\bar{p}}\alpha_0 e^{\beta\alpha_0\cdot\phi}) = 0$, which is obtained by the same conformal transformation but a different field redefinition $\phi \to \phi - \frac{1}{4\beta}\alpha_2\log(p\bar{p})$. This modified equation is related to (2.1) by a field redefinition.

for an integer k. This also acts on the solution $\Psi(z, \bar{z})$, which is denoted as $\Psi_k(z, \bar{z}) := \hat{\Omega}_k \Psi(z, \bar{z})$. The linear problem is also invariant under the transformation:

$$\hat{\Pi} : (\lambda, A_z, A_{\bar{z}}, \Psi) \rightarrow (\lambda - \frac{2\pi i}{4}, SA_zS^{-1}, SA_{\bar{z}}S^{-1}, S\Psi) \tag{2.8}$$

where $S = \exp(\frac{2\pi i}{4}\rho^\vee \cdot H)$.

We now consider the solutions of the linear differential equations (2.3) in the basic $B_2^{(1)}$-module $V^{(a)}$ ($a = 1, 2$) associated with the highest weight ω_a. Let $\mathbf{e}_j^{(a)}$ be the orthonormal basis of $V^{(a)}$ with H^i eigenvalue $(h_j^{(a)})^i$, where $i, j = 1, \cdots, \dim V^{(a)}$. For the Lie algebra B_2, $V^{(1)}$ is 5-dimensional vector representation, whose matrix representation is given by

$$E_{\alpha_1} = e_{1,2} + e_{4,5}, \quad E_{\alpha_2} = \sqrt{2}(e_{2,3} + e_{3,4}), \quad E_{\alpha_0} = -(e_{4,1} + e_{5,2}) \tag{2.9}$$

and $E_{-\alpha_i} = E_{\alpha_i}^T$. Here e_{ab} denotes the matrix whose (i, j)-element is $\delta_{ia}\delta_{jb}$. Similarly, $V^{(2)}$ is a 4-dimensional spinor representation. Its matrix representation is given by

$$E_{\alpha_1} = e_{23}, \quad E_{\alpha_2} = e_{12} + e_{34}, \quad E_{\alpha_0} = e_{41} \tag{2.10}$$

and $E_{-\alpha_i} = E_{\alpha_i}^T$.

We are interested in the small (or subdominant) solution $\Psi^{(a)}$, which decays fastest along the positive real axis. This was studied in [13] for $\hat{\mathfrak{g}}^\vee$ for an untwisted affine Lie algebra $\mathfrak{g}$. In general, the small solution $\Psi^{(a)}$ at large ρ is given by

$$\Psi^{(a)}(z, \bar{z}|\lambda, g) = C^{(a)} \exp\left(-2\mu^{(a)} \frac{\rho^{M+1}}{M+1} m \cosh(\lambda + i\theta(M+1))\right) e^{-i\theta M\rho^\vee \cdot H} \boldsymbol{\mu}^{(a)}, \tag{2.11}$$

with $C^{(a)}$ being a normalization constant. Here $\boldsymbol{\mu}^{(a)}$ and $\mu^{(a)}$ denote the eigenvector and its eigenvalue of the matrix $\Lambda_+ = E_{\alpha_0} + E_{\alpha_1} + E_{\alpha_2}$ with the eigenvalue of the largest real part. Applying the Symanzik rotation $\hat{\Omega}_k$ ($k \in \mathbf{Z}$), one obtains the small solution $\Psi_k^{(a)}$ in the Stokes sector

$$\mathcal{S}_{-k} : \left|\theta + \frac{2\pi k}{4(M+1)}\right| < \frac{\pi}{4(M+1)} \tag{2.12}$$

For the vector representation (2.9), the eigenvalues of Λ_+ are $\sqrt{2}e^{\frac{i\pi}{4}(2k+1)}$ ($k = 0, 1, 2, 3$) and 0. For the spinor representation (2.10), they are ± 1 and $\pm i$. For $V^{(1)}$, one has two eigenvalues with the largest real part and the corresponding solutions in $V^{(1)}$ are not subdominant along the real axis. So we introduce the $\frac{1}{2}$-rotated Symanzik solution $\Psi_{\frac{1}{2}}^{(1)}$. This is a solution of the linear problem with the $\frac{1}{2}$-rotated connection $(A_{\frac{1}{2}})_z$ and $(\bar{A}_{\frac{1}{2}})_{\bar{z}}$ which is obtained by replacing $E_{\pm\alpha_0} \rightarrow -E_{\pm\alpha_0}$ in (2.4). Then the $\Psi_{\frac{1}{2}}^{(1)}$ behaves along the real positive axis as (2.11) with $\mu^{(1)} = \sqrt{2}$ and $\boldsymbol{\mu}^{(1)} = (1, \sqrt{2}, \sqrt{2}, \sqrt{2}, 1)^T$.

We define the basis of the solutions around $\rho = 0$ behaves as $\rho \rightarrow 0$:

$$\mathcal{X}_i^{(a)}(z, \bar{z}|\lambda, g) = e^{-(\lambda+i\theta)\beta g \cdot h_i^{(a)}} \mathbf{e}_i^{(a)} + O(\rho), \quad i = 1, \cdots, \dim V^{(a)} \tag{2.13}$$

which are invariant under $\hat{\Omega}_k$ [13]. The small solution $\Psi_{\frac{1}{2}}^{(1)}$ and $\Psi^{(2)}$ can be expanded in this basis as

$$\Psi^{(1)}_{\frac{1}{2}}(z, \bar{z}|\lambda, g) = \sum_{i=1}^{5} Q^{(1)}_i(\lambda, g) \mathcal{X}^{(1)}_i(z, \bar{z}|\lambda, g),$$

$$\Psi^{(2)}(z, \bar{z}|\lambda, g) = \sum_{i=1}^{4} Q^{(2)}_i(\lambda, g) \mathcal{X}^{(2)}_i(z, \bar{z}|\lambda, g). \tag{2.14}$$

We call $Q^{(a)}_i(\lambda, g)$ the Q-functions. From the relation $\hat{\Omega}_1 \hat{\Pi} \Psi^{(a)} = \Psi^{(a)}$, the coefficients $Q^{(a)}_i(\lambda, g)$ satisfy the quasi-periodicity condition:

$$Q^{(a)}_i(\lambda - \frac{2\pi i}{4M}(M+1), g) = \exp(-\frac{2\pi i}{4}(\rho^\vee + \beta g) \cdot h^{(a)}_i) Q^{(a)}_i(\lambda, g). \tag{2.15}$$

Note that we can rescale z and $\bar{z}$ such that the mass parameter m is fixed to be an arbitrary non-zero constant. Then the Q-functions depend on the mass parameter through s/m.

3. ψ-System and the Bethe ansatz equations

The linear problem in the basic $B^{(1)}_2$-modules $V^{(a)}$ can be also defined in other B_2-modules corresponding to the (anti-)symmetrized tensor product of $V^{(a)}$'s. The inclusion maps between the modules induce the relation between the small solutions, which is called the ψ-system [7]. For example, we consider the inclusion map

$$\iota_1 : V^{(1)} \wedge V^{(1)} \hookrightarrow V^{(2)} \otimes V^{(2)}, \tag{3.1}$$

$$\iota_2 : V^{(2)} \wedge V^{(2)} \hookrightarrow V^{(1)}. \tag{3.2}$$

By these maps the highest weight state $\mathbf{e}^{(1)}_1 \wedge \mathbf{e}^{(1)}_2$ is mapped to $\sqrt{2}\mathbf{e}^{(2)}_1 \otimes \mathbf{e}^{(2)}_1$ and $\mathbf{e}^{(2)}_1 \wedge \mathbf{e}^{(2)}_2$ to $\mathbf{e}^{(1)}_1$. We use this map to relate the solutions of the linear problem defined on the different modules. $\Psi^{(1)}_1 \wedge \Psi^{(1)}_0$ is a solution of the linear problem (2.3) on $V^{(1)} \wedge V^{(1)}$ due to invariance of (2.3) under the Symanzik rotation $\hat{\Omega}_1$. This solution is mapped into the module $V^{(2)} \otimes V^{(2)}$ by ι_1. Now $\Psi^{(2)} \otimes \Psi^{(2)}$ is the unique solution in $V^{(2)} \otimes V^{(2)}$ with the same asymptotic behavior at large ρ. In a similar way we can identify $\Psi^{(2)}_{\frac{1}{2}} \wedge \Psi^{(2)}_{-\frac{1}{2}}$ with $\Psi^{(1)}_{\frac{1}{2}}$. Thus we obtain the ψ-system:

$$\iota_1(\Psi^{(1)}_1 \wedge \Psi^{(1)}_0) = \Psi^{(2)} \otimes \Psi^{(2)}, \tag{3.3}$$

$$\iota_2(\Psi^{(2)}_{\frac{1}{2}} \wedge \Psi^{(2)}_{-\frac{1}{2}}) = \Psi^{(1)}_{\frac{1}{2}} \tag{3.4}$$

Expanding the small solutions in the basis $\{\mathcal{X}^{(a)}_i\}$ and substituting them into the ψ-system, one obtains the functional relation for the Q-functions $Q^{(a)}_1$ and $Q^{(a)}_2$:

$$Q^{(1)}_1(\lambda - \frac{2\pi i}{8M}) Q^{(1)}_2(\lambda + \frac{2\pi i}{8M}) - Q^{(1)}_2(\lambda - \frac{2\pi i}{8M}) Q^{(1)}_1(\lambda + \frac{2\pi i}{8M}) = 2Q^{(2)}_1(\lambda) Q^{(2)}_1(\lambda),$$
$$\tag{3.5}$$

$$Q^{(2)}_1(\lambda - \frac{2\pi i}{8M}) Q^{(2)}_2(\lambda + \frac{2\pi i}{8M}) - Q^{(2)}_2(\lambda - \frac{2\pi i}{8M}) Q^{(2)}_1(\lambda + \frac{2\pi i}{8M}) = Q^{(1)}_1(\lambda). \tag{3.6}$$

Denoting the zeros of the Q-functions $Q^{(a)}_1(\lambda)$ by $\lambda^{(a)}_{1n}$ ($n = 1, 2, \ldots$), one obtains the Bethe ansatz equations

$$\frac{Q_1^{(2)}(\lambda_{1n}^{(1)} - \frac{\pi i}{4M})^2}{Q_1^{(2)}(\lambda_{1n}^{(1)} + \frac{\pi i}{4M})^2} \frac{Q_1^{(1)}(\lambda_{1n}^{(1)} + \frac{\pi i}{2M})}{Q_1^{(1)}(\lambda_{1n}^{(1)} - \frac{\pi i}{2M})} = -1, \tag{3.7}$$

$$\frac{Q_1^{(2)}(\lambda_{1n}^{(2)} - \frac{\pi i}{2M})}{Q_1^{(2)}(\lambda_{1n}^{(2)} + \frac{\pi i}{2M})} \frac{Q_1^{(1)}(\lambda_{1n}^{(2)} + \frac{\pi i}{4M})}{Q_1^{(1)}(\lambda_{1n}^{(2)} - \frac{\pi i}{4M})} = -1. \tag{3.8}$$

Note that these differ from those of the integrable model based on the $U_q(A_3^{(2)})$ [24,25], which is expected from the Langlands duality between $A_3^{(2)}$ and $B_2^{(1)}$. The Bethe ansatz equations for $U_q(A_3^{(2)})$ do not include the squared Q-functions. It would be interesting to study the solutions of the Bethe ansatz equations (3.7) and (3.8) in the conformal limit and explore the corresponding integrable model.

4. Quantum Wronskian and T–Q relations

4.1. Spinor representation and discrete symmetries

The ψ-system in the previous section has been obtained by investigating the asymptotic solution in a single Stokes sector, $\mathcal{S}_0$ for example. Now we consider the solutions of the linear problem in the whole complex plane. We focus on $V^{(2)}$ because this is the minimal dimensional representation and the solution in the vector representation can be constructed via the inclusion map ι_2.

Since we are considering a $SO(5)$ spinor, it is natural to introduce the charge conjugation. Associated with the linear problem (2.3) in the spinor representation, we define the transposed linear problem:

$$(\partial - A_z^T)\bar{\Psi} = 0, \quad (\bar{\partial} - A_{\bar{z}}^T)\bar{\Psi} = 0. \tag{4.1}$$

The solution $\bar{\Psi}(z, \bar{z}|\lambda, g)$ of these equations are related to $\Psi(z, \bar{z}|\lambda, g)$ by the charge conjugation:

$$\bar{\Psi}(z, \bar{z}|\lambda, g) = F\Psi(z, \bar{z}|\lambda, g), \tag{4.2}$$

where

$$F = \begin{pmatrix} 0 & 0 & 0 & 1 \\ 0 & 0 & -1 & 0 \\ 0 & 1 & 0 & 0 \\ -1 & 0 & 0 & 0 \end{pmatrix}. \tag{4.3}$$

This is a $\mathbf{Z}_2$ symmetry of the linear problem. Note that $\bar{\bar{\Psi}} = -\Psi$.

One can define the inner product $\langle \bar{\Psi}, \Psi \rangle := \sum_{\alpha=1}^4 \bar{\Psi}^\alpha \Psi^\alpha$ between $\Psi = (\Psi^\alpha)$ and $\bar{\Psi} = (\bar{\Psi}^\alpha)$. The inner product is independent of z and $\bar{z}$ when Ψ ($\bar{\Psi}$) is a solution of the (transposed) linear problem. The Wronskian of any four linearly independent solutions Ψ_i ($i = 1, 2, 3, 4$)

$$\langle \Psi_1, \Psi_2, \Psi_3, \Psi_4 \rangle := \det(\Psi_1, \Psi_2, \Psi_3, \Psi_4), \tag{4.4}$$

is also independent of z and $\bar{z}$.

We define the $(-k)$-rotated solution $s_k := \Psi_{-k}^{(2)}$ in the module $V^{(2)}$. This is the subdominant solution in the Stokes sector $\mathcal{S}_k$ but it gives a divergent solution in the sectors $\mathcal{S}_{k-2}$ and $\mathcal{S}_{k+2}$. One can choose $\{s_{k-1}, s_k, s_{k+1}, s_{k+2}\}$ as a basis of the solutions. We normalize the solution s_k such that

$$\langle s_{k-1}, s_k, s_{k+1}, s_{k+2} \rangle = 1, \tag{4.5}$$

by choosing the normalization constant $C^{(2)}$ in (2.11) as $(-16)^{-\frac{1}{4}}$. From the asymptotic behavior of s_k and $\bar{s}_k$ at large ρ, we find $\langle \bar{s}_k, s_k \rangle = \langle \bar{s}_k, s_{k\pm1} \rangle = 0$ and $\langle \bar{s}_k, s_{k+2} \rangle = \frac{1}{16}$. Then from the condition (4.5) we find

$$\bar{s}_k^\alpha = -\frac{1}{16} \epsilon^{\alpha \beta_1 \beta_2 \beta_3} s_{k-1}^{\beta_1} s_k^{\beta_2} s_{k+1}^{\beta_3}. \tag{4.6}$$

We write it in the form $\bar{s}_k = -\frac{1}{16} s_{k-1} \wedge s_k \wedge s_{k+1}$. Since the basis $\mathbf{e}_i^{(2)}$ is orthonormal, we can fix the normalization of $\mathcal{X}_i^{(2)}$ as

$$\det(\mathcal{X}_1^{(2)}, \mathcal{X}_2^{(2)}, \mathcal{X}_3^{(2)}, \mathcal{X}_4^{(2)}) = 1, \tag{4.7}$$

which simplifies the functional relations described below.

4.2. T–Q relation

Now we take $\{s_{-2}, s_{-1}, s_0, s_1\}$ as the basis of the solutions of the linear system. We introduce a set of functions $\mathcal{T}_{a,m}(\lambda)$ $(a = 1, 2, 3, m \in \mathbf{Z})$ by

$$\mathcal{T}_{1,m}(\lambda) = \langle s_{-2}, s_{-1}, s_0, s_{m+1} \rangle^{[-m]}, \tag{4.8}$$

$$\mathcal{T}_{2,m}(\lambda) = \langle s_{-2}, s_{-1}, s_1, s_{m+1} \rangle^{[-m]}, \tag{4.9}$$

$$\mathcal{T}_{3,m}(\lambda) = \langle s_{-2}, s_0, s_1, s_{m+1} \rangle^{[-m]}, \tag{4.10}$$

where $f^{[m]}(\lambda) \equiv f(\lambda + \frac{m}{2} \frac{2\pi i}{4M})$. A solution s_k $(k \in \mathbf{Z})$ is expanded in terms of this basis as

$$s_k = -\mathcal{T}_{1,k-2}^{[k]} s_{-2} + \mathcal{T}_{3,k-1}^{[k-1]} s_{-1} - \mathcal{T}_{2,k-1}^{[k-1]} s_0 + \mathcal{T}_{1,k-1}^{[k-1]} s_1. \tag{4.11}$$

The coefficients of s_{-1}, s_0 and s_1 follow from the definition of $\mathcal{T}_{a,m}$ directly. The coefficient of s_{-2} is evaluated as $\langle s_k, s_{-1}, s_0, s_1 \rangle$. Using the identity:

$$\langle s_{i_1}, s_{i_2}, s_{i_3}, s_{i_4} \rangle^{[2]} = \langle s_{i_1+1}, s_{i_2+1}, s_{i_3+1}, s_{i_4+1} \rangle, \tag{4.12}$$

which follows from the Symanzik rotation, it is shown to be equal to $\langle s_{-2}, s_{-1}, s_0, s_{k-1} \rangle^{[2]} = -\mathcal{T}_{1,k-2}^{[k]}$.

We expand s_{-k} in terms of the basis $\mathcal{X}_i^{(2)}$:

$$s_{-k}(z, \bar{z}) = \sum_{i=1}^{4} Q_i(\lambda - k \frac{2\pi i}{4M}, g) \mathcal{X}_i^{(2)}(z, \bar{z}|\lambda, g), \tag{4.13}$$

where $Q_i := Q_i^{(2)}$. The exterior product $s_{-i_1} \wedge s_{-i_2} \cdots \wedge s_{-i_p}$ in $\wedge^p V^{(2)}$ is also expanded in the basis $\mathcal{X}_i^{(2)}$. The coefficient of the highest weight vector is evaluated as

$$s_{-i_1} \wedge s_{-i_2} \cdots \wedge s_{-i_p} = W_{i_1 i_2 \ldots i_p}^{(p)} \mathcal{X}_1^{(2)} \wedge \cdots \wedge \mathcal{X}_p^{(2)} + \cdots, \tag{4.14}$$

where we introduce the determinant

$$W_{i_1 i_2 \ldots i_p}^{(p)} := \det \begin{pmatrix} Q_1(\lambda - i_1 \frac{2\pi i}{4M}) & Q_1(\lambda - i_2 \frac{2\pi i}{4M}) & \cdots & Q_1(\lambda - i_p \frac{2\pi i}{4M}) \\ Q_2(\lambda - i_1 \frac{2\pi i}{4M}) & Q_2(\lambda - i_2 \frac{2\pi i}{4M}) & \cdots & Q_2(\lambda - i_p \frac{2\pi i}{4M}) \\ \vdots & \vdots & & \vdots \\ Q_p(\lambda - i_1 \frac{2\pi i}{4M}) & Q_p(\lambda - i_2 \frac{2\pi i}{4M}) & \cdots & Q_p(\lambda - i_p \frac{2\pi i}{4M}) \end{pmatrix}. \tag{4.15}$$

For $p = 1$ we have $W_k^{(1)} = Q_1^{(2)[-2k]}$. For $p = 4$, we obtain $W_{i_1i_2i_3i_4}^{(4)} = \langle s_{-i_1}, s_{-i_2}, s_{-i_3}, s_{-i_4} \rangle$. In particular, from the normalization condition (4.7) we find that

$$W_{k-1,k,k+1,k+2}^{(4)} = 1. \tag{4.16}$$

The relation (4.16) can be regarded as the quantum Wronskian relation [9]. Let us consider two more examples. For $p = 2$ with $i_1 = -k$ and $i_2 = -k + 1$, using the ψ-system (3.4), we find

$$W_{k,k-1}^{(2)} = Q_1^{(1)}(\lambda - k\frac{2\pi i}{4M}). \tag{4.17}$$

For $W_{k+1,k,k-1}^{(3)}$, using (4.6), we have

$$\langle \bar{s}_{-k}, \mathcal{X}_4^{(2)} \rangle = -\frac{1}{16} W_{k+1,k,k-1}^{(3)}, \tag{4.18}$$

which becomes $\langle s_{-k}, F^T \mathcal{X}_4^{(2)} \rangle$ by the formula (4.2). We then get

$$W_{k+1,k,k-1}^{(3)} = 16 Q_1^{(2)}(\lambda - k\frac{2\pi i}{4M}). \tag{4.19}$$

We note that the determinants (4.15) satisfy the Plücker relations

$$W_{i_0i_2\cdots i_{p-1}}^{(p-1)} W_{i_1i_2\cdots i_p}^{(p)} - W_{i_1i_2\cdots i_{p-1}}^{(p-1)} W_{i_0i_2\cdots i_p}^{(p)} + W_{i_2\cdots i_{p-1}i_p}^{(p-1)} W_{i_0i_1\cdots i_{p-1}}^{(p)} = 0. \tag{4.20}$$

In particular one finds

$$\begin{aligned}
0 &= W_0^{(1)} W_{12}^{(2)} - W_1^{(1)} W_{02}^{(2)} + W_2^{(1)} W_{01}^{(2)}, \\
0 &= W_{02}^{(2)} W_{123}^{(3)} - W_{12}^{(2)} W_{023}^{(3)} - W_{32}^{(2)} W_{012}^{(3)}, \\
0 &= W_{023}^{(3)} W_{1234}^{(4)} - W_{123}^{(3)} W_{0234}^{(4)} + W_{423}^{(3)} W_{0123}^{(4)}.
\end{aligned} \tag{4.21}$$

From these equations, we can solve $W_{0234}^{(4)}$ as

$$W_{0234}^{(4)} = \frac{W_0^{(1)}}{W_1^{(1)}} + \frac{W_2^{(1)} W_{01}^{(2)}}{W_1^{(1)} W_{12}^{(2)}} + \frac{W_{23}^{(2)} W_{012}^{(3)}}{W_{123}^{(3)} W_{12}^{(2)}} + \frac{W_{234}^{(3)}}{W_{123}^{(3)}}. \tag{4.22}$$

This equation is the T–Q relation of the A_3-type quantum integrable models [26]. Now using (4.17) and (4.19), (4.22) becomes

$$\begin{aligned}
T_{1,1}^{[-1]} Q_1^{(2)} Q_1^{(1)[-1]} Q_1^{(2)[-2]} &= Q_1^{(2)[2]} Q_1^{(1)[-1]} Q_1^{(2)[-2]} + Q_1^{(2)[-4]} Q_1^{(1)[1]} Q_1^{(2)[-2]} \\
&\quad + Q_1^{(1)[-3]} Q_1^{(2)} Q_1^{(2)} + Q_1^{(2)[-4]} Q_1^{(2)} Q_1^{(1)[-1]}.
\end{aligned} \tag{4.23}$$

This is the T–Q relation for the $A_3/\mathbb{Z}_2$-type. From this relation we obtain the Bethe equations, which was also derived in the previous section by using the ψ-system.

One can also derive a set of the relations:

$$\begin{aligned}
0 &= W_{013}^{(3)} W_{2134}^{(4)} - W_{213}^{(3)} W_{0134}^{(4)} + W_{134}^{(3)} W_{0213}^{(4)}, \\
0 &= -W_{12}^{(2)} W_{013}^{(3)} + W_{01}^{(2)} W_{123}^{(3)} + W_{13}^{(2)} W_{012}^{(3)}, \\
0 &= W_{23}^{(2)} W_{134}^{(3)} - W_{13}^{(2)} W_{234}^{(3)} - W_{34}^{(2)} W_{123}^{(3)}, \\
0 &= W_2^{(1)} W_{13}^{(2)} - W_1^{(1)} W_{23}^{(2)} - W_3^{(1)} W_{12}^{(2)}, \\
0 &= W_1^{(1)} W_{23}^{(2)} - W_2^{(1)} W_{13}^{(2)} + W_3^{(1)} W_{12}^{(2)}.
\end{aligned} \tag{4.24}$$

From these equations $W_{0134}^{(4)}$ is solved as

$$W_{0134}^{(4)} = \frac{W_{01}^{(2)}}{W_{12}^{(2)}} + \frac{W_{012}^{(3)}}{W_{123}^{(3)}}\left(\frac{W_1^{(1)}W_{23}^{(2)}}{W_2^{(1)}W_{12}^{(2)}} + \frac{W_3^{(1)}}{W_2^{(1)}}\right) + \frac{W_{234}^{(3)}}{W_{123}^{(3)}}\left(\frac{W_1^{(1)}}{W_2^{(1)}} + \frac{W_3^{(1)}W_{12}^{(2)}}{W_{23}^{(2)}W_2^{(1)}}\right) + \frac{W_{34}^{(2)}}{W_{23}^{(2)}}. \quad (4.25)$$

From (4.17), (4.19), (4.25) and $W_{0134}^{(4)} = T_{2,1}^{[-3]}$, we get the T–Q relation for $T_{2,1}$, $Q_1^{(1)}$ and $Q_1^{(2)}$:

$$T_{2,1}^{[1]}Q_1^{(1)[-1]}Q_1^{(1)[1]}(Q_1^{(2)})^2 = (Q_1^{(2)})^2\left(Q_1^{(1)[-1]}Q_1^{(1)[3]} + Q_1^{(1)[-3]}Q_1^{(1)[1]}\right)$$

$$+ \left(Q_1^{(2)[2]}Q_1^{(1)[-1]} + Q_1^{(2)[-2]}Q_1^{(1)[1]}\right)^2 \quad (4.26)$$

At the zeros $\lambda_{1n}^{(2)}$ of $Q_1^{(2)}(\lambda)$, $T_{2,1}^{[1]}$ might have a double pole. Absence of the double pole in $T_{2,1}^{[1]}$ leads to eq. (3.8). By the shift of λ and evaluating (4.26) at zeros of $Q_1^{(1)}(\lambda)$, we obtain eq. (3.7). Thus one obtains the Bethe ansatz equations again.

5. T-system and Y-system

Now we study the functional relations which are satisfied by $T_{a,m}$. First we calculate the product of $T_{a,1}$ and $T_{1,m}$. From the Plücker relation

$$\langle s_{j_1}, s_{j_2}, s_{j_3}, s_{j_4}\rangle\langle s_{i_1}, s_{i_2}, s_{i_3}, s_{i_4}\rangle - \langle s_{i_1}, s_{j_2}, s_{j_3}, s_{j_4}\rangle\langle s_{j_1}, s_{i_2}, s_{i_3}, s_{i_4}\rangle$$
$$+ \langle s_{i_4}, s_{j_2}, s_{j_3}, s_{j_4}\rangle\langle s_{j_1}, s_{i_2}, s_{i_3}, s_{i_1}\rangle = 0. \quad (5.1)$$

we get the identities

$$T_{1,1}^{[+1]}T_{1,m-1}^{[m+1]} = T_{1,m}^{[m]} + T_{2,m-1}^{[m+1]},$$
$$T_{2,1}^{[+1]}T_{1,m-1}^{[m+1]} = T_{2,m}^{[m]} + T_{3,m-1}^{[m+1]},$$
$$T_{3,1}^{[+1]}T_{1,m-1}^{[m+1]} = T_{3,m}^{[m]} + T_{1,m-2}^{[m+2]}. \quad (5.2)$$

These relations are a generalization of the fusion relation of the modified sinh-Gordon equation [9] to the modified $B_2^{(1)}$ affine Toda field equation. However, one finds

$$T_{1,m}^{[+1]}T_{1,m}^{[-1]} = T_{1,m+1}T_{1,m-1} + \langle s_{-1}, s_0, s_{m+1}, s_{m+2}\rangle^{[-m-1]}, \quad (5.3)$$

where the second term in the r.h.s. is not the form of the $T_{a,m}$ functions. We add this function to a member of the T-functions and define

$$T_{1,m}(\lambda) = T_{1,m}(\lambda) = \langle s_{-2}, s_{-1}, s_0, s_{m+1}\rangle^{[-m]}, \quad (5.4)$$
$$T_{2,m}(\lambda) = \langle s_{-1}, s_0, s_{m+1}, s_{m+2}\rangle^{[-m-1]}, \quad (5.5)$$

for $m \in \mathbf{Z}$. The new function $T_{2,m}$ satisfies the identity

$$T_{2,m}^{[+1]}T_{2,m}^{[-1]} = T_{2,m-1}T_{2,m+1} + \langle s_{-1}, s_m, s_{m+1}, s_{m+1}\rangle^{[-m]}T_{1,m}. \quad (5.6)$$

We then introduce

$$T_{3,m}(\lambda) = \langle s_{-1}, s_m, s_{m+1}, s_{m+2}\rangle^{[-m]}. \quad (5.7)$$

But this is not new. Using (4.2) and (4.6), we can show that $T_{3,m} = T_{1,m}$. Finally we obtain the T-system of $A_3/\mathbf{Z}_2$ type:

$$T^{[+1]}_{1,m}T^{[-1]}_{1,m} = T_{1,m-1}T_{1,m+1} + T_{2,m}$$
$$T^{[+1]}_{2,m}T^{[-1]}_{2,m} = T_{2,m+1}T_{2,m-1} + T_{1,m}T_{1,m},$$ (5.8)

which is obtained by the reduction of A_3 T-system with the identification $T_{1,m} = T_{3,m}$. Other functions $\mathcal{T}_{2,m}$, $\mathcal{T}_{3,m}$ can be expressed in terms of $T_{a,m}$ by using $\mathcal{T}_{2,1} = T^{[-1]}_{2,1}$, $\mathcal{T}_{3,1} = T^{[-2]}_{3,1}$ and (5.2). They also satisfy the identities:

$$\mathcal{T}_{3,m+1}\mathcal{T}_{1,m-1} = \mathcal{T}^{[-1]}_{3,m}\mathcal{T}^{[2]}_{1,m} - \mathcal{T}_{2,m},$$
$$\mathcal{T}^{[+1]}_{2,m+1}\mathcal{T}_{1,m} = \mathcal{T}^{[+1]}_{1,m+1}\mathcal{T}_{2,m} + \mathcal{T}_{2,m+1}.$$ (5.9)

We next introduce the Y-functions by

$$Y_{1,m} = \frac{T_{1,m+1}T_{1,m-1}}{T_{2,m}}, \quad Y_{2,m} = \frac{T_{2,m+1}T_{2,m-1}}{T_{1,m}T_{1,m}}.$$ (5.10)

They satisfy the Y-system of $A_3/\mathbf{Z}_2$ type

$$\frac{Y^{[+1]}_{a,m}Y^{[-1]}_{a,m}}{Y_{a+1,m}Y_{a-1,m}} = \frac{(1 + Y_{a,m+1})(1 + Y_{a,m-1})}{(1 + Y_{a+1,m})(1 + Y_{a-1,m})},$$ (5.11)

where $a = 1,2$ and $Y_{3,m} = Y_{1,m}$. The T-system (5.8) and the Y-system (5.11) imply that the Langlands duality between the modified $B_2^{(1)}$ affine Toda equation and the functional equations of the $A_3/\mathbf{Z}_2$ quantum integrable system.

We now discuss the boundary condition of the T-system and the Y-system. It is easy to see that $T_{a,-1} = 0$ and $T_{a,0} = 1$. In order to determine the boundary conditions $T_{1,m}$ for large m, we need to study the small solutions s_m in the whole complex plane. When $4(M + 1)$ is not a rational number, the Stokes sectors cover the complex plane infinitely many times. So the T-functions $T_{a,m}$ are defined independently for arbitrary positive integer m.

In this paper we will consider the case $4(M + 1) = n$ with $n \geq 6$ being a positive integer in detail.[2] In this case there are n Stokes sectors in the complex plane. When we go around the origin, the solution $s_k(ze^{-2\pi i})$ is defined in the sector $\mathcal{S}_{k+n}$, which is the same as $\mathcal{S}_k$. Then the small solution $s_{k+n}(z)$ is proportional to $s_k(ze^{-2\pi i})$:

$$s_{k+n}(z) \propto s_k(ze^{-2\pi i}).$$ (5.12)

For $g = 0$, the linear system has no simple pole at the origin. The solution has no monodromy around it. Then we have $s_k(ze^{-2\pi i}) = s_k(z)$, which implies

$$s_{k+n}(z, \lambda) \propto s_k(z, \lambda).$$ (5.13)

The condition (5.13) leads to the boundary conditions for the T-/Y-functions: $T_{a,n-3} = 0$ and $Y_{a,n-4} = 0$. The truncated T-/Y-system becomes the same as the one for the n-point gluon scattering amplitudes in AdS$_4$ at strong coupling [19].

For $g \neq 0$, the solutions of the linear system have monodromy around the origin. We introduce a monodromy matrix $\Omega(\lambda)$ by

$$\begin{pmatrix} s_1 \\ s_0 \\ s_{-1} \\ s_{-2} \end{pmatrix}(ze^{-2\pi i}, \lambda) = \Omega(\lambda)\begin{pmatrix} s_1 \\ s_0 \\ s_{-1} \\ s_{-2} \end{pmatrix}(z, \lambda).$$ (5.14)

---

[2] When n is a rational number, we can do similar arguments. But it is not discussed in this paper.

From the normalization condition (4.5) we find $\det\Omega(\lambda) = 1$. We also introduce the proportion-ality factor $B(\lambda)$ in (5.13) for $k = 1$ by

$$s_{n+1}(z, \lambda) = B(\lambda)s_1(ze^{-2\pi i}, \lambda). \tag{5.15}$$

Let us expand the solution $s_0(z, \lambda)$ in the basis $\mathcal{X}_i(z, \bar{z}|\lambda, g)$ whose coefficient has been defined as $Q_i(\lambda, g)$. Then we substitute its Symanzik rotation into (5.15). In the basis $\mathcal{X}_i$, the monodromy matrix becomes diagonal and takes the form $diag(e^{2\pi i\beta g \cdot h_1^{(2)}}, \ldots, e^{2\pi i\beta g \cdot h_4^{(2)}})$. Moreover from the quasi-periodicity condition (2.15) one finds that $B(\lambda) = -1$. Plugging (5.15) into (5.14), we get the relation

$$\begin{pmatrix} s_{n+1} \\ s_n \\ s_{n-1} \\ s_{n-2} \end{pmatrix}(z, \lambda) = -\Omega(\lambda) \begin{pmatrix} s_1 \\ s_0 \\ s_{-1} \\ s_{-2} \end{pmatrix}(z, \lambda), \tag{5.16}$$

which generalizes the condition (5.13) and determines the boundary condition for the T-system. It is convenient to use the (multi-)trace of the monodromy matrix Ω: $\text{tr}\Omega$ and $\text{tr}^{(2)}\Omega \equiv \frac{1}{2}((\text{tr}\Omega)^2 - \text{tr}\Omega^2)$, which are basis independent quantities. These traces can be also expressed using the Wron-skians:

$$\text{tr}\Omega = -\langle s_{-2}, s_{-1}, s_0, s_{n+1}\rangle + \langle s_{-2}, s_{-1}, s_1, s_n\rangle - \langle s_{-2}, s_0, s_1, s_{n-1}\rangle$$
$$+ \langle s_{-1}, s_0, s_1, s_{n-2}\rangle,$$
$$\text{tr}^{(2)}\Omega = \langle s_{-2}, s_{-1}, s_n, s_{n+1}\rangle + \langle s_0, s_1, s_{n-2}, s_{n-1}\rangle + \langle s_{-2}, s_{n-1}, s_0, s_{n+1}\rangle$$
$$+ \langle s_{n-2}, s_{-1}, s_0, s_{n+1}\rangle + \langle s_{-2}, s_{n-1}, s_n, s_1\rangle + \langle s_{n-2}, s_{-1}, s_n, s_1\rangle. \tag{5.17}$$

Here the r.h.s. of these equations are expressed by $T_{2,m}$, $T_{3,m}$ and the auxiliary T-functions $W_{1,m}$, $W_{2,m}$, $\bar{W}_{2,m}$ defined in [23], in addition to the T-functions (5.5). In Appendix A, we will sum-marize these auxiliary T-functions and their recursion relations. In the diagonal basis, they are evaluated as

$$\text{tr}\Omega = 4\cos(\beta g_1 \pi)\cos(\beta g_2 \pi) \tag{5.18}$$

$$\text{tr}^{(2)}\Omega = 2 + 4\cos[\beta(g_1 - g_2)\pi]\cos[\beta(g_1 + g_2)\pi] \tag{5.19}$$

where $g_i \equiv g \cdot e_i$ $(i = 1, 2)$. For $g = 0$, one finds that $\text{tr}\Omega = 4$ and $\text{tr}^{(2)}\Omega = 6$. The monodromy conditions (5.17) determine $T_{a,n}$ $(a = 1, 2)$. Then the T-system extends up to $m = n - 1$ and the T-functions $T_{a,m}$ for $m \geq n - 1$ are determined by the T-system and the monodromy conditions. Concerning the Y-system (5.11), it also extends up to $m = n - 2$. It is convenient to introduce new Y-functions $\bar{Y}_a$ $(a = 1, 2, 3)$ by

$$\bar{Y}_1 = \bar{Y}_3 = -\frac{T_{1,n-2}}{T_{2,n-1}}, \quad \bar{Y}_2 = \frac{T_{2,n-2}}{T_{1,n-1}T_{1,n-1}}. \tag{5.20}$$

whose functional relations are given by

$$\frac{\bar{Y}_a^{[+1]}\bar{Y}_a^{[-1]}}{\bar{Y}_{a+1}\bar{Y}_{a-1}} = \frac{1 + Y_{a,n-2}}{(1 + Y_{a+1,n-1})(1 + Y_{a-1,n-1})}. \tag{5.21}$$

The Y-system (5.11) for $m = n - 2$ and (5.21) contains $Y_{a,n-1}$ in the r.h.s. of the equations. $Y_{a,n-1}$ are expressed as

$$Y_{1,n-1} = -T_{1,n}\bar{Y}_1, \quad Y_{2,n-1} = T_{2,n}\bar{Y}_2, \tag{5.22}$$

and $T_{a,n}$ are expressed in terms of the lower T-functions. For the $n \neq 4\ell$ ($\ell = 1, 2, \cdots$) case, they are also expressed in terms of the lower Y-functions by solving (5.10). Then (5.11) and (5.21) with (5.22) become the closed functional relations. Note that the present T- and Y-systems are the same as those of form factors in AdS$_4$ [23]. However the function $p(z)$ has different pole structure from the present one.

In the case of even n and $g_1 = 0$ (or $g_2 = 0$), one can consider the limit to the modified sinh-Gordon equation [9] (or gluon scattering amplitudes in AdS$_3$ [19]), where in this limit the $SO(5)$ spinor is decomposed into left and right-handed spinors. In this reduction the T-functions $T_{a,m}$ reduce to the functions T_k ($k = 1, \cdots, \frac{n}{2} - 2$), which are defined by the inner product of the left-handed spinors. They satisfy

$$T_{1,2k+1} = 0, \quad T_{1,2k} = -T_k^{[2]}, \quad T_{2,2k} = T_k^{[3]} T_k^{[+1]}, \quad T_{2,2k+1} = -T_k^{[2]} T_{k+1}^{[2]},$$
$$\langle s_{-2}, s_0, s_1, s_{n-1} \rangle = T_{\frac{n}{2}-2}^{[n+2]}, \quad \langle s_{-2}, s_{-1}, s_1, s_n \rangle = T_{\frac{n}{2}}^{[n]}. \tag{5.23}$$

Here T_k obey the functional relations

$$T_k^{[2]} T_k^{[-2]} = 1 + T_{k-1} T_{k+1}. \tag{5.24}$$

Using (5.23), we can rewrite trΩ in terms of the left-handed part and right-handed part. Decomposing these two parts we obtain $T_{\frac{n}{2}} - T_{\frac{n}{2}-2} = 2\cos \pi \beta g_2$, which is the trace of monodromy in left-handed part. The Y-functions $Y_{a,m}$ and $\bar{Y}_a$ reduce to $Y_k = T_{k+1} T_{k-1}$ ($k = 1, \cdots, n/2 - 2$) and $\bar{Y} = -T_{\frac{n}{2}-2}$ as

$$Y_{1,2k} = 0, \quad Y_{1,2k+1} = -1, \quad Y_{2,2k+1} = \infty, \quad Y_{2,2k} = Y_k,$$
$$\bar{Y}_1 Y_{2,n-2} = \bar{Y}^{[2]}, \quad \bar{Y}_2 = \infty. \tag{5.25}$$

Here the Y-functions Y_k and $\bar{Y}$ satisfy the $D_{n/2}$-type Y-system [9,22]

$$Y_k^{[2]} Y_k^{[-2]} = (1 + Y_{k-1})(1 + Y_{k+2}), \quad (k = 1, \ldots, \frac{n}{2} - 3),$$
$$\bar{Y}^{[2]} \bar{Y}^{[-2]} = 1 + Y_{\frac{n}{2}-2},$$
$$Y_{\frac{n}{2}-2}^{[2]} Y_{\frac{n}{2}-2}^{[-2]} = (1 + Y_{\frac{n}{2}-3})(1 - 2\cos \pi \beta g_2 \bar{Y} + \bar{Y}^2). \tag{5.26}$$

6. High-temperature limit of the Y-system

In the previous section we have seen that the T-/Y-system becomes the extended one in the presence of monodromy. The standard approach to analyze the (extended) Y-system is to derive the Thermodynamic Bethe ansatz (TBA) equations and investigate their free energy. The IR (or low-temperature) limit of the TBA system are characterized by the WKB approximation, whereas in the UV (or high-temperature) limit is characterized by the spectral parameter independent Y-functions and the free energy is determined by the dilog formulas [27–29]. Since the present Y-system is very complicated, we leave the detailed TBA analysis to the subsequent paper. Instead we will study the high-temperature limit of the Y-system and their solutions explicitly for the simplest case $n = 6$.

For $n = 6$. Using the T-system, $T_{a,m}$ ($a = 1, 2$, $m = 1, \ldots, 6$) are solved in terms of $T_{1,1} = x$ and $T_{2,1} = y$ using (5.8). Substituting them into the monodromy conditions we obtain two equations for x and y

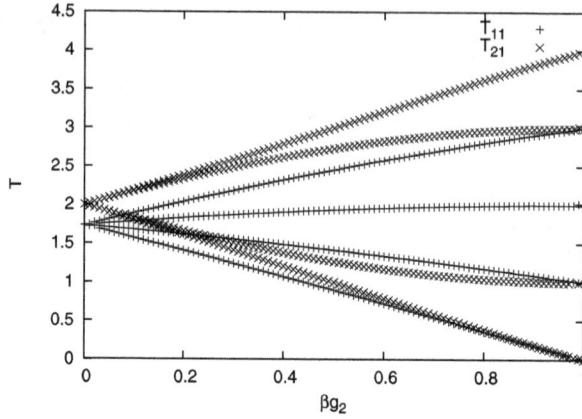

Fig. 1. Plots of $T_{1,1}$ and $T_{2,1}$ at $g_1 = 0$.

$$x^6 - 6x^4(-1+y) - 2y(-3+y^2) + 3x^2(-1-4y+3y^2) + 4\cos[\beta g_1\pi]\cos[\beta g_2\pi] = 0,$$
$$-2x^6 + y^2(-3+y^2)^2 + 3x^4(8-8y+3y^2) - 6x^2(3-2y^3+y^4)$$
$$-4\cos[\beta(g_1-g_2)\pi]\cos[\beta(g_1+g_2)\pi] = 0. \tag{6.1}$$

For $g_1 = g_2 = 0$, *i.e.* when there is no monodromy around the origin, we find the solutions of the above algebraic equations are given by $(x, y) = (0, -1), (0, 2), (\pm 2\sqrt{3}, 5)$ and $(\pm\sqrt{3}, 2)$. For $(x, y) = (0, -1)$, we get $Y_{1,2k-1} = -1$, $Y_{1,2k} = 0$, $Y_{2,2k-1} = \infty$ $(k \geq 1)$, $Y_{2,2} = 0$ and $Y_{2,4} = -1$. This solution corresponds to the AdS_3 limit of the Y-system. For $(x, y) = (\pm\sqrt{3}, 2)$, we get $Y_{1,1} = \frac{1}{2}$, $Y_{2,2} = \frac{1}{3}$ and $Y_{1,2} = Y_{2,2} = 0$, which corresponds to the constant Y-system of the 6-point amplitudes [19]. For other solutions, we do not find any corresponding physical quantities.

Now we turn on g_2 with keeping $g_1 = 0$. We find the solutions of (6.1) with $x = 0$ are given by

$$y = \pm\sqrt{2 + 2\cos\frac{2\pi}{3}(1 - \beta g_2)}. \tag{6.2}$$

This gives

$$Y_{2,2} = 1 + 2\cos\frac{2\pi}{3}(1 - \beta g_2), \quad \bar{Y}_1 Y_{2,4} = -y, \tag{6.3}$$

which turns out to be a constant solution. This corresponds to a constant solution of the AdS_3 form factor with $n = 6$ gluons [22]. Note that $\bar{Y}^{[2]} = \bar{Y}_1 Y_{2,4}$ in (5.25) for $n = 6$. As for the solutions starting from $(\sqrt{3}, 2)$, we solve eqs. (6.1) numerically. The graphs of $T_{1,1} = x(g_2)$ and $T_{2,1} = y(g_2)$ are shown in Fig. 1 and the corresponding Y-functions $Y_{1,1}$ and $Y_{2,1}$ are shown in Fig. 2. There arise four branches from the point $(\sqrt{3}, 2)$, which arrive at the solutions $(x, y) = (3, 4), (2, 3), (1, 0)$ and $(0, 1)$ at $g_2 = 1$. These solutions provide a deformation of the constant T-/Y-systems by the monodromy parameter g_2.

7. Conclusions and discussion

In this paper, we studied the massive ODE/IM correspondence for modified $B_2^{(1)}$ affine Toda field equation. By investigating the solutions of the linear problem associated with the modified

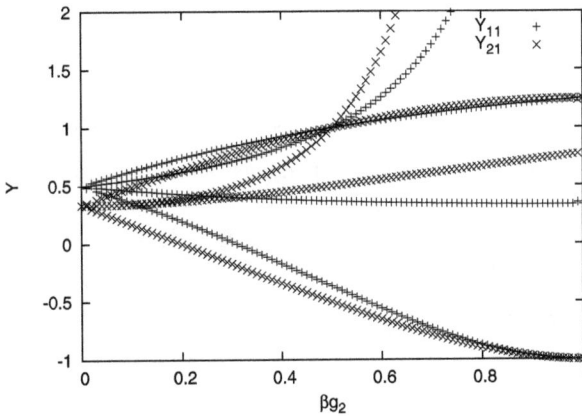

Fig. 2. Plots of $Y_{1,1}$ and $Y_{2,1}$ at $g_1 = 0$.

affine Toda equation, we derived the ψ-system. This leads to the Bethe ansatz equations corresponding to the integrable model which is not identified yet. We also derived the same Bethe ansatz equations from the T–Q relations. We constructed the T-system and Y-system from the Wronskians of the solutions of the linear problem. These systems have non-trivial boundary conditions due to the presence of monodromy around the origin. It would be interesting to generalize the present approach to modified affine Toda field equations associated with other affine Lie algebras which are not of the Langlands dual of an untwisted affine Lie algebra [30]. It is also interesting to study the massless limit of this linear problem and investigate the description by using the free field realization of conformal field theory [31–34].

For the linear system associated the null-polygonal minimal surface in AdS_4, we have seen that the corresponding integrable system is the homogeneous sine-Gordon model [20,21]. When the solution has monodromy around the origin, we have seen the T-system and Y-system are extended and they take the form that appears in the strong-coupling limit of the form factor in $\mathcal{N} = 4$ super Yang–Mills theory. For a general polynomial $p(z)$ and the appropriate boundary conditions for the solutions of the linear problem, one can describe the minimal surface problem using the massive ODE/IM correspondence. In particular it is interesting to explore the ODE/IM correspondence for the minimal surface in AdS_5, where the corresponding quantum integrable model is not known yet.

Acknowledgements

We would like to thank Yuji Satoh and Christopher Locke for valuable discussion and collaboration in an early stage of this work. This work is supported in part by Grant-in-Aid for Scientific Research 15K05043 from Japan Society for the Promotion of Science (JSPS).

Appendix A. Auxiliary T-functions

In this appendix we summarize the auxiliary T-functions and their recursion relations [23]. From these relations we express $\mathrm{tr}\,\Omega$ and $\mathrm{tr}^{(2)}\Omega$ in terms of $T_{a,m}$ ($m \leq n$). Furthermore we can express $Y_{a,n-1}$ in the lower Y-functions and get a closed Y-system. We define the functions $U_{1,m}$, $V_{1,m}$, $W_{1,m}$, $W_{2,m}$ $\bar{W}_{2,m}$ ($m \in \mathbf{Z}$) by

$$U_{1,m} = \langle s_{-2}, s_{-1}, s_m, s_{m+2} \rangle^{[-m]}, \quad V_{1,m} = \langle s_{-2}, s_0, s_{m+1}, s_{m+2} \rangle^{[-m]}, \tag{A.1}$$

$$W_{1,m} = \langle s_{-2}, s_0, s_{m+1}, s_{m+3} \rangle^{[-m-1]}, \tag{A.2}$$

$$W_{2,m} = \langle s_{-1}, s_0, s_{m+1}, s_{m+4} \rangle^{[-m-2]}, \quad \bar{W}_{2,m} = \langle s_{-2}, s_1, s_{m+2}, s_{m+3} \rangle^{[-m-2]}. \tag{A.3}$$

From the Plücker relation (5.1), we can show that these auxiliary T-functions satisfy

$$U_{1,m} T_{1,m} = T_{1,m-1}^{[-1]} T_{2,m+1} + T_{1,m+1}^{[+1]} T_{2,m}^{[-1]}, \tag{A.4}$$

$$V_{1,m} T_{3,m} = T_{1,m-1}^{[+1]} T_{2,m+1} + T_{3,m+1}^{[-1]} T_{2,m}^{[+1]}, \tag{A.5}$$

$$W_{1,m} T_{2,m} = V_{1,m}^{[-1]} U_{1,m}^{[+1]} - T_{1,m}^{[-1]} T_{3,m}^{[+1]}, \tag{A.6}$$

$$W_{2,m} T_{2,m+1} = U_{1,m+1}^{[+1]} U_{1,m} - T_{2,m}^{[-1]} T_{2,m+2}^{[+1]}, \tag{A.7}$$

$$\bar{W}_{2,m} T_{2,m+1} = V_{1,m+1}^{[-1]} V_{1,m} - T_{2,m}^{[+1]} T_{2,m+2}^{[-1]}. \tag{A.8}$$

Then $U_{1,m}$, $V_{1,m}$, $W_{1,m}$, $W_{2,m}$ and $\bar{W}_{2,m}$ are expressed in terms of $T_{a,s}$. The (symmetrized) trace of the monodromy matrix becomes

$$\mathrm{tr}\,\Omega = -\left(T_{1,n}^{[n]} - \mathcal{T}_{2,n-1}^{[n-1]} + \mathcal{T}_{3,n-2}^{[n-2]} - T_{1,n-4}^{[n-2]} \right), \tag{A.9}$$

$$\mathrm{tr}^{(2)}\,\Omega = T_{2,n}^{[n-1]} + T_{2,n-4}^{[n-1]} - W_{1,n-2}^{[n-1]} + W_{2,n-3}^{[n-1]} + \bar{W}_{2,n-3}^{[n-1]} - W_{1,n-4}^{[n-1]}. \tag{A.10}$$

Using $\mathcal{T}_{3,1} = T_{1,1}^{[-2]}$, $\mathcal{T}_{2,1} = T_{2,1}^{[-1]}$ and the identities (5.2), we get

$$\mathcal{T}_{3,n-2} = T_{1,1}^{[-n+1]} T_{1,n-3}^{[+1]} - T_{1,n-4}^{[+2]} \tag{A.11}$$

$$\mathcal{T}_{2,n-1} = T_{2,1}^{[-n+1]} T_{1,n-2}^{[+1]} - \{ T_{1,1}^{[-n+1]} T_{1,n-3}^{[+1]} - T_{1,n-4}^{[+2]} \}. \tag{A.12}$$

The (symmetrized) traces are expressed in terms of $T_{a,s}$ as

$$\mathrm{tr}\,\Omega = -\left(T_{1,n}^{[n]} - T_{1,n-4}^{[n-2]} - \left(T_{2,1}^{[-n+1]} T_{1,n-2}^{[+1]} - (T_{1,1} T_{1,n-3}^{[n]} - T_{1,n-4}^{[n+1]}) \right) \right. $$
$$\left. + (T_{1,1}^{[-1]} T_{1,n-3}^{[n-1]} - T_{1,n-4}^{[n]}) \right),$$

$$\mathrm{tr}^{(2)}\,\Omega = T_{2,n}^{[n-1]} + T_{2,n-4}^{[n-1]}$$
$$- \left(-\frac{T_{3,n-2}^{[n]} T_{1,n-2}^{[n-2]}}{T_{2,n-2}^{[n-1]}} \right.$$
$$+ \frac{1}{T_{2,n-2}^{[n-1]}} \frac{T_{1,n-1}^{[n+1]} T_{2,n-2}^{[n-1]} + T_{2,n-1}^{[n]} T_{1,n-3}^{[n-1]}}{T_{3,n-2}^{[n]}} \frac{T_{2,n-2}^{[n-1]} T_{1,n-1}^{[n-3]} + T_{1,n-3}^{[n-1]} T_{2,n-1}^{[n-2]}}{T_{3,n-2}^{[n-2]}} \right)$$
$$+ \left(-\frac{T_{2,n-1}^{[n]} T_{2,n-3}^{[n-2]}}{T_{2,n-2}^{[n-1]}} \right.$$
$$+ \frac{1}{T_{2,n-2}^{[n-1]}} \frac{T_{1,n-3}^{[n-1]} T_{2,n-1}^{[n]} + T_{1,n-1}^{[n+1]} T_{2,n-2}^{[n-1]}}{T_{1,n-2}^{[n]}} \frac{T_{1,n-4}^{[n-2]} T_{2,n-2}^{[n-1]} + T_{1,n-2}^{[n]} T_{2,n-3}^{[n-2]}}{T_{1,n-3}^{[n-1]}} \right)$$
$$+ \left(-\frac{T_{2,n-3}^{[n]} T_{2,n-1}^{[n-2]}}{T_{2,n-2}^{[n-1]}} \right.$$

$$+ \frac{1}{T_{2,n-2}^{[n-1]}} \frac{T_{2,n-3}^{[n]} T_{1,n-2}^{[n-2]} + T_{1,n-4}^{[n]} T_{2,n-2}^{[n-1]}}{T_{3,n-3}^{[n-1]}} \frac{T_{2,n-2}^{[n-1]} T_{1,n-1}^{[n-3]} + T_{1,n-3}^{[n-1]} T_{2,n-1}^{[n-2]}}{T_{3,n-2}^{[n-2]}} \Bigg)$$

$$- \Bigg(-\frac{T_{3,n-4}^{[n]} T_{1,n-4}^{[n-2]}}{T_{2,n-4}^{[n-1]}}$$

$$+ \frac{1}{T_{2,n-4}^{[n-1]}} \frac{T_{1,n-3}^{[n+1]} T_{2,n-4}^{[n-1]} + T_{2,n-3}^{[n]} T_{1,n-5}^{[n-1]}}{T_{3,n-4}^{[n]}} \frac{T_{2,n-4}^{[n-1]} T_{1,n-3}^{[n-3]} + T_{1,n-5}^{[n-1]} T_{2,n-3}^{[n-2]}}{T_{3,n-4}^{[n-2]}} \Bigg). \quad (A.13)$$

From these equations we can write $T_{1,n}$ and $T_{2,n}$ in terms of lower T-functions. In the case of $n \neq 4\ell$ with $\ell = 1, 2, \cdots$, the T-functions can be further expressed in terms of the Y-functions by solving (5.10). We then obtain a closed Y-system.

References

[1] D. Gaiotto, G.W. Moore, A. Neitzke, arXiv:0907.3987 [hep-th].
[2] N.A. Nekrasov, S.L. Shatashvili, arXiv:0908.4052 [hep-th].
[3] L.F. Alday, D. Gaiotto, J. Maldacena, J. High Energy Phys. 1109 (2011) 032, arXiv:0911.4708 [hep-th].
[4] P. Dorey, R. Tateo, J. Phys. A 32 (1999) L419, arXiv:hep-th/9812211.
[5] V.V. Bazhanov, S.L. Lukyanov, A.B. Zamolodchikov, J. Stat. Phys. 102 (2001) 567, arXiv:hep-th/9812247.
[6] P. Dorey, C. Dunning, R. Tateo, J. Phys. A 40 (2007) R205, arXiv:hep-th/0703066.
[7] P. Dorey, C. Dunning, D. Masoero, J. Suzuki, R. Tateo, Nucl. Phys. B 772 (2007) 249, arXiv:hep-th/0612298.
[8] J. Sun, SIGMA 8 (2012) 028, arXiv:1201.1614 [math.QA].
[9] S.L. Lukyanov, A.B. Zamolodchikov, J. High Energy Phys. 1007 (2010) 008, arXiv:1003.5333 [math-ph].
[10] P. Dorey, S. Faldella, S. Negro, R. Tateo, Philos. Trans. R. Soc. Lond. A 371 (2013) 20120052, arXiv:1209.5517 [math-ph].
[11] K. Ito, C. Locke, Nucl. Phys. B 885 (2014) 600, arXiv:1312.6759 [hep-th].
[12] P. Adamopoulou, C. Dunning, J. Phys. A 47 (2014) 205205, arXiv:1401.1187 [math-ph].
[13] K. Ito, C. Locke, Nucl. Phys. B 896 (2015) 763, arXiv:1502.00906 [hep-th].
[14] D. Masoero, A. Raimondo, D. Valeri, arXiv:1501.07421 [math-ph].
[15] D. Masoero, A. Raimondo, D. Valeri, arXiv:1511.00895 [math-ph].
[16] H.J. De Vega, N.G. Sanchez, Phys. Rev. D 47 (1993) 3394.
[17] B.A. Burrington, P. Gao, J. High Energy Phys. 1004 (2010) 060, arXiv:0911.4551 [hep-th].
[18] B.A. Burrington, J. High Energy Phys. 1109 (2011) 002, arXiv:1105.3227 [hep-th].
[19] L.F. Alday, J. Maldacena, A. Sever, P. Vieira, J. Phys. A 43 (2010) 485401, arXiv:1002.2459 [hep-th].
[20] Y. Hatsuda, K. Ito, K. Sakai, Y. Satoh, J. High Energy Phys. 1004 (2010) 108, arXiv:1002.2941 [hep-th].
[21] Y. Hatsuda, K. Ito, K. Sakai, Y. Satoh, J. High Energy Phys. 1104 (2011) 100, arXiv:1102.2477 [hep-th].
[22] J. Maldacena, A. Zhiboedov, J. High Energy Phys. 1011 (2010) 104, arXiv:1009.1139 [hep-th].
[23] Z. Gao, G. Yang, J. High Energy Phys. 1306 (2013) 105, arXiv:1303.2668 [hep-th].
[24] N.Y. Reshetikhin, P.B. Wiegmann, Phys. Lett. B 189 (1987) 125.
[25] A. Kuniba, J. Suzuki, J. Phys. A 28 (1995) 711, arXiv:hep-th/9408135.
[26] P. Dorey, C. Dunning, R. Tateo, J. Phys. A 33 (2000) 8427, arXiv:hep-th/0008039.
[27] A.N. Kirillov, Zap. Nauč. Semin. POMI 164 (1987) 121, J. Math. Sci. 47 (1989) 2450.
[28] A. Kuniba, T. Nakanishi, J. Suzuki, Int. J. Mod. Phys. A 9 (1994) 5215, arXiv:hep-th/9309137.
[29] A. Kuniba, T. Nakanishi, J. Suzuki, Int. J. Mod. Phys. A 9 (1994) 5267, arXiv:hep-th/9310060.
[30] C. Locke, ODE/IM Correspondence and Affine Toda Field Equations, Doctoral thesis, Tokyo Institute of Technology, 2015.
[31] V.V. Bazhanov, S.L. Lukyanov, A.B. Zamolodchikov, Commun. Math. Phys. 177 (1996) 381, arXiv:hep-th/9412229.
[32] V.V. Bazhanov, S.L. Lukyanov, A.B. Zamolodchikov, Commun. Math. Phys. 190 (1997) 247, arXiv:hep-th/9604044.
[33] V.V. Bazhanov, A.N. Hibberd, S.M. Khoroshkin, Nucl. Phys. B 622 (2002) 475, arXiv:hep-th/0105177.
[34] T. Kojima, J. Phys. A 41 (2008) 355206, arXiv:0803.3505 [nlin.SI].

13

Topological defects in a deformed gauge theory

Mir Faizal [a,b,*], Tsou Sheung Tsun [c]

[a] *Irving K. Barber School of Arts and Sciences, University of British Columbia - Okanagan, Kelowna, British Columbia V1V 1V7, Canada*
[b] *Department of Physics and Astronomy, University of Lethbridge, Lethbridge, Alberta, T1K 3M4, Canada*
[c] *Mathematical Institute, University of Oxford, Andrew Wiles Building, Radcliffe Observatory Quarter, Woodstock Road, Oxford OX2 6GG, United Kingdom*

Editor: Stephan Stieberger

Abstract

In this paper, we will analyse the topological defects in a deformation of a non-abelian gauge theory using the Polyakov variables. The gauge theory will be deformed by the existence of a minimum measurable length scale in the background spacetime. We will construct the Polyakov loops for this deformed non-abelian gauge theory, and use these deformed loop space variables for obtaining a deformed loop space curvature. It will be demonstrated that this curvature will vanish if the deformed Bianchi identities are satisfied. However, it is possible that the original Bianchi identities are satisfied, but the deformed Bianchi identities are violated at the leading order in the deformation parameter, due to some topological defects. Thus, topological defects could be produced purely from a deformation of the background geometry.

1. Introduction

Topological defects can be analysed using the Polyakov variables, and these Polyakov variables are defined using the holonomies of the gauge fields [1–4]. In this paper, we will call these holonomies of the gauge fields as Polyakov loops, as they were introduced by Polyakov [1]. So, these Polyakov loops would be constructed using the gauge fields as the holonomies of closed

---

* Corresponding author.
 E-mail address: mirfaizalmir@gmail.com (M. Faizal).

loops in spacetime. They are also called as the Dirac phase factors in the physics literature. They do not depend on the parameterization chosen, and they capture some interesting topological properties of the gauge theory. In fact, they resemble the Wilson's loops, but unlike the Wilson loops, no trace is taken over the gauge group for Polyakov loops. Thus, in this paper, there is a difference between these Polyakov loops, and the usual Wilson's loops. This is because Wilson's loops only represented by a number, but Polyakov loops are gauge group-valued functions of the infinite-dimensional loop space [1]. So, they can be used to analyse various interesting structures in the gauge theory, and this includes topological defects produced by the existence of non-abelian monopoles. It may be noted that recently Polyakov loops have been used for analysing various interesting physical systems including fractional M2-branes [5]. They have also been used for analysing three dimensional supersymmetric gauge theory [6], and these theories are important to study systems like M2-branes and D2-branes. The four dimensional supersymmetric gauge theories have also been analysed using this formalism [7]. In fact, this formalism was used to analyse the non-abelian monopoles in four dimensional supersymmetric gauge theories. Thus, it is possible to use this formalism to analysing various interesting generalizations of the usual gauge theories. So, in this paper, we will analyse the effect of topological defects on a deformed non-abelian gauge theory. This gauge theory will be deformed by the existence of a minimum measurable length scale in the background geometry.

Such a deformation of the gauge theories by the existence of a minimum measurable length scale in the background geometry is in turn motivated from low energy effects of quantum gravity. This is because almost all the approaches to quantum gravity restrict the measurement of spacetime below the Planck scale. The string theory is one of the most important approaches for analysing quantum gravity, and the fundamental string is the smallest probe available in perturbative string theory, and so it is not possible to probe spacetime below the string length scale in string theory [8–12]. Thus, the string length acts, which is given by $l_s = \alpha'$ as a minimum measurable length in string theory. Furthermore, if non-perturbative effects are taken into consideration, then it is possible to have $D0$-brane which is a point like object. However, it has been argued that even in presence of such brane, there is an intrinsic minimal length of the order of $l_{min} = l_s g_s^{1/3}$, where g_s is the string coupling constant [13,14]. The total energy of the quantized string depends on the excitation n and winding number w, and under T-duality the n and w gets interchanged, as $R \to l_s^2/R$ and $n \to w$. So, a description of string theory below l_s is the same as the description above it, and so it can be argued from T-duality that the string theory cannot be described below the string length scale [13]. The T-duality has been used to construct an effective path integral for the center of mass of the string, and analyse the corresponding Green's function [15,16]. This has been done by analysing strings propagating in spacetime with compactified additional dimensions. It has been demonstrated that this Green's function also has an minimal length associated with it [15,16]. So, string theory due to T-duality has a minimal length associated with it. It may be noted that this minimal length can be different from Planck length [13]. This is because the Planck length l_{PL} can be expressed as $l_{PL} = g_s^{1/4} l_s$ [13]. It has also been argued that a minimal length may exist in models of quantum gravity, such as the loop quantum gravity [17]. The physics of black holes restricts the measurement to scales larger than the Planck scale. This is because the energy needed to probe a region of spacetime below Planck scale is greater than the energy needed to form a mini black hole in the region of spacetime [18,19]. So, if we try to probe the spacetime at a scale smaller than the Planck scale, a mini black hole will form in that region of spacetime, and this will in turn restrict our ability to analyse that region of spacetime. It has also been argued that this length can be much larger than the Planck length, and its scale would be fixed by present experimental data [20,21]. So, it may be possible to have such effects observed in future experiments, and thus it would be interesting to study different aspects of such effects.

However, the problem with the existence of such a minimum measurable length scale is that it is not consistent with the foundations principles of ordinary quantum mechanics. This is because the ordinary quantum mechanics is based on the Heisenberg uncertainty principle, and according to this principle it is possible to detect the position of a particle with arbitrary accuracy, if the momentum is not measured. Thus, according to the Heisenberg uncertainty principle there is no bound on the accuracy to which the length can be measured as long as the momentum is not measured. Thus, in principle, according to the Heisenberg uncertainty principle, we can analyse the spacetime at a length scale smaller than the Planck scale, and so no minimum measurable length scale exists. However, the Heisenberg uncertainty principle can be modified to incorporate the existence of a minimum measurable length scale. This can be done by deforming the usual Heisenberg uncertainty principle $\Delta x \Delta p \geq 1/2$, to $\Delta x \Delta p \geq 1/2(1 + \beta(\Delta p)^2)$, where β is a parameter in the theory. This modified Heisenberg uncertainty principle is called the generalized uncertainty principle (GUP).

As the Heisenberg uncertainty principle is closely related to the Heisenberg algebra, such a deformation of the Heisenberg uncertainty principle will generate a deformation of the Heisenberg algebra. In this GUP deformed Heisenberg algebra, the commutator of momentum and position operators is a function of momentum, $[x_i, p_j] = i(\delta_{ij} + \beta(p^2\delta_{ij} + 2p_i p_j))$ [12,22–26]. This deformation of the Heisenberg algebra will also produce a deformation of the coordinate representation of the momentum operator. In fact, it is possible to write the deformed momentum operator, to the first order in β, as $p_i = -i\partial_i(1 - \beta\partial^j\partial_j)$. In this paper, we will analyse a relativistic version of this deformation, and the corresponding gauge theory using the Polyakov loop formalism.

2. Loop space

In this section, we will construct the Polyakov loops for a deformed gauge theory, which will be deformed by the deformation of the Heisenberg algebra by GUP. It is also possible to define a relativistic version of the GUP deformed Heisenberg algebra, and study the quantum field theory corresponding to such a deformed algebra [27–32]. Thus, the full covariant algebra can be written as, $[\hat{x}^\mu, \hat{p}_\nu] = i\delta^\mu_\nu[1 + \beta\hat{p}^\rho\hat{p}_\rho] + 2i\beta\hat{p}^\mu\hat{p}_\nu$. The generalized uncertainty for this deformed algebra can be expressed as $\Delta x^\mu \Delta p_\mu \geq 1/2(1 + 3\beta\Delta p^\mu \Delta p_\mu + 3\beta\langle p^\rho\rangle\langle p_\rho\rangle)$, and this generalized uncertainty can be used to obtain the following bound $\Delta x^\mu_{min} = \sqrt{3\beta}\sqrt{1 + 3\beta\langle p^\rho\rangle\langle p_\rho\rangle}$ [27]. So, there exists a minimum length l_s and a minimum time t_s in this algebra, such that $l_s = \sqrt{3\beta}$, and $t_s = \sqrt{3\beta}$. To the first order in β, we can write the deformed momentum operator as $p_\mu = -i\partial_\mu(1 - \beta\partial^\rho\partial_\rho) + \mathcal{O}(\beta^2)$. It is possible to define a gauge covariant derivative which is consistent with the existence of a minimum length scale as

$$\mathcal{D}_\mu = (1 - \beta D^\rho D_\rho) D_\mu, \tag{1}$$

where $D_\mu = \partial_\mu + iA^A_\mu T_A$. Here T_A are the generators of the Lie algebra $[T_A, T_B] = if^C_{AB}T_C$. Now as the covariant derivative transform $D_\mu \to UD_\mu U^{-1}$, so the deformed covariant derivative transform as [28]

$$\mathcal{D}_\mu \to -i\left(1 - \beta UD^\rho U^{-1}UD_\rho U^{-1}\right)UD_\mu U^{-1}$$

$$= -iU \left(1 - \beta D^\rho D_\rho \right) D_\mu U^{-1}$$
$$= U \mathcal{D}_\mu U^{-1}. \tag{2}$$

So, the GUP deformed covariant derivative still transforms like a regular covariant derivative. It is possible to show that the Bianchi identity will hold, but the algebraic manipulations are long (we used the package Quantum Mathematica to prove this result),

$$[\mathcal{D}_\lambda, [\mathcal{D}_\mu, \mathcal{D}_\nu]] + [\mathcal{D}_\mu, [\mathcal{D}_\nu, \mathcal{D}_\lambda]] + [\mathcal{D}_\nu, [\mathcal{D}_\lambda, \mathcal{D}_\mu]]$$
$$= [(1 - \beta D^\rho D_\rho)D_\lambda, [(1 - \beta D^\tau D_\tau)D_\mu, (1 - \beta D^\sigma D_\sigma)D_\nu]]$$
$$+ [(1 - \beta D^\tau D_\tau)D_\mu, [(1 - \beta D^\sigma D_\sigma)D_\nu, (1 - \beta D^\rho D_\rho)D_\lambda]]$$
$$+ [(1 - \beta D^\sigma D_\sigma)D_\nu, [(1 - \beta D^\rho D_\rho)D_\lambda, (1 - \beta D^\tau D_\tau)D_\mu]]$$
$$= 0. \tag{3}$$

Motivated from the definition of the usual field tensor $F_{\mu\nu} = -i[D_\mu, D_\nu]$, the deformed field tensor is defined as,

$$\mathcal{F}_{\mu\nu} = -i \left[\mathcal{D}_\mu, \mathcal{D}_\nu \right]$$
$$= -i \left[\left(1 - \beta D^\rho D_\rho \right) D_\mu, \left(1 - \beta D^\rho D_\rho \right) D_\nu \right]$$
$$= \left(1 - \beta D^\rho D_\rho \right) \left[\left(1 - \beta D^\rho D_\rho \right) F_{\mu\nu} - \beta \left(D^\rho F_{\mu\rho} D_\nu - D^\rho F_{\nu\rho} D_\mu \right) \right.$$
$$\left. - \beta \left(F_{\mu\rho} D^\rho D_\nu - F_{\nu\rho} D^\rho D_\mu \right) \right],$$
$$= F_{\mu\nu} - 2\beta D^\rho D_\rho F_{\mu\nu} - \beta \left(D^\rho F_{\mu\rho} D_\nu - D^\rho F_{\nu\rho} D_\mu \right)$$
$$- \beta \left(F_{\mu\rho} D^\rho D_\nu - F_{\nu\rho} D^\rho D_\mu \right)$$
$$= F_{\mu\nu} + \beta \tilde{F}_{\mu\nu}, \tag{4}$$

where $F_{\mu\nu} = -i[D_\mu, D_\nu]$ is the un-deformed field tensor, and

$$\tilde{F}_{\mu\nu} = -2D^\rho D_\rho F_{\mu\nu} - \left(D^\rho F_{\mu\rho} D_\nu - D^\rho F_{\nu\rho} D_\mu \right)$$
$$- \left(F_{\mu\rho} D^\rho D_\nu - F_{\nu\rho} D^\rho D_\mu \right). \tag{5}$$

Writing out equation (1) in terms of the undeformed potential A_μ, we have

$$\mathcal{D}_\mu = [1 - \beta(\partial^\rho + iA_\rho)(\partial_\rho + iA_\rho)](\partial_\mu + iA_\mu). \tag{6}$$

It is now clear that we can consider a modified potential $\mathcal{A}_\mu$ differing from the undeformed A_μ by a term proportional to β:

$$\mathcal{A}_\mu = A_\mu + \beta \tilde{A}_\mu \tag{7}$$

with the extra term obtained from equation (6)

$$\tilde{A}_\mu = (\partial^\rho + iA_\rho)(\partial_\rho + iA_\rho)(\partial_\mu + iA_\mu). \tag{8}$$

We now wish to define loop space variables based on this deformed field tensor $\mathcal{F}_{\mu\nu}$ to study topological obstructions in this GUP spacetime.

First we consider the space of loops in undeformed spacetime, with a fixed base point. A loop is parameterized by the coordinates $\xi^\mu(s)$,

$$C : \{\xi^\mu(s) : s = 0 \to 2\pi, \ \xi^\mu(0) = \xi^\mu(2\pi)\}, \tag{9}$$

where $\xi^\mu(0) = \xi^\mu(2\pi)$ is the chosen (but arbitrary) base point [1–4]. Next we define the loop space variable

$$\Phi[\xi] = P_s \exp i \int_0^{2\pi} A^\mu(\xi(s)) \frac{d\xi_\mu}{ds}, \tag{10}$$

where P_s denotes ordering in s increasing from right to left. From this we can define its logarithmic derivative as a kind of loop space connection

$$F_\mu[\xi|s] = i\,\Phi^{-1}[\xi] \frac{\delta}{\delta\xi^\mu(s)} \Phi[\xi]. \tag{11}$$

The derivative in s is taken from below. It may be noted as the loop variable $\Phi[\xi]$ only depends on C and not the manner in which C is parameterized, so labeling it with a fixed point is over complete. In fact, any other parameterization of C will only change the variable in the integration and not the loop space variable $\Phi[\xi]$.

We can obtain a formula relating $F_\mu[\xi|s]$ to the spacetime curvature $F_{\mu\nu}$ by first defining a parallel transport from a point $\xi(s_1)$ to a point $\xi(s_2)$ as [1–4]

$$\Phi[\xi : s_1, s_2] = P_s \exp i \int_{s_1}^{s_2} A^\mu(\xi(s)) \frac{d\xi_\mu}{ds}. \tag{12}$$

Thus

$$F^\mu[\xi|s] = \Phi^{-1}[\xi : s, 0] F^{\mu\nu} \Phi[\xi : s, 0] \frac{d\xi_\nu(s)}{ds}. \tag{13}$$

This formula can be understood as follows. We parallel transport from a fixed point along a fixed path to another fixed point. After reaching that point, we will take a detour then turn back along the same path till we reach the original point. Thus, the phase factor generated by going along the path from the original point to final point will be canceled by the phase factor generated by going from the final point back to the original point. However, there will be a contribution generated by the transport along the infinitesimal circuit along the final point, which is proportional to the spacetime curvature at that point.

We can repeat the same construction using our deformed variables. So, we can define a deformed loop variable with a deformed connection. As this deformed connection, is a connection in the deformed theory, we can write

$$\mathbf{\Phi}[\xi] = P_s \exp i \int_0^{2\pi} \mathcal{A}^\mu(\xi(s)) \frac{d\xi_\mu}{ds}. \tag{14}$$

Here again P_s denotes ordering in s increasing from right to left. Now we can define the logarithmic derivative of this deformed variable as a deformed loop space connection

$$\mathcal{F}_\mu[\xi|s] = i\,\mathbf{\Phi}^{-1}[\xi] \frac{\delta}{\delta\xi^\mu(s)} \mathbf{\Phi}[\xi]. \tag{15}$$

We can also deformed a parallel transport from a point $\xi(s_1)$ to a point $\xi(s_2)$ as

$$\mathbf{\Phi}[\xi : s_1, s_2] = P_s \exp i \int_{s_1}^{s_2} \mathcal{A}^\mu(\xi(s)) \frac{d\xi_\mu}{ds}. \tag{16}$$

Now we can use this deformed parallel transport to go from a fixed point to another fixed point, along a fixed path. We can also take a detour from that final fixed point and go back to the initial fixed point along the same path. So, the phase generated by going to the final fixed point will exactly cancel the phase generated by going back to the initial fixed point. However, to take a detour, we will have to produce an infinitesimal circuit along the final point. This infinitesimal circuit will produce a contribution, and as we are using the deformed parallel transport, we can write this contribution as

$$\mathcal{F}^{\mu}[\xi|s] = \Phi^{-1}[\xi : s, 0]\mathcal{F}^{\mu\nu}(\xi(s))\Phi[\xi : s, 0]\frac{d\xi_{\nu}(s)}{ds}. \tag{17}$$

Now since the GUP deformation in the covariant derivative is first order in β, in this expression we can actually replace the deformed Φ by the undeformed Φ

$$\begin{aligned}
\mathcal{F}^{\mu}[\xi|s] &= \Phi^{-1}[\xi : s, 0]\mathcal{F}^{\mu\nu}(\xi(s))\Phi[\xi : s, 0]\frac{d\xi_{\nu}(s)}{ds} \\
&= \Phi^{-1}[\xi : s, 0]\big[F^{\mu\nu} + \beta\tilde{F}^{\mu\nu}\big](\xi(s))\Phi[\xi : s, 0]\frac{d\xi_{\nu}(s)}{ds} \\
&= F^{\mu}[\xi|s] + \beta\tilde{F}^{\mu}[\xi|s].
\end{aligned} \tag{18}$$

Now this is important to note that if the original $F^{\mu\nu} = 0$, then $F^{\mu}[\xi|s] = 0$. However, it is possible that even if $F^{\mu\nu} = 0$, we can have $\tilde{F}^{\mu\nu} \neq 0$, and so $\mathcal{F}^{\mu\nu} \neq 0$. This would mean that even if $F^{\mu}[\xi|s] = 0$, we can still have $\tilde{F}^{\mu}[\xi|s] \neq 0$, and so $\mathcal{F}^{\mu}[\xi|s] \neq 0$. Thus, there could be a contribution to the Polyakov loop produced solely from the deformation of the background geometry.

3. Topological defects

We can regard $\mathcal{F}_{\mu}[\xi|s]$ as the connection in the loop space as it represents the change in phase of $\Phi[\xi]$ as one moves in the loop space. It is interesting to note that the connection is loop space $\mathcal{F}_{\mu}[\xi|s]$ is proportional to the field strength in spacetime $\mathcal{F}_{\mu}(\xi(s))$. As $\mathcal{F}_{\mu}[\xi|s]$ acts as a connection in the loop space, we can define covariant derivative in loop space $\Delta_{\mu}(s) = \delta/\delta\xi^{\mu}(s) + i\mathcal{F}_{\mu}[\xi|s]$. This covariant derivative can be used to define the curvature of the loop space $-i\mathcal{G}_{\mu\nu}[\xi, s_1, s_2]$ as the commutator of these covariant derivatives $[\Delta_{\mu}[\xi(s_1)], \Delta_{\nu}[\xi(s_2)]]$. So, we obtain the following expression for the curvature of the deformed loop space

$$\begin{aligned}
\mathcal{G}_{\mu\nu}[\xi(s_1, s_2)] = &\frac{\delta}{\delta\xi^{\mu}(s_2)}\mathcal{F}_{\nu}[\xi|s_1] - \frac{\delta}{\delta\xi^{\nu}(s_1)}\mathcal{F}_{\mu}[\xi|s_2] \\
&+ i[\mathcal{F}_{\mu}[\xi|s_1], \mathcal{F}_{\nu}[\xi|s_2]] \\
&G_{\mu\nu}[\xi(s_1, s_2)] + \beta\tilde{G}_{\mu\nu}[\xi(s_1, s_2)].
\end{aligned} \tag{19}$$

Here the original curvature in loop space is given by [2]

$$\begin{aligned}
G_{\mu\nu}[\xi(s_1, s_2)] = &\frac{\delta}{\delta\xi^{\mu}(s_2)}F_{\nu}[\xi|s_1] - \frac{\delta}{\delta\xi^{\nu}(s_1)}F_{\mu}[\xi|s_2] \\
&+ i[F_{\mu}[\xi|s_1], F_{\nu}[\xi|s_2]],
\end{aligned} \tag{20}$$

and $\tilde{G}_{\mu\nu}[\xi(s_1, s_2)]$ is the correction to the original loop space curvature. It may be noted that the deformed loop connection, $\mathcal{F}_{\mu}[\xi|s]$ represents a change in phase Φ as one moves in the deformed loop space. So, for a deformed gauge connection $\mathcal{A}_{\mu}$, it is possible to construct the holonomy using the deformed field tensor $\mathcal{F}^{\mu\nu}$. However, now $\mathcal{F}^{\mu}[\xi|s]$ is also a connection in the deformed loop space, and so we can construct the corresponding holonomy. Thus, we can go from a fixed point in the deformed loop space to another fixed point, and then take a detour back to the initial point. We will go back along the same path we initially took, and so the contribution of going to the final point will exactly cancel the contribution of going back to the initial point. However, to take a detour, we will have to make an infinitesimal circuit, and this will have a contribution. As we are moving in the deformed loop space, this contribution would be equal to $\mathcal{G}_{\mu\nu}[\xi(s_1, s_2)]$. It may be noted that in spacetime, this would appear as sweeping out an infinitesimal two dimensional surface enveloping a three dimensional volume. Now the value of this deformed loop space curvature will depend on what is inside this volume. This loop space curvature can be used to analyse the presence of a topological defect in the original theory. This is because if a monopole is not present in the spacetime, then this deformed loop space curvature term vanishes. This can be seen by showing that this deformed loop space curvature is proportional to left-hand side of the Bianchi identity (3).

In fact, following closely similar arguments for the usual gauge theories [2], we consider variations of the curve in two orthogonal directions λ and κ. Now first we define three displaced curves,

$$(\xi_1^\mu(s))_\lambda = (\xi^\mu(s))_\lambda + \Delta\delta_\lambda^\mu\delta(s - s_1)$$
$$(\xi_2^\mu(s))_\kappa = (\xi^\mu(s))_\kappa + \Delta'\delta_\kappa^\mu\delta(s - s_2)$$
$$(\xi_3^\mu(s))_\kappa = (\xi_1^\mu(s))_\kappa + \Delta'\delta_\kappa^\mu\delta(s - s_2), \tag{21}$$

where the Kronecker delta δ_λ^μ means that the variation is zero if $\mu \neq \lambda$, and similarly for δ_κ^μ. Then by definition

$$\frac{\delta}{\delta\xi^\kappa(s_2)}\mathcal{F}_\lambda[\xi|s_1] = \lim_{\Delta\to 0}\lim_{\Delta'\to 0}\frac{1}{\Delta\Delta'}\frac{i}{g}\left\{\Phi^{-1}[\xi_2]\Phi[\xi_3] - \Phi^{-1}[\xi]\Phi[\xi_1]\right\}. \tag{22}$$

It may be noted that the right-hand side usually has the implicit indices λ and κ as noted above.

Next we calculate the value of $\Phi^{-1}[\xi_2]\Phi[\xi_3] - \Phi^{-1}[\xi]\Phi[\xi_1]$. Using parallel transport along these paths, we obtain

$$\Phi[\xi_1] = \Phi[\xi] - i\int ds\,\Phi[\xi:2\pi,s]\mathcal{F}(\xi(s))\Phi(\xi:s,0), \tag{23}$$

where

$$\mathcal{F}(\xi(s)) = \mathcal{F}^{\mu\nu}(\xi(s))\frac{d\xi_\nu(s)}{ds}\Delta\delta_\mu^\lambda\delta(s - s_1). \tag{24}$$

Furthermore, we also obtain,

$$\Phi[\xi_2] = \Phi[\xi] - i\int ds\,\Phi[\xi:2\pi,s]\mathcal{F}(\xi(s))\Phi[\xi:s,0], \tag{25}$$

where

$$\mathcal{F}(\xi(s)) = \mathcal{F}^{\mu\nu}(\xi(s))\frac{d\xi_\nu(s)}{ds}\Delta'\delta_\mu^\kappa\delta(s - s_2). \tag{26}$$

Finally, we obtain

$$\Phi[\xi_3] = \Phi[\xi_1] - i\int ds\,\Phi[\xi_1:2\pi,s]\mathcal{F}(\xi_1(s))\Phi[\xi_1:s,0], \tag{27}$$

where

$$\mathcal{F}(\xi_1(s)) = \mathcal{F}^{\mu\nu}(\xi_1(s)) \frac{d\xi_{1\nu}(s)}{ds} \Delta' \delta_\mu^\kappa \delta(s - s_2). \tag{28}$$

We can also write similar expressions for $\Phi[\xi : 2\pi, s]$ and $\Phi[\xi_1 : s, 0]$. Now collecting all these, we obtain the following expression,

$$\frac{\delta}{\delta\xi_\mu(s_2)} \mathcal{F}_\nu[\xi|s_1] = \Phi^{-1}[\xi : s_1, 0] \mathcal{D}^\nu \mathcal{F}^{\mu\rho}(\xi(s_2))$$

$$\times \frac{d\xi_\rho(s_1)}{ds_1} \Phi[\xi : s_1, 0] \delta(s_2 - s_1)$$

$$+ \Phi^{-1}[\xi : s_2, 0] \mathcal{F}_{\mu\nu}(\xi(s_2)) \Phi[\xi : s_2, 0]$$

$$\times \frac{d}{ds_1} \delta(s_2 - s_1)$$

$$+ i[\mathcal{F}_\mu[\xi|s_2], \mathcal{F}_\nu[\xi|s_1]] \theta(s_1 - s_2),$$

$$\frac{\delta}{\delta\xi_\nu(s_1)} \mathcal{F}_\mu[\xi|s_2] = \Phi^{-1}[\xi : s_2, 0] \mathcal{D}^\mu \mathcal{F}^{\nu\tau}(\xi(s_1))$$

$$\times \frac{d\xi_\tau(s_2)}{ds_2} \Phi[\xi : s_2, 0] \delta(s_1 - s_2)$$

$$+ \Phi^{-1}[\xi : s_1, 0] \mathcal{F}_{\nu\mu}(\xi(s_1)) \Phi[\xi : s_1, 0]$$

$$\times \frac{d}{ds_2} \delta(s_1 - s_2)$$

$$+ i[\mathcal{F}_\nu[\xi|s_1], \mathcal{F}_\mu[\xi|s_2]] \theta(s_2 - s_1). \tag{29}$$

So, the loop space curvature can be written as,

$$\mathcal{G}_{\mu\nu}[\xi(s_1, s_2)] = \frac{\delta}{\delta\xi^\mu(s_2)} \mathcal{F}_\nu[\xi|s_1] - \frac{\delta}{\delta\xi^\nu(s_1)} \mathcal{F}_\mu[\xi|s_2]$$

$$+ i[\mathcal{F}_\mu[\xi|s_1], \mathcal{F}_\nu[\xi|s_2]]$$

$$= \Phi^{-1}[\xi : s_1, 0] \Big[[\mathcal{D}_\mu, \mathcal{F}_{\nu\tau}] + [\mathcal{D}_\nu, \mathcal{F}_{\tau\mu}] + [\mathcal{D}_\tau, \mathcal{F}_{\mu\nu}] \Big]$$

$$\times \Phi[\xi : s_1, 0] \frac{d\xi^\tau(s_1)}{ds} \delta(s_1 - s_2). \tag{30}$$

Thus, the deformed loop space curvature is proportional to the deformed Bianchi identity in the spacetime. It is known that the Bianchi identity are satisfied in absence of a topological defect in spacetime, $[\mathcal{D}_\mu, \mathcal{F}_{\nu\tau}] + [\mathcal{D}_\nu, \mathcal{F}_{\tau\mu}] + [\mathcal{D}_\tau, \mathcal{F}_{\mu\nu}] = 0$, and so the loop space curvature vanishes in absence of a topological defect in spacetime, $\mathcal{G}_{\mu\nu}[\xi(s_1, s_2)] = 0$. However, if a monopole exists in spacetime, then Bianchi identity are not satisfied $[\mathcal{D}_\mu, \mathcal{F}_{\nu\tau}] + [\mathcal{D}_\nu, \mathcal{F}_{\tau\mu}] + [\mathcal{D}_\tau, \mathcal{F}_{\mu\nu}] \neq 0$. Now if the world-line of a monopole goes through the point represented by s_1, then the loop space curvature does not vanish $\mathcal{G}_{\mu\nu}[\xi(s_1, s_2)] \neq 0$. However, if the topological defect only contributes at the order β, then the original Bianchi identity will be satisfied, and the deformed Bianchi identity will be violated at the order β. Thus, the loop space curvature is proportional will also have a contribution at the order β, and it will not vanish. So, it is possible to produce topological defects in the gauge theory from the deformation of the background geometry by minimum measurable length scale. It would be interesting to analyse the consequences of such a deformation further.

4. Monopole charge

Now we will finally obtain an expression for the non-abelian monopole charge in such deformed field theories. It is possible to obtain the non-abelian monopole charge for the usual gauge theories using the concept of loop of loops [2]. In this section, we will generalize this construction to deformed gauge theories, and thus obtain a generalized monopole charge for deformed gauge theories. It is also possible to construct a loop in the loop space by using the connection in the loop space, $\mathcal{F}^{\mu}[\xi|s]$. In order to do that, we define Σ as

$$\Sigma : \{\xi^{\mu}(s) : s = 0 \to 2\pi, \ t = 0 \to 2\pi\}, \tag{31}$$

where

$$\xi^{\mu}(t:0) = \xi^{\mu}(t:2\pi), \ t = 0 \to 2\pi,$$
$$\xi^{\mu}(0:s) = \xi(2\pi:s), \ s = 0 \to 2\pi. \tag{32}$$

So, for each t, we have $\xi^{\mu}(t:s)$ and this represents a closed loop $C(t) \ s = 0 \to 2\pi$,

$$C(t) : \{\xi^{\mu}(t:s), s = 0 \to 2\pi\}. \tag{33}$$

Here $C(t)$ traces out a closed loop as t varies, and it shrinks to a point for $t = 0$ and $t = 2\pi$. Now using Σ, we can construct a loop in the loop space. Thus, for the usual un-deformed gauge theories, this will be given by

$$\Theta(\Sigma) = P_t \exp i \int_0^{2\pi} dt \int_0^{2\pi} F_{\mu}[\xi|t,s] \frac{\partial \xi^{\mu}[\xi|t,s]}{\partial t}. \tag{34}$$

This loop in the loop space is a parameterized surface in spacetime. Thus, this loop in the loop space encloses a volume. So, it can be used to measure the monopole inside such a volume. We will now generalize this construction to deformed gauge theories, and then apply that deformed formalism to analyse the monopole charge for the deformed gauge theory.

However, as the deformation by the generalized uncertainty principle, deforms $F_{\mu}[\xi|t,s]$ to $\mathcal{F}_{\mu}[\xi|t,s]$, we can construct a loop in the loop space of deformed theories using

$$\Theta(\Sigma) = P_t \exp i \int_0^{2\pi} dt \int_0^{2\pi} \mathcal{F}_{\mu}[\xi|t,s] \frac{\partial \xi^{\mu}[\xi|t,s]}{\partial t}$$

$$= P_t \exp i \int_0^{2\pi} dt \int_0^{2\pi} \left[F_{\mu}[\xi|t,s] + \beta \tilde{F}_{\mu}[\xi|t,s] \right] \frac{\partial \xi^{\mu}[\xi|t,s]}{\partial t}$$

$$= \Theta(\Sigma) + \beta \tilde{\Theta}(\Sigma). \tag{35}$$

This Θ measures the charge of a non-abelian monopole for a deformed gauge theory, since $\Theta(\Sigma) = \zeta$, where ζ is the generalized monopole charge enclosed by the surface Σ. Note that $\Theta(\Sigma) = I$ the group identity represents the vacuum, i.e. no topological charge is enclosed by the surface. As an example, let us consider a gauge theory with $SO(3)$ as its gauge group. In this case, monopole charges are $+1$ for no monopole, and -1 for a monopole. If a monopole is not present, then $\Theta(\Sigma)$ will wind fully around the gauge group and will equal to the identity. However, in presence of a monopole, $\Theta(\Sigma)$ cannot wind fully around the gauge group and will equal the identity. It may be noted that the expression $\Theta(\Sigma) = \zeta$ is interesting as it can be used to evaluate the monopole charge. It is possible to demonstrate that this result holds for all non-abelian Yang–Mills theories with gauge group is $SU(N)/Z_N$. The monopole charge for such a gauge group is given by $\zeta = \exp i2\pi r/N$, where $r = 0, 1, 2, \cdots, (N-1)$. Thus, with this modification this result can be applied to Yang–Mills theory with any gauge group. It is interesting to note that even the change has a β contribution coming from deformation. This occurs because the topological defects can occur at the order β, even if they do not occur in the original theory. Thus, even in deformed gauge theories, the Polyakov loops space formalism can be used to analyse the topological defects.

It may be noted as $G_{\mu\nu}[\xi(s_1, s_2)]$ is analogous to $\mathcal{F}^{\mu\nu}$ in deformed loop space, it can be constructed using a logarithmic derivative of $\Theta(\Sigma)$. So, basically, we can argue that the logarithmic derivative of $\Theta(\Sigma)$ would produce a connection in this loop of loop space. In fact, this has been done for ordinary loop space [2], and the same argument can be used for deformed loop space by using deformed quantities. Now for $s_1 \neq s_2$, $G_{\mu\nu}[\xi(s_1, s_2)]$ does not enclose any volume, and so for this $\Theta(\Sigma) = I$, which is the group identity, and its logarithmic derivative vanishes. This also occurs for $s_1 = s_2$, of $\xi(s)$ does not intersect with a monopole worldline, which we can represent by $Y^\rho(\tau)$. So, in that case again $\theta(\Sigma) = I$. However, when $s_1 = s_2$, and $x_1(s)$ intersects a monopole worldline $Y^\rho(\tau)$, $G_{\mu\nu}[\xi(s_1, s_2)]$ corresponds to Σ enclosing a monopole. Now for original un-deformed loop space variable we have [2],

$$G_{\mu\nu}[\xi(s_1, s_2)] = \frac{-\pi}{g} \int d\tau \kappa[\xi|s] \frac{d\xi_\sigma(s)}{ds} \frac{dY_\rho(\tau)}{d\tau} \delta(\xi(s) - Y(\tau))\delta(s_1 - s_2), \tag{36}$$

where $\exp i\pi\kappa = \zeta$. However, in deformed loop space, we can write $G_{\mu\nu}[\xi(s_1, s_2)] = G_{\mu\nu}[\xi(s_1, s_2)] + \beta \tilde{G}_{\mu\nu}[\xi(s_1, s_2)]$, so we obtain

$$G_{\mu\nu}[\xi(s_1, s_2)] = \frac{-\pi}{g} \int d\tau \kappa[\xi|s] \frac{d\xi_\sigma(s)}{ds} \frac{dY_\rho(\tau)}{d\tau} \delta(\xi(s) - Y(\tau))\delta(s_1 - s_2)$$
$$- \beta \tilde{G}_{\mu\nu}[\xi(s_1, s_2)]. \tag{37}$$

So, even if $G_{\mu\nu}[\xi(s_1, s_2)] = 0$, due to the original Bianchi identity being satisfied, we still have

$$\frac{-\pi}{g} \int d\tau \kappa[\xi|s] \frac{d\xi_\sigma(s)}{ds} \frac{dY_\rho(\tau)}{d\tau} \delta(\xi(s) - Y(\tau))\delta(s_1 - s_2) \neq 0, \tag{38}$$

and so the deformation of the loop space produces a topological defect in spacetime. Thus, we will have demonstrated that a monopole contribution can generated from deformation of loop space variables. It may be noted that monopoles in general have been analysed in loop space using a duality which reduces to electromagnetic Hodge duality for abelian theories [35–38]. However, as far as we know, all such constructions use the loop space formalism, and we are not aware of any proof of this duality using space–time variables alone. Therefore we are restricted at present to such a discussion in loop space only.

We would like to point out that solitonic solutions of the 't Hooft–Polyakov type are sometimes called non-abelian monopoles, but the magnetic charge carried by them is usually an abelian magnetic charge (with symmetry breaking into a $U(1)$ subgroup). As far as we know, no solutions of the pure Yang–Mills equation i.e., (without the introduction of symmetry breaking), with a non-abelian monopole charge has been constructed, either using spacetime variables or loop variables. Furthermore, the monopoles (with symmetry breaking) are solutions only outside of a sphere of a finite radius, usually interpreted as the size of the monopole. Inside of this sphere, not much is known, since the interactions due to the original non-abelian forces become non-negligible. Here we are interested in studying the genuinely non-abelian magnetic charge, without involving the Higgs fields. Their existence in ordinary spacetime is governed by topology. In spacetimes with a minimum length scale, as we study here, the obstruction to the vanishing of the relevant loop space curvature indicates also the topological nature of this obstruction, which by analogy we think of as generalized monopoles. As no non-abelian monopole solutions in ordinary spacetime are known, to construct one for GUP spacetime would really be interesting, but perhaps not feasible at present. In this paper we have demonstrated that minimal length in spacetime can give rise to a certain topological charge. However, we would like to point out that it is possible that an object would not exist even if such a charge is allowed to exist [39]. So, we only demonstrate that such an object can exist due to the existence of a topological charge produced by minimal length.

5. Validity of the approximation

In this section, we will argue that the higher order contributions cannot cancel the topological defects produced at a certain order in β. Thus, we will be able to demonstrate that the results obtain in this paper are not a consequence of the approximation that we have used. This is because if we had considered the deformation to the next order, then we would get higher order contribution to the field strength, which would produce higher order contributions to the loop space variables. Thus, if we analysed the theory to the order β^2, then the corrected field strength would be given by

$$\mathcal{F}_{\mu\nu} = F_{\mu\nu} + \beta \tilde{F}_{\mu\nu} + \beta^2 \bar{\tilde{F}}_{\mu\nu}. \tag{39}$$

This would occur because $\mathcal{D}_\mu$ will also have a β^2 contribution to it. This will in turn produce a β^2 contribution to the connection,

$$\mathcal{A}_\mu = A_\mu + \beta \tilde{A}_\mu + \beta^2 \bar{\tilde{A}}_\mu. \tag{40}$$

Now by using this new connection in the loop space formalism, we can obtain the β^2 contribution to all the loop space variables. Thus, by using the expression of the connection to the order β^2, we obtain

$$\Phi[\xi] = P_s \exp i \int_0^{2\pi} \mathcal{A}^\mu(\xi(s)) \frac{d\xi_\mu}{ds}$$

$$= P_s \exp i \int_0^{2\pi} [A^\mu + \beta \tilde{A}^\mu + \beta^2 \bar{\tilde{A}}^\mu](\xi(s)) \frac{d\xi_\mu}{ds}. \tag{41}$$

We can also write, to the β^2 order,

$$\Phi[\xi : s_1, s_2] = P_s \exp i \int_{s_1}^{s_2} \mathcal{A}^\mu(\xi(s)) \frac{d\xi_\mu}{ds}$$

$$= P_s \exp i \int_{s_1}^{s_2} [A^\mu + \beta \tilde{A}^\mu + \beta^2 \bar{\tilde{A}}^\mu](\xi(s)) \frac{d\xi_\mu}{ds}. \tag{42}$$

Finally, we can also obtain $\mathcal{F}^{\mu}[\xi|s]$ to the order β^2 as

$$\mathcal{F}^{\mu}[\xi|s] = \Phi^{-1}[\xi:s,0]\mathcal{F}^{\mu\nu}(\xi(s))\Phi[\xi:s,0]\frac{d\xi_{\nu}(s)}{ds}. \tag{43}$$

Thus, we can demonstrate that to the order β^2,

$$\mathcal{F}^{\mu}[\xi|s] = \Phi^{-1}[\xi:s,0]\left[F^{\mu\nu} + \beta\tilde{F}^{\mu\nu} + \beta^2\bar{\tilde{F}}_{\mu\nu}\right](\xi(s))$$
$$\times \Phi[\xi:s,0]\frac{d\xi_{\nu}(s)}{ds}. \tag{44}$$

Thus by repeating this argument we have used for in this paper, to the order β^2, we can demonstrate that to the order β^2,

$$\mathcal{G}_{\mu\nu}[\xi(s_1,s_2)] = \frac{\delta}{\delta\xi^{\mu}(s_2)}\mathcal{F}_{\nu}[\xi|s_1] - \frac{\delta}{\delta\xi^{\nu}(s_1)}\mathcal{F}_{\mu}[\xi|s_2]$$
$$+ i[\mathcal{F}_{\mu}[\xi|s_1],\mathcal{F}_{\nu}[\xi|s_2]]$$
$$= \Phi^{-1}[\xi:s_1,0]\left[[\mathcal{D}_{\mu},\mathcal{F}_{\nu\tau}] + [\mathcal{D}_{\nu},\mathcal{F}_{\tau\mu}] + [\mathcal{D}_{\tau},\mathcal{F}_{\mu\nu}]\right]$$
$$\times \Phi[\xi:s_1,0]\frac{d\xi^{\tau}(s_1)}{ds}\delta(s_1-s_2), \tag{45}$$

where we have considered all the covariant derivatives and the field strengths deformed to the order β^2. This is because the $\mathcal{F}^{\mu}[\xi|s]$ also contain the β^2 terms. It could be demonstrated by repeating the calculations we did to the order β, that the Bianchi identity also holds iteratively for higher order β deformations, and thus it would hold for β^2 deformation. Thus, we can argue that it would be possible for the $\mathcal{G}_{\mu\nu}[\xi(s_1,s_2)]$ not to be zero at the order β^2, even if it is zero at the order β. Thus, topological defects can occur at higher order, even if they do not occur at lower order. However, if $\mathcal{G}_{\mu\nu}[\xi(s_1,s_2)] \neq 0$ at the order β, then it cannot vanish at any higher order. This is because at higher order say β^2, the contribution to $\mathcal{G}_{\mu\nu}[\xi(s_1,s_2)]$ will come from $\bar{\tilde{F}}_{\mu\nu}$ which is of order β^2, and no additional contribution will come at the order β. Now as $\beta < 1$, the β^2 contribution cannot cancel the β contributions to the loop space variables. Thus, the topological defect which is present at the order β cannot be eliminated by considering by considering higher order corrections to the loop space. It may be noted that at β^2 order $\Theta(\Sigma) = \Theta + \beta\tilde{\Theta} + \beta^2\bar{\tilde{\Theta}}$.

In fact, this argument can be made iteratively for the loop space variables at any order. Thus, if a topological defect exists at the order β^n, it cannot be eliminated at the order β^{n+m}, when $m \geq 1$. This is because the $\mathcal{D}_{\mu}$ will have an contribution to the order β^n at the order n and β^{n+m} at the order $n+m$. So, the field strength $\mathcal{F}_{\mu\nu}$ at the order n will also contain terms proportional to $1\cdots\beta^n$, and the field strength $\mathcal{F}_{\mu\nu}$ at the order $n+m$ will contain terms proportional to $1\cdots\beta^{n+m}$. So, the connection $\mathcal{A}_{\mu}$ will also contain terms proportional to β^n and β^{n+m} at the orders β^n and β^{n+m}, respectively. Now repeating the argument used in this section, we can define the loop space variables for each the deformation at any order, and $\mathcal{G}_{\mu\nu}[\xi(s_1,s_2)]$ would also be given in terms of the Bianchi identity at the corresponding order of the deformation parameter. This implies that $\mathcal{G}_{\mu\nu}[\xi(s_1,s_2)]$ will contain terms proportional to β^n at the order n and β^{n+m} at the order $n+m$. Now if $\mathcal{G}_{\mu\nu}[\xi(s_1,s_2)] \neq 0$ at β^n, then this contribution cannot be canceled at the order β^{n+m}, because $\beta < 1$. Thus, the topological defects produced at any order cannot be eliminated by considering higher order contributions in the deformation parameter. We would also like to point out that the $\Theta(\Sigma)$ will also contributions proportional to $1\cdots\beta^n$ at n order, and $1\cdots\beta^{n+m}$ at $n+m$ order.

6. Conclusion

In this paper, we were able to analyse the deformation of a gauge theory by the existence of a minimum measurable length scale. This was done using the loop space formalism. We explicitly constructed the loop space variable for this deformed theory. This loop space variable was then used for constructing the loop space curvature. This curvature did not vanish in presence of a non-abelian monopole. Hence, we were able to demonstrate that the non-vanishing of the loop space curvature indicates the existence of a topological obstruction even for deformed gauge theories. We have also constructed an explicit expression for the charge of a non-abelian monopole using the loop in the loop space. However, it was possible to consider configurations, for which the original field strength vanished, but the deformation did not vanish. Using these field configurations, it was possible to demonstrate that the Polyakov connection can get contributions purely from the deformation, and the loop space curvature can also get β order contributions, even if originally it vanished. Thus, it is possible the deformation of gauge theories by the deformation of the background geometry can give rise to topological defects. We have also demonstrated that higher order corrections cannot cancel the topological defects produced at a certain order in the loop space formalism. So, the presence of a minimum length actually may create topological obstructions like monopoles. It thus does not seem to be an artifact of any approximation, but what may happen if quantum effects are taken into account, in the way we propose. It may be noted that the production of magnetic monopoles and even electric charge by quantum gravitational effects is not a new idea, and such charges have been constructed using Wheeler–DeWitt approach [33,34]. However, all such work was done only for abelian gauge theories, and this is the first time it has been proposed that quantum gravity may produce monopoles in non-abelian gauge theories. We would like to point out that we have only used the deformed gauge theories to obtain such results, however, such a deformation of gauge theories occurs due to quantum gravity. This is because a deformation of quantum mechanics can occur due to a low energy effects from quantum gravity [20,21], and the corresponding deformation of quantum field theories (including gauge theories) can also occur from such quantum gravitational effects [27–32]. This deformation of gauge theory, from quantum gravitational effects, is the deformation we have used to obtain the results of this paper. Thus, it is possible that the topological defects produced from the deformation studied in this paper, could occurs due to quantum gravitational effects because of the existence of minimal length in spacetime.

The loop space formalism has been used to construct loop space duality for ordinary Yang–Mills theories [35–38]. This duality reduces to the usual electromagnetic Hodge duality for abelian gauge theories. So, even though the Hodge duality cannot be generalized to non-abelian gauge theories, this loop space duality can be used to construct a dual potential even in case of non-abelian gauge theories. This dual potential has also been used for constructing a Dualized Standard Model [40–44], and which has in turn been used for explaining the difference of masses between different generations of fermions [45,46]. This model has also been used for analysing the Neutrino oscillations [47], Lepton transmutations [48], and off-diagonal elements of the CKM matrix [49]. The dual potential used for obtaining the Dualized Standard Model has also been used for constructing the 't Hooft's order-disorder parameters [50–52]. It would be interesting to repeat this analysis for a gauge theory deformed by a minimum measurable length. Thus, we can use the results of this paper to construct a dual potential for gauge theories deformed by generalized uncertainty principle. This dual potential can in turn be used for constructing a deformed version of Dualized Standard Model. This deformed Dualized Standard Model can be used for analysing the effect on generalized uncertainty principle on the off-diagonal elements of the CKM matrix, difference of masses between different generations of fermions, Neutrino oscillations and Lepton transmutations. It would also be interesting to analyse the 't Hooft's order-disorder parameters for gauge theories deformed by generalized uncertainty principle.

Acknowledgements

We would like to thank Mohammed Khalil for proving the Bianchi identity for deformed gauge theories.

References

[1] A.M. Polyakov, Nucl. Phys. B 164 (1980) 171.
[2] H.M. Chan, S.T. Tsou, Some Elementary Gauge Theory Concepts, World Scientific, 1993.
[3] H.M. Chan, P. Scharbach, S.T. Tsou, Ann. Phys. 167 (1986) 454.
[4] H.M. Chan, S.T. Tsou, Acta Phys. Pol. B 17 (1986) 259.
[5] M. Faizal, S.T. Tsou, Int. J. Theor. Phys. 54 (2015) 896.
[6] M. Faizal, Europhys. Lett. 103 (2013) 21003.
[7] M. Faizal, S.T. Tsou, Europhys. Lett. 107 (2014) 20008.
[8] D. Amati, M. Ciafaloni, G. Veneziano, Phys. Lett. B 216 (1989) 41.
[9] A. Kempf, G. Mangano, R.B. Mann, Phys. Rev. D 52 (1995) 1108.
[10] L.N. Chang, D. Minic, N. Okamura, T. Takeuchi, Phys. Rev. D 65 (2002) 125027.
[11] L.N. Chang, D. Minic, N. Okamura, T. Takeuchi, Phys. Rev. D 65 (2002) 125028.
[12] S. Benczik, L.N. Chang, D. Minic, N. Okamura, S. Rayyan, T. Takeuchi, Phys. Rev. D 66 (2002) 026003.
[13] S. Hossenfelder, Living Rev. Relativ. 16 (2013) 2.
[14] M.R. Douglas, D.N. Kabat, P. Pouliot, S.H. Shenker, Nucl. Phys. B 485 (1997) 85.
[15] A. Smailagic, E. Spallucci, T. Padmanabhan, arXiv:hep-th/0308122.
[16] M. Fontanini, E. Spallucci, T. Padmanabhan, Phys. Lett. B 633 (2006) 627.
[17] P. Dzierzak, J. Jezierski, P. Malkiewicz, W. Piechocki, Acta Phys. Pol. B 41 (2010) 717.
[18] M. Maggiore, Phys. Lett. B 304 (1993) 65.
[19] M.I. Park, Phys. Lett. B 659 (2008) 698.
[20] S. Das, E.C. Vagenas, Phys. Rev. Lett. 101 (2008) 221301.
[21] I. Pikovski, M.R. Vanner, M. Aspelmeyer, M. Kim, C. Brukner, Nat. Phys. 8 (2012) 393.
[22] L.J. Garay, Int. J. Mod. Phys. A 10 (1995) 145.
[23] C. Bambi, F.R. Urban, Class. Quantum Gravity 25 (2008) 095006.
[24] K. Nozari, Phys. Lett. B 629 (2005) 41.
[25] A. Kempf, J. Phys. A 30 (1997) 2093.
[26] A.F. Ali, S. Das, E.C. Vagenas, Phys. Rev. D 84 (2011) 44013.
[27] M. Kober, Phys. Rev. D 82 (2010) 085017.
[28] M. Kober, Int. J. Mod. Phys. A 26 (2011) 4251.
[29] M. Faizal, Int. J. Geom. Methods Mod. Phys. 12 (2015) 1550022.
[30] M. Faizal, S.I. Kruglov, Int. J. Mod. Phys. D 25 (2016) 1650013.
[31] M. Faizal, B. Majumder, Ann. Phys. 357 (2015) 49.
[32] V. Husain, D. Kothawala, S.S. Seahra, Phys. Rev. D 87 (2013) 025014.
[33] R. Garattini, B. Majumder, Nucl. Phys. B 883 (2014) 598.
[34] R. Garattini, Phys. Lett. B 666 (2008) 189.
[35] H.M. Chan, J. Faridani, S.T. Tsou, Phys. Rev. D 52 (1995) 6134.
[36] M. Faizal, S.T. Tsou, Found. Phys. 45 (2015) 1421.
[37] M. Faizal, S.T. Tsou, Eur. Phys. J. C 75 (2015) 316.
[38] H.M. Chan, J. Faridani, S.T. Tsou, Phys. Rev. D 53 (1996) 7293.
[39] O. Aharony, N. Seiberg, Y. Tachikawa, J. High Energy Phys. 1308 (2013) 115.
[40] H.M. Chan, J. Bordes, S.T. Tsou, Int. J. Mod. Phys. A 14 (1999) 2173.
[41] H.M. Chan, S.T. Tsou, Acta Phys. Pol. B 28 (1997) 3027.
[42] H.M. Chan, S.T. Tsou, Acta Phys. Pol. B 33 (2002) 4041.
[43] H.M. Chan, Int. J. Mod. Phys. A 16 (2001) 163.

[44] H.M. Chan, S.T. Tsou, Acta Phys. Pol. B 28 (1997) 3041.
[45] J. Bordes, H.M. Chan, J. Faridani, J. Pfaudler, S.T. Tsou, Phys. Rev. D 58 (1998) 013004.
[46] J. Bordes, H.M. Chan, J. Faridani, J. Pfaudler, S.T. Tsou, Phys. Rev. D 60 (1999) 013005.
[47] J. Bordes, H.M. Chan, J. Pfaudler, S.T. Tsou, Phys. Rev. D 58 (1998) 053003.
[48] J. Bordes, H.M. Chan, S.T. Tsou, Phys. Rev. D 65 (2002) 093006.
[49] M. Kobayashi, T. Maskawa, Prog. Theor. Phys. 49 (1973) 652–657.
[50] G. 't Hooft, Nucl. Phys. B 138 (1978) 1.
[51] H.M. Chan, S.T. Tsou, Phys. Rev. D 57 (1998) 2507.
[52] H.M. Chan, S.T. Tsou, Phys. Rev. D 56 (1997) 3646.

14

Holographic entanglement entropy close to crossover/phase transition in strongly coupled systems

Shao-Jun Zhang

Institute for Advanced Physics and Mathematics, Zhejiang University of Technology, Hangzhou 310023, China

Editor: Stephan Stieberger

Abstract

We investigate the behavior of entanglement entropy in the holographic QCD model proposed by Gubser et al. By choosing suitable parameters of the scalar self-interaction potential, this model can exhibit various types of phase structures: crossover, first order and second order phase transitions. We use entanglement entropy to probe the crossover/phase transition, and find that it drops quickly/suddenly when the temperature approaches the critical point which can be seen as a signal of confinement. Moreover, the critical behavior of the entanglement entropy suggests that we may use it to characterize the corresponding phase structures.

1. Introduction

In the past two decades, AdS/CFT [1–3] or the more generic gauge/gravity duality has attracted lots of attention and efforts, which relates a quantum field theory (QFT) in $(d + 1)$ dimensions to some gravitational theory in $(d + 2)$ dimensions. As a strong/weak duality, it provides a powerful tool to deal with strongly coupled field systems for which traditional methods of perturbative QFT confront great challenge or even break down. It has been applied on various areas of modern theoretical physics, including QCD [4–6], condensed matter physics [7–11] and cosmology [12], and achieved great successes.

E-mail address: sjzhang84@hotmail.com.

On the other hand, experiments of heavy ions collisions on RHIC [13–16] have opened a novel window into the physics of strongly interacting hadronic matters. The existing data suggest the following evolution picture: After collision, the hot QCD matters undergo a very fast thermalization process to reach thermal equilibrium where a ball of quark–gluon plasma (QGP) forms; And subsequently, the QGP expands to cool down until the temperature falls below the QCD transition (or crossover) point where it finally hadronizes. The QGP can be described very well by relativistic hydrodynamics with a very small η/s [17], where η/s is the ratio of shear viscosity to entropy density. This implies that the QGP is strongly coupled, and thus a treatment beyond the perturbative QCD is called for. With the help of the most well-established example of the AdS/CFT correspondence, namely the duality between $N = 4$ superconformal Yang–Mills (SYM) theory in four-dimensional Minkowski spacetime and type IIB supergravity in $AdS_5 \times S^5$, η/s of $N = 4$ SYM is produced which is very close to the hydrodynamic result [18,19]. This remarkable result shows the validity and powerfulness of the holographic method. However, $N = 4$ SYM is a conformal field theory and thus does not exhibit crossover behavior or any kind of phase transition which is very different from QCD. Moreover, lattice data indicates that the QGP is not a fully conformal fluid in the relevant RHIC energy range $1 \leq T/T_c \leq 3$, and the deviation from conformality may play an important role near the crossover/phase transition [20–22]. Therefore, it is interesting to seek more realistic holographic models to model non-conformal field theories such as QCD.

Now there are various ways to construct holographic models dual to non-conformal field theories, either following a top–down approach by studying a specific gravitational theory which has a string theory construction or a bottom–up approach in which the gravitational background is phenomenologically fixed to fit the lattice QCD (lQCD) data. Recently, in Refs. [23,24] Gubser and his collaborators proposed an interesting bottom–up holographic model intending to mimic the equation of state of QCD. In their model, beyond the Einstein gravity sector, a nontrivial massive scalar field as well as a judicious choice of its self-interaction potential are introduced in the bulk to break the conformal symmetry. The scalar potential has several parameters. By choosing appropriate values of these parameters, the model can exhibit a crossover behavior at some critical temperature and the equation of state generated agrees well with the result from lQCD [25]. Moreover, this simple model can also realize various types of phase transitions by choosing other values of the parameters in the scalar potential, for example the first and second order phase transitions. Thus, in additional to mimic properties of QCD, this model provides us a good background to study various phase structures of strongly coupled field systems. Many efforts have been devoted to investigate properties of this model. Various first and second order hydrodynamic transport coefficients have been calculated in Ref. [26]. In Refs. [27,28], quasinormal modes are used to probe the crossover/phase transition and a number of novel features are observed which were not present in the conformal case. In Ref. [29], this model is extended to include the effect of finite chemical potential and a more complete phase diagram of QCD is thus studied. Other models are also proposed, see Refs. [30–34] for the improved holographic QCD model (IHQCD), Refs. [35,36] for a top–down model and Refs. [37,38] for a semi-analytical holographic QCD model.

In this paper, based on Gubser's model, we aim to use one of non-local observables, the entanglement entropy, to probe the crossover/phase transition. Typically, there are three important non-local observables—the two-point function, the Wilson loop and the entanglement entropy— one can consider as probes to track the number of degrees of freedom and reflect non-local information (correlations between parts for example) of the system. However, they are difficult to calculate in the field theory side. AdS/CFT correspondence makes the calculations easier by relating them to some geometric quantities in the bulk. These non-local observables have been extensively used as probes in the study of the holographic thermalization process of strongly coupled field systems (QGP for example). For reviews, see Refs. [39,40] and references therein. The time evolution of these non-local observables can reflect thermalization process of different regions of the system and explicitly explain the very short thermalization time. They are also used to describe phase transitions in holographic models of condensed matter systems (holographic superconductor models for example) [41–63]. In this paper, we will mainly focus on the study of the entanglement entropy, which, according to the AdS/CFT dictionary, is related to the area of some extremal codimension-two surface in the bulk. We will consider three sets of parameters in the scalar potential, which exhibit respectively a crossover, first order and second order phase transitions. Our goal is to investigate the behavior of the entanglement entropy close to the crossover/phase transitions, and to see if this non-local observable can give us some information characterizing the phase structures.

The paper is organized as follows. In the next section, we will give a brief introduction of the holographic QCD model proposed by Guber et al. Then, in Sec. 3, thermodynamics of the model is discussed. In Sec. 4, behavior of the entanglement entropy close to the crossover/phase transitions is investigated. We will consider two shapes of the entanglement region: the strip one and the ball one. The last section is devoted to summary and discussions.

2. Review of the holographic QCD model

In this section, we give a brief introduction of the holographic model proposed by Gubser et al. in Refs. [23,24] intending to mimic the equation of state of QCD. The bulk action is a Einstein–dilaton action,

$$S = \int d^5x \sqrt{-g} \left[R - \frac{1}{2}(\partial\phi)^2 - V(\phi) \right],$$
(1)

where the scalar potential assumes the form [27,28]

$$V(\phi) = -12\cosh(\gamma\phi) + b_2\phi^2 + b_4\phi^4 + b_6\phi^6.$$
(2)

This potential is parameterized by four constants, γ, b_2, b_4 and b_6, whose values we can choose. It has the following small ϕ expansion

$$V(\phi) \sim -12 + \frac{1}{2}m^2\phi^2 + \mathcal{O}(\phi^4).$$
(3)

The first term is the negative cosmological constant (note that we have chosen the unit to set the AdS radius to be one), and the second term is the mass term with $m^2 \equiv 2(b_2 - 6\gamma^2)$. According to the AdS/CFT dictionary, the scalar field ϕ in the bulk is dual to a scalar operator O_ϕ in the dual boundary field theory. The conformal dimension of the scalar operator is related to the mass parameter of the scalar field as $\Delta(\Delta - 4) = m^2$. The mass square m^2 can be negative and is constrained by the Breitenloner–Freedman (BF) bound $m^2 \geq -4$ [64,65]. Holographically, this gravity model is dual to a deformation of the boundary conformal field theory

$$\mathcal{L} = \mathcal{L}_{\text{CFT}} + \Lambda^{4-\Delta} O_\phi,$$
(4)

where Λ is an energy scale. In this paper, we consider $2 \leq \Delta < 4$ which corresponds to relevant deformations of the CFT.

Table 1
Parameters for the three scalar potentials [28].

Potential	γ	b_2	b_4	b_6	Δ
V_{QCD}	0.606	1.4	−0.1	0.0034	3.55
V_{2nd}	$1/\sqrt{2}$	1.958	0	0	3.38
V_{1st}	$\sqrt{7/12}$	2.5	0	0	3.41

By choosing suitable values of the parameters (γ, b_2, b_4, b_6), this model can produce an equation of state which agrees well with the lQCD data. Moreover, by choosing other values of the parameters, this model can also realize various types of phase transitions. In this work, as in Refs. [27,28] we consider three sets of parameters, labeled by V_{QCD}, V_{1st} and V_{2nd} respectively, which are summarized in Table 1. The parameters for V_{QCD} have been chosen to fit the lQCD data from Ref. [25], and the system is known to possess a crossover behavior at zero baryon chemical potential as we will show later. Parameters of potentials V_{1st} and V_{2nd} were chosen so that the corresponding dual field systems exhibit respectively the 1st, and the 2nd order phase transitions.

As we want to study properties of the dual field system at finite temperature, in the gravity side we need black hole solutions. To seek these solutions, we take the following ansatz as in Refs. [23,24],

$$ds^2 = e^{2A}(-h dt^2 + d\vec{x}^2) + \frac{e^{2B}}{h}dr^2.$$
$$\phi = r, \tag{5}$$

where A, B and h are only functions of r (or, equivalently ϕ). The above ansatz takes a gauge $\phi = r$ which greatly simplifies the solving of the field equations. Then the field equations of motion are

$$A'' - A'B' + \frac{1}{6} = 0, \tag{6}$$
$$h'' + (4A' - B')h' = 0, \tag{7}$$
$$6A'h' + h(24A'^2 - 1) + 2e^{2B}V = 0, \tag{8}$$
$$4A' - B' + \frac{h'}{h} - \frac{e^{2B}}{h}V' = 0, \tag{9}$$

where the prime denotes a derivative with respect to ϕ. The horizon $\phi = \phi_H$ is determined by the zero point of the blackening function h:

$$h(\phi_H) = 0. \tag{10}$$

We follow the method proposed in Refs. [23,24] to solve the field equations, in which by defining a function $G(\phi) \equiv A'(\phi)$ the solution of field equations can be expressed as:

$$A(\phi) = A_H + \int_{\phi_H}^{\phi} d\tilde{\phi}\, G(\tilde{\phi}), \tag{11}$$

$$B(\phi) = B_H + \ln\left(\frac{G(\phi)}{G(\phi_H)}\right) + \int_{\phi_H}^{\phi} \frac{d\tilde{\phi}}{6G(\tilde{\phi})}, \tag{12}$$

$$h(\phi) = h_H + h_1 \int\limits_{\phi_H}^{\phi} d\tilde{\phi} e^{-4A(\tilde{\phi})+B(\tilde{\phi})}, \tag{13}$$

where the integration constants A_H, B_H, h_H and h_1 are determined by requiring the appropriate boundary conditions at the horizon Eq. (10) and the infinite boundary,

$$A_H = \frac{\ln \phi_H}{\Delta - 4} + \int\limits_0^{\phi_H} d\phi \left[G(\phi) - \frac{1}{(\Delta - 4)\phi} \right], \tag{14}$$

$$B_H = \ln \left(-\frac{4V(\phi_H)}{V(0)V'(\phi_H)} \right) + \int\limits_0^{\phi_H} \frac{d\phi}{6G(\phi)}, \tag{15}$$

$$h_H = 0, \tag{16}$$

$$h_1 = \frac{1}{\int_{\phi_H}^0 d\phi e^{-4A(\phi)+B(\phi)}}. \tag{17}$$

So, once we get the solution of $G(\phi)$, the full solution can be generated. As in Ref. [23], by manipulating Eqs. (6)–(9), it is found that $G(\phi)$ satisfies the following "master equation"

$$\frac{G'}{G + V/3V'} = \frac{d}{d\phi} \ln \left(\frac{G'}{G} + \frac{1}{6G} - 4G - \frac{G'}{G + V/3V'} \right). \tag{18}$$

From it, the series expansion of $G(\phi)$ near the horizon $\phi = \phi_H$ can be obtained,

$$G(\phi) = -\frac{V(\phi)}{3V'(\phi)} + \frac{1}{6} \left(\frac{V(\phi_H)V''(\phi_H)}{V'(\phi_H)^2} - 1 \right) (\phi - \phi_H) + \mathcal{O}(\phi - \phi_H)^2, \tag{19}$$

which can be used as the appropriate boundary conditions to solve $G(\phi)$. Note that it is hard to solve the "master equation" analytically, so we rely on numerical method.

From the above expressions, we can see that given one value of the horizon ϕ_H we can obtain one unique black hole solution. In this paper, we will vary the value of ϕ_H and obtain a family of black hole solutions numerically.

3. Thermodynamics

In this section, we study the thermodynamics of the dual field system. We will focus on the temperature-dependence of the entropy density and the speed of sound.

From the ansatz Eq. (5), the Hawking temperature and the entropy density can be obtained,

$$T = \frac{e^{A_H - B_H} |h'(\phi_H)|}{4\pi}, \qquad s = \frac{e^{3A_H}}{4}. \tag{20}$$

From them, we can get the square of the speed of sound

$$c_s^2 = \frac{d \ln T / d\phi_H}{d \ln s / d\phi_H}. \tag{21}$$

In the following three subsections, we will discuss the three cases listed in Table 1 respectively.

To make a comparison with the conformal case, here we also show the results for the five-dimensional Schwarzschild–AdS black hole with the metric

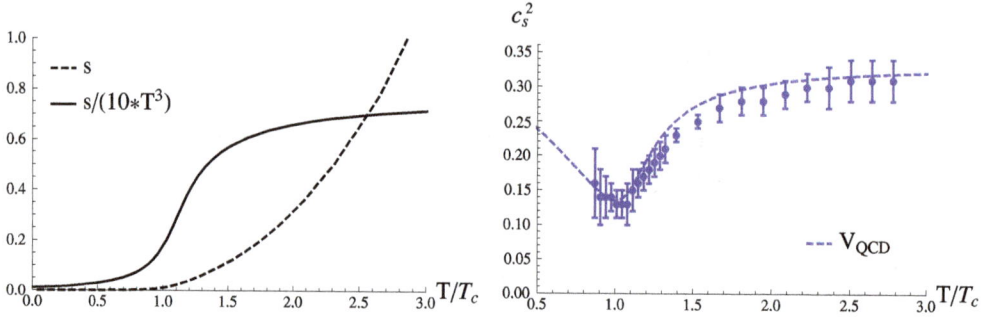

Fig. 1. (Color online.) Entropy density s and speed of sound c_s^2 for V_{QCD}. In the left panel, the two curves are for s (dashed) and $s/(10 * T^3)$ (solid) respectively. In the right panel, the lattice data points with error bar are from Ref. [25].

$$ds_{SAdS}^2 = \frac{1}{z^2}\left[-h(z)dt^2 + \frac{dz^2}{h(z)} + d\vec{x}^2 \right],$$

$$h(z) = 1 - \left(\frac{z}{z_H}\right)^4.$$
(22)

The horizon is at $z = z_H$ and the infinite boundary $z = 0$. From the above metric and using the AdS/CFT dictionary, it is easy to derive the following relations

$$s/T^3 \propto \text{(number of degrees of freedom)}, \qquad c_s^2 = \frac{1}{3},$$
(23)

which are expected for a CFT.

3.1. V_{QCD}

In Fig. 1, we show the dependence of the entropy density s and the square of the speed of sound c_s^2 on the temperature for V_{QCD}. From the right panel, we can see that there is a critical temperature T_c which corresponds to the lowest dip of c_s^2. According to Ref. [26], this value should be 143.8 MeV for QCD, while in our present units it is $T_c = 0.181033$. As shown in the left panel, the entropy density and its derivative to the temperature are both continuous at the critical temperature which means a crossover. When the temperature is beyond the critical point, s/T^3 approaches constant suggesting that even in the non-conformal case s/T^3 may be considered to approximately count the effective number of degrees of freedom. This claim is also true for the other two cases, V_{1st} and V_{2nd}, as we will see later. Moreover, as the temperature approaching T_c, the entropy density s/T^3 drops quickly indicating that the number of underlying degrees of freedom is largely suppressed, which can be understood as a signal of confinement. The dependence of c_s^2 on the temperature agrees well with the lattice result [25,27]. Moreover, for high temperature, c_s^2 approaches its conformal value $1/3$ as expected.

3.2. V_{1st}

In Fig. 2, the dependence of the entropy density and the square of the speed of sound on the temperature for V_{1st} are shown. From the left panel, we can see that In some range of the temperature, there are three branches of solutions, in which two of them are stable (shown in blue curves) and one is unstable (shown in red curve). And there is a minimum temperature

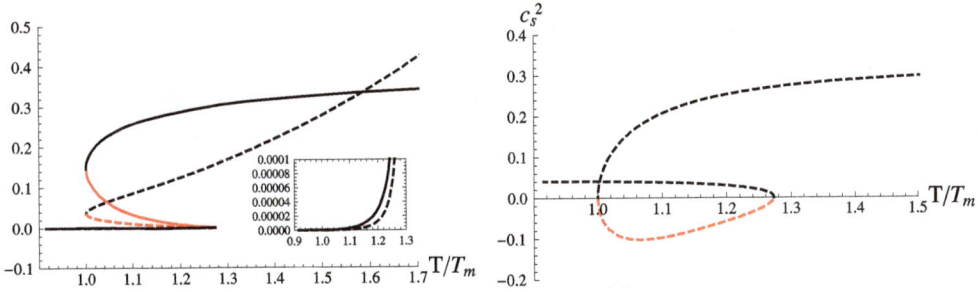

Fig. 2. (Color online.) Entropy density s and speed of sound c_s^2 for V_{1st}. In the left panel, the two curves are respectively s (dashed) and $s/(20*T^3)$ (solid). In a certain range of temperature, there are three branches of solutions, two of which are stable (blue curves) and one is unstable (red curve).

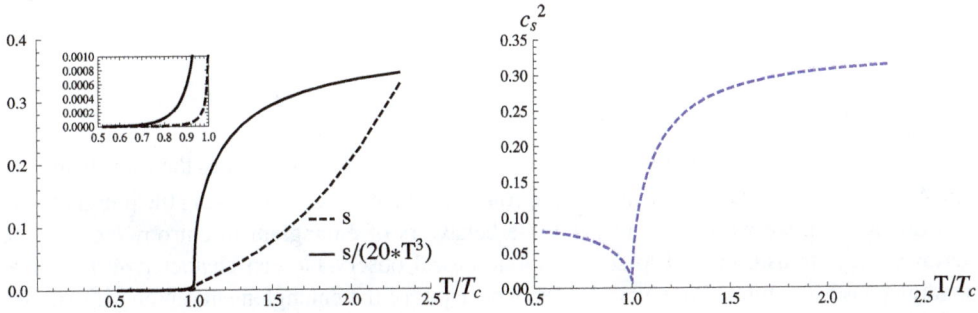

Fig. 3. (Color online.) Entropy density s and speed of sound c_s^2 for V_{2nd}. In the left panel, the two curves are for s (dashed) and $s/(20*T^3)$ (solid) respectively.

T_m below which no unstable black hole solutions exist. This can be seen more clearly from the right panel, where for the unstable branch of solutions, c_s^2 becomes negative. In our present units, $T_m = 0.232287$. There is a phase transition between the two stable branches of solutions with critical temperature $T_c \simeq 1.05T_m$ [28], which can be read off by comparing the free energy of the two stable branches of solutions. As the entropy density is discontinuous at the critical temperature, the phase transition is first order. Moreover, from the left panel, we can see that s/T^3 drops suddenly as the temperature approaching T_c which also indicates that the number of underlying degrees of freedom is largely suppressed, and can also be understood as a kind of confinement. However, we should note that in the cases V_{1st} and V_{2nd}, we do not intend to mimic the equation of state of QCD, but rather realize various types of phase structures within the same framework.

3.3. V_{2nd}

In Fig. 3, the dependence of the entropy density and the square of the speed of sound on the temperature for V_{2nd} are shown. From the figure, we can see that there is a critical temperature T_c at which the speed of sound c_s vanishes. The entropy density s is continuous at T_c but not its derivative with respect to the temperature, thus the phase transition is second order. In our present units, $T_c = 0.156841$. Once again, we can see that s/T^3 drops quickly as the temperature approaching T_c.

4. Behaviors of holographic entanglement entropy

Suppose the system of the boundary field system living in Minkowski spacetime is divided into two parts $\mathcal{A}$ and $\mathcal{A}^c$, with $\mathcal{A}^c$ being the complement of $\mathcal{A}$. Then the entanglement entropy of the subregion $\mathcal{A}$ is defined as the Von Neuman entropy,

$$S_{\text{EE}} \equiv -\text{tr}_{\mathcal{A}}\rho_{\mathcal{A}}\ln\rho_{\mathcal{A}}, \tag{24}$$

where ρ_A is the reduced density matrix of $\mathcal{A}$.

Generally, it is difficult to calculate the entanglement entropy directly from the field theory side. However, the calculation can be made easier with the help of the holographic method, which relates this entropy to some geometric quantity of the dual bulk geometry. According to the conjecture proposed in Refs. [66,67], in Einstein gravity, the holographic entanglement entropy formulae is

$$S_{\text{EE}} = \frac{1}{4G_N^{(d+1)}}\text{ext}[\text{Area}(\gamma_{\mathcal{A}})], \tag{25}$$

where $\gamma_{\mathcal{A}}$ is the extremal codimensional-two surface in the bulk which shares the same boundary with $\mathcal{A}$, i.e. $\partial\gamma_{\mathcal{A}} = \partial\mathcal{A}$. $G_N^{(d+1)}$ is the Newton constant which we set to be one in the present work.

In this section, we would like to study the behaviors of entanglement entropy close to the crossover/phase transition, and to see if this non-local observable can characterize the corresponding phase structures. We will consider two shapes of the entanglement region $\mathcal{A}$: strip one and ball one.

4.1. Coordinate redefinition

Before diving into the calculations of the three non-local observables, we would like to transform the metric into the form as

$$ds^2 = e^{2\tilde{A}(z)}\left[-\tilde{h}(z)dt^2 + \frac{dz^2}{\tilde{h}(z)} + d\vec{x}^2\right], \tag{26}$$

which can be achieved by redefining the radial coordinate as

$$\phi = \phi(z). \tag{27}$$

The form of $\phi(z)$ can be obtained by matching the line elements before and after the transformation. The result is

$$z(\phi) = \int_0^\phi d\tilde{\phi}e^{B(\tilde{\phi})-A(\tilde{\phi})}, \tag{28}$$

and then $\tilde{A}(z) = A(\phi)$ and $\tilde{h}(z) = h(\phi)$. In this new radial coordinate, the horizon lies at $z = z_H \equiv z(\phi_H)$ and the infinite boundary at $z = 0$. We found that it is more convenient to work with the z-coordinate to do the calculations. In the following, without causing confusion, we will omit the tilde of the metric functions in Eq. (26) for the sake of simplicity.

4.2. Entanglement entropy: strip shape

On the boundary with coordinates $(t, \vec{x})$, we consider the entanglement region $\mathcal{A}$ to be a strip:

$$\mathcal{A}: \qquad x_1 \in [-l/2, l/2], \quad x_2, x_3 \in [-L/2, +L/2], \tag{29}$$

which has a width l in the x_1 direction and length L in the other two spatial directions. We consider the case $L \gg l$ so that $\mathcal{A}$ preserves translation invariance in the x_2 and x_3 directions. According to the holographic entanglement entropy formulae Eq. (25), the entanglement entropy of the entanglement region $\mathcal{A}$ can be calculated by computing the area of the extremal surface which starts from the boundary of the region and extended into the bulk. Taking into account of the symmetry, the extremal surface $\gamma_{\mathcal{A}}$ can be parameterized by only one function $z = z(x_1)$, and thus the induced metric on the surface is

$$ds_{\gamma_{\mathcal{A}}}^2 = e^{2A}\left(1 + \frac{z'^2}{h}\right)dx_1^2 + e^{2A}(dx_2^2 + dx_3^2). \tag{30}$$

The entanglement entropy then is

$$S_{EE} = \frac{1}{4}\int dx_1 dx_2 dx_3 \sqrt{\gamma},$$

$$= \frac{V_2}{2}\int_0^{l/2} dx_1 Q^{1/2} e^{3A}, \tag{31}$$

$$Q \equiv 1 + \frac{z'^2}{h}, \tag{32}$$

where $V_2 \equiv \int dx_2 dx_3$ and γ is the determinant of the induced metric on $\gamma_{\mathcal{A}}$. The equation for $z(x_1)$ can be obtained by extremizing S_{EE}, and the result is

$$2hz'' - \left(6h\frac{dA}{dz} + \frac{dh}{dz}\right)z'^2 - 6h^2\frac{dA}{dz} = 0. \tag{33}$$

To solve $z(x_1)$, boundary conditions are needed, which, taking into account of the symmetry, are

$$z(0) = z_*, \qquad z'(0) = 0, \qquad z(\pm l/2) = 0. \tag{34}$$

z_* is the deepest position the extremal surface can reach in the z-direction.

From the expression of S_{EE}, we can see that it does not depend on x_1 explicitly which leads to a conservation equation

$$Q^{1/2} = \frac{e^{3A}}{e^{3A(z_*)}}. \tag{35}$$

Using it, S_{EE} can be expressed as

$$S_{EE} = \frac{V_2}{2}\int_0^{l/2} dx_1 \frac{e^{6A(z)}}{e^{3A(z_*)}}. \tag{36}$$

To study the behavior of the entanglement entropy close to the crossover/phase transition, we define a renormalized entanglement entropy density $s_{EE}^{re} \equiv \frac{S_{EE} - S_{EE}^0}{lV_2}$ (lV_2 is the volume of

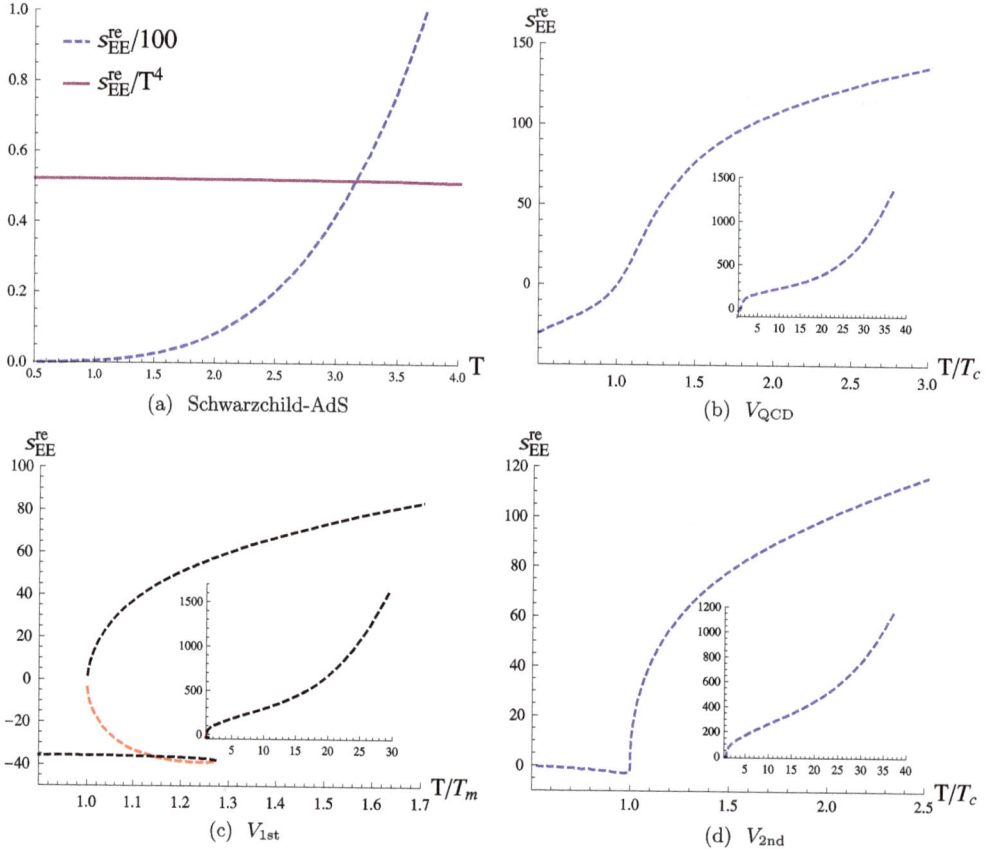

Fig. 4. (Color online.) Entanglement entropy density for Schwarzschild–AdS black hole solution and the three scalar potentials. The width of the entanglement strip is fixed to be $l = 0.04$. The red curve in the panel (c) corresponds to the unstable branch of solutions. The insets of panels (b), (c) and (d) show the behaviors of renormalized entanglement entropy density at sufficiently high temperature.

the strip) and study its dependence on the temperature. S_{EE}^0 is the entanglement entropy of the strip in some reference spacetime. Generally, the integration in the expression of the holographic entanglement entropy Eq. (36) diverges, and by introducing S_{EE}^0 the divergence can be renormalized. We choose the reference spacetimes as follows: For the Schwarzschild–AdS case, it is the pure AdS spacetime; For the V_{QCD} and V_{2nd} cases, they are the critical black hole solutions with the critical temperature T_c; For the V_{1st} case, it is the critical black hole solution with the temperature T_m. Without losing generality and for simplicity, we fix the width of the strip $l = 0.04$ and change the temperature while keeping in the regime $Tl \ll 1$.

In Fig. 4, the dependence of the renormalized entanglement entropy density on the temperature for the three scalar potentials are shown. To make a comparison, we also show the result for the Schwarzschild–AdS black hole which corresponds to the conformal case.

From the panel (a), we can see that for the Schwarzschild–AdS case, $s_{EE}^{re}/T^4 = const$. This numerical result agrees with the analytical result in Ref. [68] where it is shown that in the regime $Tl \ll 1$, $s_{EE}^{re}/T^4 \propto N_c^2$ (N_c denotes the color number of the dual conformal field theory and N_c^2 is thus the effective number of degrees of freedom). However, from panels (b), (c) and (d), we can see that for the three non-conformal cases, the entanglement entropy shows very different behavior. Different from the observable of the thermal entropy density, here we can not expect s_{EE}^{re}/T^4 to count the effective number of degrees of freedom in the non-conformal cases. This difference can be understood as the effect of the additional energy scale l beyond the mass m of the scalar field (or equivalently the conformal dimension Δ of the scalar operator), which is absent for the observable of thermal entropy density.

We can also see that as T approaching T_c, similar to the thermal entropy density, the renormalized entanglement entropy density drops quickly/suddenly. Physically, we can understand this behavior as follows: as the temperature approaching T_c, the field system undergoes confinement so that the number of degrees of freedom contributing to the entanglement is largely suppressed. This behavior has also been seen in holographic superconductor models [50], where condensation decreases the entanglement entropy. Moreover, from the insets of panels (b), (c) and (d) and by fitting the data, we can see that at sufficiently high temperature (while keeping in regime $Tl \ll 1$), the behaviors of renormalized entanglement entropy density are restored to that of the conformal case, i.e., $s_{EE}^{re} \sim T^4$. This indicates that the effects of conformal symmetry breaking become insignificant at sufficiently high temperature.

Moreover, for the three non-conformal cases, the renormalized entanglement entropy density exhibit behavior characterizing the corresponding phase structures close to T_c: For the crossover case V_{QCD}, s_{EE}^{re} and its derivative with respect to the temperature are both continuous at T_c; For the 1st order case V_{1st}, s_{EE}^{re} is discontinuous at T_c; For the 2nd order case V_{2nd}, s_{EE}^{re} is continuous at T_c but not its derivative with respect to the temperature. These behaviors suggest that, as the thermal entropy, the entanglement entropy may also be used to characterize the type of phase transition.

4.3. Entanglement entropy: ball shape

In this subsection, we consider the entanglement region $\mathcal{A}$ to take a ball shape with radius R, i.e. $\sum_{i=1}^3 x_i^2 \le R^2$. It is more convenient to work with spherical coordinates (ρ, Ω_2), under which the bulk metric takes the form

$$ds^2 = -e^{2A}\left[-hdt^2 + \frac{dz^2}{h} + d\rho^2 + \rho^2 d\Omega_2^2\right]. \tag{37}$$

Then, in the spherical coordinates (ρ, Ω_2), the entanglement region is parameterized as $\rho \le R$. Taking into account of the symmetry, the extremal surface γ_A can be parameterized by only one function, i.e. $z = z(\rho)$, and with boundary conditions

$$z(0) = z_*, \quad z'(0) = 0, \quad z(R) = 0, \tag{38}$$

where z_* is the tip of the extremal surface denoting the deepest position the extremal surface can reach in the z-direction. The induced metric on the extremal surface is

$$ds_{\gamma_A} = e^{2A}\left(1 + \frac{z'^2}{h}\right)d\rho^2 + e^{2A}\rho^2 d\Omega_2^2. \tag{39}$$

The entanglement entropy then is

$$S_{EE} = \frac{1}{4}\int d\rho d\Omega_2 \sqrt{\gamma} \tag{40}$$

$$= \pi \int\limits_0^R d\rho \sqrt{1 + \frac{z'^2}{h} e^{3A} \rho^2}. \tag{41}$$

The equation for $z(\rho)$ can be obtained by extremizing S_{EE}, and the result is

$$2h\rho z'' + 4z'^3 - \left(6h\rho \frac{dA}{dz} + \rho \frac{dh}{dz}\right) z'^2 + 4hz' - 6h^2 \rho \frac{dA}{dz} = 0. \tag{42}$$

Note that this equation has a singular point at $\rho = 0$, so practically in doing numerical calculations we impose the boundary conditions near the central point $\rho = 0$ to avoid numerical problems,

$$z(\epsilon) = z_* + \mathcal{O}(\epsilon^2), \quad z'(\epsilon) = \mathcal{O}(\epsilon), \tag{43}$$

where ϵ is a small parameter with typical order of 10^{-3}. The higher order terms can be obtained by solving the equation Eq. (42) near $\rho = 0$ order by order.

To study the behavior of the entanglement entropy close to the crossover/phase transition, similarly we define a renormalized entanglement entropy density $s_{EE}^{re} \equiv \frac{S_{EE} - S_{EE}^0}{4\pi R^3/3}$ ($4\pi R^3/3$ is the volume of the entanglement ball) and study its dependence on the temperature. S_{EE}^0 is the entanglement entropy of the ball in the reference spacetime with the same choices as in the strip case. Without losing generality and for simplicity, we fix the radius of the entanglement ball to be $R = 0.04$ and change the temperature while keeping in the regime $TR \ll 1$. The results are shown in Fig. 5.

From panel (a), we can see that for the conformal case, in the regime $TR \ll 1$ once again we have the relation $s_{EE}^{re}/T^4 = const$. For the three non-conformal cases, the renormalized entanglement entropy density shows a similar dependence on the temperature as in the strip case. These results suggest that the behavior of entanglement entropy close to T_c does not depend on the specific shape of the entanglement region. This implies that our previous suggestion, that the entanglement entropy may be applied to characterize the type of the phase transition, does not depend on the specific shape of the entanglement region.

5. Summary and discussions

In this paper, we discuss the behavior of entanglement entropy close to crossover/phase transition in the holographic QCD model proposed by Gubser et al. [23,24]. This holographic model is proposed intending to mimic the equation of state of QCD by introducing a nontrivial scalar field in the bulk to break the conformal symmetry. The scalar self-interaction potential is parameterized by four constants whose values we can choose to fit the lQCD results. Moreover, by choosing other values of the four parameters, this simple model can also realize various types of phase structures. In this paper, we consider three sets of parameters V_{QCD}, V_{1st} and V_{2nd} possessing respectively crossover, 1st and 2nd phase transitions, which can be seen by studying their thermodynamic properties, such as the entropy and the square of the speed of sound.

Our results show that, similar to the thermal entropy, the entanglement entropy drops quickly/suddenly as the temperature approaching the critical value. This can be understood as a signal of confinement. Moreover, at the critical temperature, it is found that the entanglement entropy shows behavior characterizing the type of the phase transition which is also similar to the thermal entropy. These results suggest that we may apply the entanglement entropy to characterize the phase structures of strongly coupled field systems. Moreover, by studying two cases

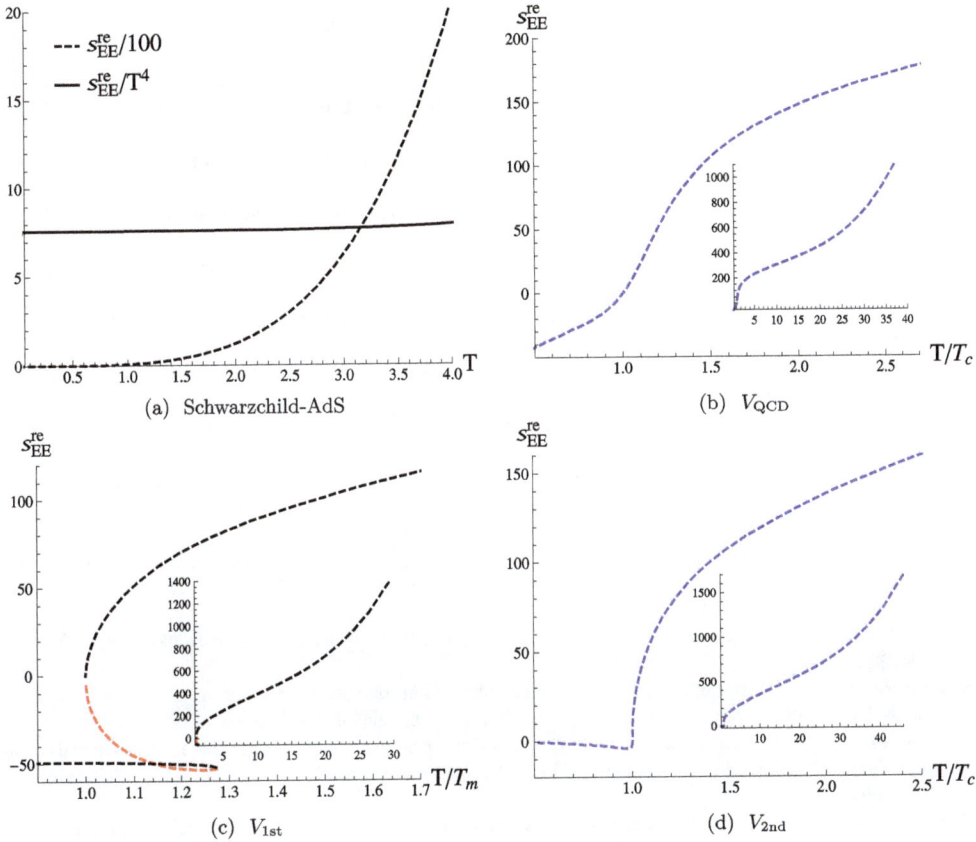

Fig. 5. (Color online.) Entanglement entropy density for Schwarzschild–AdS black hole solution and the three scalar potentials. The radius of the entanglement ball is fixed to be $R = 0.04$. The red curve in the panel (c) corresponds to the unstable branch of solutions.

with different shapes of entanglement region, the strip one and the ball one, we show that our claim does not depend on the specific shape of entanglement region.

We only consider the holographic QCD model proposed by Guber et al. [23,24]. Whether our claim is still hold for other holographic models, for example the improved holographic QCD model proposed in Refs. [30–34], the top–down model in Refs. [35,36] and for the semi-analytical holographic QCD model in Refs. [37,38], needs further investigations.

Acknowledgement

This work was supported in part by National Natural Science Foundation of China (No. 11605155).

References

[1] J.M. Maldacena, The large N limit of superconformal field theories and supergravity, Int. J. Theor. Phys. 38 (1999) 1113, Adv. Theor. Math. Phys. 2 (1998) 231, arXiv:hep-th/9711200.

[2] S.S. Gubser, I.R. Klebanov, A.M. Polyakov, Gauge theory correlators from noncritical string theory, Phys. Lett. B 428 (1998) 105, arXiv:hep-th/9802109.

[3] E. Witten, Anti-de Sitter space and holography, Adv. Theor. Math. Phys. 2 (1998) 253, arXiv:hep-th/9802150.

[4] D. Mateos, String theory and quantum chromodynamics, Class. Quantum Gravity 24 (2007) S713, arXiv:0709.1523 [hep-th].

[5] S.S. Gubser, A. Karch, From gauge-string duality to strong interactions: a Pedestrian's guide, Annu. Rev. Nucl. Part. Sci. 59 (2009) 145, arXiv:0901.0935 [hep-th].

[6] J. Casalderrey-Solana, H. Liu, D. Mateos, K. Rajagopal, U.A. Wiedemann, Gauge/string duality, hot QCD and heavy ion collisions, arXiv:1101.0618 [hep-th].

[7] S.A. Hartnoll, Lectures on holographic methods for condensed matter physics, Class. Quantum Gravity 26 (2009) 224002, arXiv:0903.3246 [hep-th].

[8] C.P. Herzog, Lectures on holographic superfluidity and superconductivity, J. Phys. A 42 (2009) 343001, arXiv:0904.1975 [hep-th].

[9] J. McGreevy, Holographic duality with a view toward many-body physics, Adv. High Energy Phys. 2010 (2010) 723105, arXiv:0909.0518 [hep-th].

[10] G.T. Horowitz, Introduction to holographic superconductors, Lect. Notes Phys. 828 (2011) 313, arXiv:1002.1722 [hep-th].

[11] R.G. Cai, L. Li, L.F. Li, R.Q. Yang, Introduction to holographic superconductor models, Sci. China, Phys. Mech. Astron. 58 (6) (2015) 060401, arXiv:1502.00437 [hep-th].

[12] T. Banks, W. Fischler, The holographic approach to cosmology, arXiv:hep-th/0412097.

[13] J. Adams, et al., STAR Collaboration, Experimental and theoretical challenges in the search for the quark gluon plasma: the STAR collaboration's critical assessment of the evidence from RHIC collisions, Nucl. Phys. A 757 (2005) 102, arXiv:nucl-ex/0501009.

[14] B.B. Back, et al., The PHOBOS perspective on discoveries at RHIC, Nucl. Phys. A 757 (2005) 28, arXiv:nucl-ex/0410022.

[15] I. Arsene, et al., BRAHMS Collaboration, Quark gluon plasma and color glass condensate at RHIC? The perspective from the BRAHMS experiment, Nucl. Phys. A 757 (2005) 1, arXiv:nucl-ex/0410020.

[16] K. Adcox, et al., PHENIX Collaboration, Formation of dense partonic matter in relativistic nucleus–nucleus collisions at RHIC: experimental evaluation by the PHENIX collaboration, Nucl. Phys. A 757 (2005) 184, arXiv:nucl-ex/0410003.

[17] M. Luzum, P. Romatschke, Conformal relativistic viscous hydrodynamics: applications to RHIC results at $s(NN)^{**}(1/2) = 200$-GeV, Phys. Rev. C 78 (2008) 034915, Erratum: Phys. Rev. C 79 (2009) 039903, arXiv:0804.4015 [nucl-th].

[18] G. Policastro, D.T. Son, A.O. Starinets, The Shear viscosity of strongly coupled $N = 4$ supersymmetric Yang–Mills plasma, Phys. Rev. Lett. 87 (2001) 081601, arXiv:hep-th/0104066.

[19] P. Kovtun, D.T. Son, A.O. Starinets, Viscosity in strongly interacting quantum field theories from black hole physics, Phys. Rev. Lett. 94 (2005) 111601, arXiv:hep-th/0405231.

[20] F. Karsch, D. Kharzeev, K. Tuchin, Universal properties of bulk viscosity near the QCD phase transition, Phys. Lett. B 663 (2008) 217, arXiv:0711.0914 [hep-ph].

[21] H.B. Meyer, A calculation of the bulk viscosity in SU(3) gluodynamics, Phys. Rev. Lett. 100 (2008) 162001, arXiv:0710.3717 [hep-lat].

[22] H.B. Meyer, Energy-momentum tensor correlators and viscosity, PoS LATTICE 2008 (2008) 017, arXiv:0809.5202 [hep-lat].

[23] S.S. Gubser, A. Nellore, Mimicking the QCD equation of state with a dual black hole, Phys. Rev. D 78 (2008) 086007, arXiv:0804.0434 [hep-th].

[24] S.S. Gubser, A. Nellore, S.S. Pufu, F.D. Rocha, Thermodynamics and bulk viscosity of approximate black hole duals to finite temperature quantum chromodynamics, Phys. Rev. Lett. 101 (2008) 131601, arXiv:0804.1950 [hep-th].

[25] S. Borsanyi, G. Endrodi, Z. Fodor, S.D. Katz, S. Krieg, C. Ratti, K.K. Szabo, QCD equation of state at nonzero chemical potential: continuum results with physical quark masses at order mu^2, J. High Energy Phys. 1208 (2012) 053, arXiv:1204.6710 [hep-lat].

[26] S.I. Finazzo, R. Rougemont, H. Marrochio, J. Noronha, Hydrodynamic transport coefficients for the non-conformal quark–gluon plasma from holography, J. High Energy Phys. 1502 (2015) 051, arXiv:1412.2968 [hep-ph].

[27] R.A. Janik, G. Plewa, H. Soltanpanahi, M. Spalinski, Linearized nonequilibrium dynamics in nonconformal plasma, Phys. Rev. D 91 (12) (2015) 126013, arXiv:1503.07149 [hep-th].

[28] R.A. Janik, J. Jankowski, H. Soltanpanahi, Quasinormal modes and the phase structure of strongly coupled matter, J. High Energy Phys. 1606 (2016) 047, arXiv:1603.05950 [hep-th].

[29] O. DeWolfe, S.S. Gubser, C. Rosen, A holographic critical point, Phys. Rev. D 83 (2011) 086005, arXiv:1012.1864 [hep-th].

[30] U. Gursoy, E. Kiritsis, Exploring improved holographic theories for QCD: part I, J. High Energy Phys. 0802 (2008) 032, arXiv:0707.1324 [hep-th].

[31] U. Gursoy, E. Kiritsis, F. Nitti, Exploring improved holographic theories for QCD: part II, J. High Energy Phys. 0802 (2008) 019, arXiv:0707.1349 [hep-th].

[32] U. Gursoy, E. Kiritsis, L. Mazzanti, F. Nitti, Holography and thermodynamics of 5D dilaton-gravity, J. High Energy Phys. 0905 (2009) 033, arXiv:0812.0792 [hep-th].

[33] E. Kiritsis, Dissecting the string theory dual of QCD, Fortschr. Phys. 57 (2009) 396, arXiv:0901.1772 [hep-th].

[34] U. Gursoy, E. Kiritsis, L. Mazzanti, F. Nitti, Improved holographic Yang–Mills at finite temperature: comparison with data, Nucl. Phys. B 820 (2009) 148, arXiv:0903.2859 [hep-th].

[35] A. Buchel, S. Deakin, P. Kerner, J.T. Liu, Thermodynamics of the $N = 2^*$ strongly coupled plasma, Nucl. Phys. B 784 (2007) 72, arXiv:hep-th/0701142.

[36] A. Buchel, M.P. Heller, R.C. Myers, Equilibration rates in a strongly coupled nonconformal quark–gluon plasma, Phys. Rev. Lett. 114 (25) (2015) 251601, arXiv:1503.07114 [hep-th].

[37] D. Li, S. He, M. Huang, Q.S. Yan, Thermodynamics of deformed AdS_5 model with a positive/negative quadratic correction in graviton–dilaton system, J. High Energy Phys. 1109 (2011) 041, arXiv:1103.5389 [hep-th].

[38] R.G. Cai, S. He, D. Li, A hQCD model and its phase diagram in Einstein–Maxwell–Dilaton system, J. High Energy Phys. 1203 (2012) 033, arXiv:1201.0820 [hep-th].

[39] V. Balasubramanian, A. Bernamonti, J. de Boer, N. Copland, B. Craps, E. Keski-Vakkuri, B. Muller, A. Schafer, et al., Thermalization of strongly coupled field theories, Phys. Rev. Lett. 106 (2011) 191601, arXiv:1012.4753 [hep-th].

[40] V. Balasubramanian, A. Bernamonti, J. de Boer, N. Copland, B. Craps, E. Keski-Vakkuri, B. Muller, A. Schafer, et al., Holographic thermalization, Phys. Rev. D 84 (2011) 026010, arXiv:1103.2683 [hep-th].

[41] T. Nishioka, T. Takayanagi, AdS bubbles, entropy and closed string tachyons, J. High Energy Phys. 0701 (2007) 090, arXiv:hep-th/0611035.

[42] I.R. Klebanov, D. Kutasov, A. Murugan, Entanglement as a probe of confinement, Nucl. Phys. B 796 (2008) 274, arXiv:0709.2140 [hep-th].

[43] A. Pakman, A. Parnachev, Topological entanglement entropy and holography, J. High Energy Phys. 0807 (2008) 097, arXiv:0805.1891 [hep-th].

[44] T. Nishioka, S. Ryu, T. Takayanagi, Holographic entanglement entropy: an overview, J. Phys. A 42 (2009) 504008, arXiv:0905.0932 [hep-th].

[45] J. de Boer, M. Kulaxizi, A. Parnachev, Holographic entanglement entropy in lovelock gravities, J. High Energy Phys. 1107 (2011) 109, arXiv:1101.5781 [hep-th].

[46] L.Y. Hung, R.C. Myers, M. Smolkin, On holographic entanglement entropy and higher curvature gravity, J. High Energy Phys. 1104 (2011) 025, arXiv:1101.5813 [hep-th].

[47] N. Ogawa, T. Takayanagi, Higher derivative corrections to holographic entanglement entropy for AdS solitons, J. High Energy Phys. 1110 (2011) 147, arXiv:1107.4363 [hep-th].

[48] T. Albash, C.V. Johnson, Holographic entanglement entropy and renormalization group flow, J. High Energy Phys. 1202 (2012) 095, arXiv:1110.1074 [hep-th].

[49] R.C. Myers, A. Singh, Comments on holographic entanglement entropy and RG flows, J. High Energy Phys. 1204 (2012) 122, arXiv:1202.2068 [hep-th].

[50] T. Albash, C.V. Johnson, Holographic studies of entanglement entropy in superconductors, J. High Energy Phys. 1205 (2012) 079, arXiv:1202.2605 [hep-th].

[51] R.G. Cai, S. He, L. Li, Y.L. Zhang, Holographic entanglement entropy in insulator/superconductor transition, J. High Energy Phys. 1207 (2012) 088, arXiv:1203.6620 [hep-th].

[52] R.G. Cai, S. He, L. Li, Y.L. Zhang, Holographic entanglement entropy on P-wave superconductor phase transition, J. High Energy Phys. 1207 (2012) 027, arXiv:1204.5962 [hep-th].

[53] R.G. Cai, S. He, L. Li, L.F. Li, Entanglement entropy and Wilson loop in Stúckelberg holographic insulator/superconductor model, J. High Energy Phys. 1210 (2012) 107, arXiv:1209.1019 [hep-th].

[54] R.E. Arias, I.S. Landea, Backreacting p-wave superconductors, J. High Energy Phys. 1301 (2013) 157, arXiv:1210.6823 [hep-th].

[55] X.M. Kuang, E. Papantonopoulos, B. Wang, Entanglement entropy as a probe of the proximity effect in holographic superconductors, J. High Energy Phys. 1405 (2014) 130, arXiv:1401.5720 [hep-th].

[56] W. Yao, J. Jing, Holographic entanglement entropy in metal/superconductor phase transition with Born–Infeld electrodynamics, Nucl. Phys. B 889 (2014) 109, arXiv:1408.1171 [hep-th].

[57] A. Dey, S. Mahapatra, T. Sarkar, Very general holographic superconductors and entanglement thermodynamics, J. High Energy Phys. 1412 (2014) 135, arXiv:1409.5309 [hep-th].

[58] O. Ben-Ami, D. Carmi, J. Sonnenschein, Holographic entanglement entropy of multiple strips, J. High Energy Phys. 1411 (2014) 144, arXiv:1409.6305 [hep-th].

[59] Y. Peng, Y. Liu, A general holographic metal/superconductor phase transition model, J. High Energy Phys. 1502 (2015) 082, arXiv:1410.7234 [hep-th].

[60] Y. Ling, P. Liu, C. Niu, J.P. Wu, Z.Y. Xian, Holographic entanglement entropy close to quantum phase transitions, J. High Energy Phys. 1604 (2016) 114, arXiv:1502.03661 [hep-th].

[61] A. Dey, S. Mahapatra, T. Sarkar, Thermodynamics and entanglement entropy with Weyl corrections, Phys. Rev. D 94 (2) (2016) 026006, arXiv:1512.07117 [hep-th].

[62] Y. Ling, P. Liu, J.P. Wu, Characterization of quantum phase transition using holographic entanglement entropy, Phys. Rev. D 93 (12) (2016) 126004, arXiv:1604.04857 [hep-th].

[63] Y. Peng, Holographic entanglement entropy in two-order insulator/superconductor transitions, arXiv:1607.08305 [hep-th].

[64] P. Breitenlohner, D.Z. Freedman, Positive energy in anti-De Sitter backgrounds and gauged extended supergravity, Phys. Lett. B 115 (1982) 197.

[65] P. Breitenlohner, D.Z. Freedman, Stability in gauged extended supergravity, Ann. Phys. 144 (1982) 249.

[66] S. Ryu, T. Takayanagi, Holographic derivation of entanglement entropy from AdS/CFT, Phys. Rev. Lett. 96 (2006) 181602, arXiv:hep-th/0603001.

[67] V.E. Hubeny, M. Rangamani, T. Takayanagi, A covariant holographic entanglement entropy proposal, J. High Energy Phys. 0707 (2007) 062, arXiv:0705.0016 [hep-th].

[68] W. Fischler, S. Kundu, Strongly coupled gauge theories: high and low temperature behavior of non-local observables, J. High Energy Phys. 1305 (2013) 098, arXiv:1212.2643 [hep-th].

Permissions

All chapters in this book were first published in NPB, by Elsevier B.V.; hereby published with permission under the Creative Commons Attribution License or equivalent. Every chapter published in this book has been scrutinized by our experts. Their significance has been extensively debated. The topics covered herein carry significant findings which will fuel the growth of the discipline. They may even be implemented as practical applications or may be referred to as a beginning point for another development.

The contributors of this book come from diverse backgrounds, making this book a truly international effort. This book will bring forth new frontiers with its revolutionizing research information and detailed analysis of the nascent developments around the world.

We would like to thank all the contributing authors for lending their expertise to make the book truly unique. They have played a crucial role in the development of this book. Without their invaluable contributions this book wouldn't have been possible. They have made vital efforts to compile up to date information on the varied aspects of this subject to make this book a valuable addition to the collection of many professionals and students.

This book was conceptualized with the vision of imparting up-to-date information and advanced data in this field. To ensure the same, a matchless editorial board was set up. Every individual on the board went through rigorous rounds of assessment to prove their worth. After which they invested a large part of their time researching and compiling the most relevant data for our readers.

The editorial board has been involved in producing this book since its inception. They have spent rigorous hours researching and exploring the diverse topics which have resulted in the successful publishing of this book. They have passed on their knowledge of decades through this book. To expedite this challenging task, the publisher supported the team at every step. A small team of assistant editors was also appointed to further simplify the editing procedure and attain best results for the readers.

Apart from the editorial board, the designing team has also invested a significant amount of their time in understanding the subject and creating the most relevant covers. They scrutinized every image to scout for the most suitable representation of the subject and create an appropriate cover for the book.

The publishing team has been an ardent support to the editorial, designing and production team. Their endless efforts to recruit the best for this project, has resulted in the accomplishment of this book. They are a veteran in the field of academics and their pool of knowledge is as vast as their experience in printing. Their expertise and guidance has proved useful at every step. Their uncompromising quality standards have made this book an exceptional effort. Their encouragement from time to time has been an inspiration for everyone.

The publisher and the editorial board hope that this book will prove to be a valuable piece of knowledge for researchers, students, practitioners and scholars across the globe.

List of Contributors

Johannes Broedel
Institut für Mathematik und Institut für Physik, Humboldt-Universität zu Berlin, IRIS Adlershof, Zum Großen Windkanal 6, 12489 Berlin, Germany

Martin Sprenger
Institut für Theoretische Physik, Eidgenössische Technische Hochschule Zürich, Wolfgang-Pauli-Strasse 27, 8093 Zürich, Switzerland

Alejandro Torres Orjuela
Institut für Mathematik und Institut für Physik, Humboldt-Universität zu Berlin, IRIS Adlershof, Zum Großen Windkanal 6, 12489 Berlin, Germany
Institut für Mathematik, Technische Universität Berlin, Straße des 17. Juni 136,10623 Berlin, Germany

M.R. Setare and H. Adami
Department of Science, University of Kurdistan, Sanandaj, Iran

Cheng-Yuan Zhang, Ya-Bo Wu, Ya-Nan Zhang, Huan-Yu Wang and Meng-Meng Wu
Department of Physics, Liaoning Normal University, Dalian, 116029, China

B. Nikolić and B. Sazdović
Institute of Physics Belgrade, University of Belgrade, Pregrevica 118, 11080 Belgrade, Serbia

Ding-fang Zeng
Theoretical Physics Division, College of Applied Sciences, Beijing University of Technology, China
State Key Laboratory of Theoretical Physics, Institute of Theoretical Physics, Chinese Academy of Sciences, Beijing, 100124, China

Jie-Xiong Mo and Gu-Qiang Li
Institute of Theoretical Physics, Lingnan Normal University, Zhanjiang, 524048, Guangdong, China

Department of Physics, Lingnan Normal University, Zhanjiang, 524048, Guangdong, China

Ze-Tao Lin
Department of Physics, Lingnan Normal University, Zhanjiang, 524048, Guangdong, China

Xiao-Xiong Zeng
School of Material Science and Engineering, Chongqing Jiaotong University, Chongqing, 400074, China
Institute of Theoretical Physics, Chinese Academy of Sciences, Beijing 100190, China

Deyou Chen
Institute of Theoretical Physics, China West Normal University, Nanchong, 637009, China

Gan qingyu and Jun Tao
Center for Theoretical Physics, College of Physical Science and Technology, Sichuan University, Chengdu, 610064, China

Binoy Krishna Patra
Department of Physics, Indian Institute of Technology Roorkee, Roorkee 247 667, India

Bhaskar Arya
Department of Mechanical and Industrial Engineering, Indian Institute of Technology Roorkee, Roorkee 247 667, India

Sota Hanazawa and Makoto Sakaguchi
Department of Physics, Ibaraki University, Mito 310-8512, Japan

Yan Peng
School of Mathematical Sciences, Qufu Normal University, Qufu, Shandong 273165, China
School of Mathematics and Computer Science, Shaanxi Sci-Tech University, Hanzhong, Shaanxi 723000, China

Qiyuan Pan
Department of Physics, Key Laboratory of Low Dimensional Quantum Structures and Quantum Control of Ministry of Education, Hunan Normal University, Changsha, Hunan 410081, China

Yunqi Liu
School of Physics, Huazhong University of Science and Technology, Wuhan, Hubei 430074, China

Sergey Koshkarev and Stefan Groote
Institute of Physics, University of Tartu, Tartu 51010, Estonia

Katsushi Ito and Hongfei Shu
Department of Physics, Tokyo Institute of Technology, Tokyo, 152-8551, Japan

Mir Faizal
Irving K. Barber School of Arts and Sciences, University of British Columbia -Okanagan, Kelowna, BritishColumbiaV1V 1V7, Canada Department of Physics and Astronomy, University of Lethbridge, Lethbridge, Alberta, T1K 3M4, Canada

Tsou Sheung Tsun
Mathematical Institute, University of Oxford, Andrew Wiles Building, Radcliffe Observatory Quarter, Woodstock Road, Oxford OX2 6GG, United Kingdom

Shao-Jun Zhang
Institute for Advanced Physics and Mathematics, Zhejiang University of Technology, Hangzhou 310023, China

Index

www.ingramcontent.com/pod-product-compliance
Lightning Source LLC
Chambersburg PA
CBHW061958190326
41458CB00009B/2906